Selected Titles in This Series

677 **Volodymyr V. Lyubashenko,** Squared Hopf algebras, 1999

676 **S. Strelitz,** Asymptotics for solutions of linear differential equations having turning points with applications, 1999

675 **Michael B. Marcus and Jay Rosen,** Renormalized self-intersection local times and Wick power chaos processes, 1999

674 **R. Lawther and D. M. Testerman,** A_1 subgroups of exceptional algebraic groups, 1999

673 **John Lott,** Diffeomorphisms and noncommutative analytic torsion, 1999

672 **Yael Karshon,** Periodic Hamiltonian flows on four dimensional manifolds, 1999

671 **Andrzej Rosłanowski and Saharon Shelah,** Norms on possibilities I: Forcing with trees and creatures, 1999

670 **Steve Jackson,** A computation of δ_5^1, 1999

669 **Seán Keel and James McKernan,** Rational curves on quasi-projective surfaces, 1999

668 **E. N. Dancer and P. Poláčik,** Realization of vector fields and dynamics of spatially homogeneous parabolic equations, 1999

667 **Ethan Akin,** Simplicial dynamical systems, 1999

666 **Mark Hovey and Neil P. Strickland,** Morava K-theories and localisation, 1999

665 **George Lawrence Ashline,** The defect relation of meromorphic maps on parabolic manifolds, 1999

664 **Xia Chen,** Limit theorems for functionals of ergodic Markov chains with general state space, 1999

663 **Ola Bratteli and Palle E. T. Jorgensen,** Iterated function systems and permutation representation of the Cuntz algebra, 1999

662 **B. H. Bowditch,** Treelike structures arising from continua and convergence groups, 1999

661 **J. P. C. Greenlees,** Rational S^1-equivariant stable homotopy theory, 1999

660 **Dale E. Alspach,** Tensor products and independent sums of $\mathcal{L}_p$-spaces, $1 < p < \infty$, 1999

659 **R. D. Nussbaum and S. M. Verduyn Lunel,** Generalizations of the Perron-Frobenius theorem for nonlinear maps, 1999

658 **Hasna Riahi,** Study of the critical points at infinity arising from the failure of the Palais-Smale condition for n-body type problems, 1999

657 **Richard F. Bass and Krzysztof Burdzy,** Cutting Brownian paths, 1999

656 **W. G. Bade, H. G. Dales, and Z. A. Lykova,** Algebraic and strong splittings of extensions of Banach algebras, 1999

655 **Yuval Z. Flicker,** Matching of orbital integrals on $GL(4)$ and $GSp(2)$, 1999

654 **Wancheng Sheng and Tong Zhang,** The Riemann problem for the transportation equations in gas dynamics, 1999

653 **L. C. Evans and W. Gangbo,** Differential equations methods for the Monge-Kantorovich mass transfer problem, 1999

652 **Arne Meurman and Mirko Primc,** Annihilating fields of standard modules of $\mathfrak{sl}(2,\mathbb{C})^\sim$ and combinatorial identities, 1999

651 **Lindsay N. Childs, Cornelius Greither, David J. Moss, Jim Sauerberg, and Karl Zimmermann,** Hopf algebras, polynomial formal groups, and Raynaud orders, 1998

650 **Ian M. Musson and Michel Van den Bergh,** Invariants under Tori of rings of differential operators and related topics, 1998

649 **Bernd Stellmacher and Franz Georg Timmesfeld,** Rank 3 amalgams, 1998

648 **Raúl E. Curto and Lawrence A. Fialkow,** Flat extensions of positive moment matrices: Recursively generated relations, 1998

647 **Wenxian Shen and Yingfei Yi,** Almost automorphic and almost periodic dynamics in skew-product semiflows, 1998

(Continued in the back of this publication)

Squared Hopf Algebras

MEMOIRS
of the
American Mathematical Society

Number 677

Squared Hopf Algebras

Volodymyr V. Lyubashenko

November 1999 • Volume 142 • Number 677 (third of 4 numbers) • ISSN 0065-9266

American Mathematical Society
Providence, Rhode Island

1991 *Mathematics Subject Classification.*
Primary 16W30, 18D15, 18E10, 17B37; Secondary 18D20, 16D90.

Library of Congress Cataloging-in-Publication Data

Lyubashenko, Volodymyr V., 1959–
Squared Hopf algebras / Volodymyr V. Lyubashenko.
p. cm. — (Memoirs of the American Mathematical Society, ISSN 0065-9266 ; no. 677)
"November 1999, volume 142, number 677 (third of 4 numbers)."
Includes bibliographical references.
ISBN 0-8218-1361-7
1. Hopf algebras. I. Title. II. Series.
QA613.8.L98 1999
512′.55–dc21 99-39861
CIP

Memoirs of the American Mathematical Society

This journal is devoted entirely to research in pure and applied mathematics.

Subscription information. The 1999 subscription begins with volume 137 and consists of six mailings, each containing one or more numbers. Subscription prices for 1999 are $448 list, $358 institutional member. A late charge of 10% of the subscription price will be imposed on orders received from nonmembers after January 1 of the subscription year. Subscribers outside the United States and India must pay a postage surcharge of $30; subscribers in India must pay a postage surcharge of $43. Expedited delivery to destinations in North America $35; elsewhere $130. Each number may be ordered separately; *please specify number* when ordering an individual number. For prices and titles of recently released numbers, see the New Publications sections of the *Notices of the American Mathematical Society.*

Back number information. For back issues see the *AMS Catalog of Publications.*

Subscriptions and orders should be addressed to the American Mathematical Society, P. O. Box 5904, Boston, MA 02206-5904. *All orders must be accompanied by payment.* Other correspondence should be addressed to Box 6248, Providence, RI 02940-6248.

Memoirs of the American Mathematical Society is published bimonthly (each volume consisting usually of more than one number) by the American Mathematical Society at 201 Charles Street, Providence, RI 02904-2294. Periodicals postage paid at Providence, RI. Postmaster: Send address changes to Memoirs, American Mathematical Society, P. O. Box 6248, Providence, RI 02940-6248.

This publication is indexed in *Science Citation Index®*, *SciSearch®*, *Research Alert®*, *CompuMath Citation Index®*, *Current Contents®/Physical, Chemical & Earth Sciences.*
Printed in the United States of America.

∞ The paper used in this book is acid-free and falls within the guidelines established to ensure permanence and durability.
Visit the AMS home page at URL: `http://www.ams.org/`

10 9 8 7 6 5 4 3 2 1 04 03 02 01 00 99

Contents

ABSTRACT. Given an abelian $\Bbbk$-linear rigid monoidal category $\mathcal{V}$, where $\Bbbk$ is a perfect field, we define *squared coalgebras* as objects of the cocompleted $\mathcal{V} \boxtimes \mathcal{V}$ (Deligne's tensor product of categories) equipped with the appropriate notion of comultiplication. Based on this, (squared) bialgebras and Hopf algebras are defined without use of braiding. If $\mathcal{V}$ is the category of $\Bbbk$-vector spaces, squared (co)algebras coincide with conventional ones. If $\mathcal{V}$ is braided, a braided Hopf algebra can be obtained from a squared one.

Reconstruction theorems give equivalence of squared co- (bi-, Hopf) algebras in $\mathcal{V}$ and corresponding fibre functors to $\mathcal{V}$ (which is not the case with the usual definitions). Finally, a squared quasitriangular Hopf coalgebra is a solution to the problem of defining quantum groups in braided categories.

Received by the editor May 19, 1997; and in revised form May 18, 1998.

Introduction

Classical reconstruction theorem (e.g. Saavedra [**26**, Section 2.3.2.1]) asserts that a $\Bbbk$-coalgebra can be reconstructed from the underlying functor from its category of comodules to vector spaces. Joyal and Street review in [**17**] the relationship of reconstruction theorems for coalgebras and bialgebras with quantum groups and braided categories. Saavedra [**26**, Section 2.6.3 a)] and later Schauenburg [**27**] also prove that an essentially small abelian $\Bbbk$-linear category equipped with an exact faithful functor ω to the category of finite dimensional $\Bbbk$-vector spaces is equivalent to the category of finite dimensional comodules over some $\Bbbk$-coalgebra. A direct attempt to generalise these results replacing the category of vector spaces by an abelian $\Bbbk$-linear rigid monoidal category $\mathcal{V}$ fails. For instance, the category of comodules over the coalgebra constructed from ω is bigger than the initial category (precise results are formulated by Pareigis [**23**]). However, if one modifies the definitions of coalgebras and comodules in a monoidal category, the reconstruction theorem will be recovered. This is the main conclusion of this work.

The new notion will be called a *squared coalgebra*. The monoidal version of the reconstruction theorem dictates the definition of a squared bialgebra. Squared Hopf (co)algebras based on $\mathcal{V}$ can also be defined, even if $\mathcal{V}$ is not braided, but satisfies a much weaker condition! If $\mathcal{V}$ is braided, a squared Hopf (co)algebra determines a braided Hopf algebra, but not vice versa. Finally, a squared quasitriangular Hopf coalgebra is a solution to the problem of defining quantum groups in braided categories.

All definitions are based on the notion of tensor product of abelian categories given by Deligne [**7**]. The squared notions (coalgebras, bialgebras, Hopf algebras) are objects of the cocompleted tensor square of the initial category $\mathcal{V}$, whence the terminology. The structure maps – comultiplication, multiplication etc. – are morphisms in tensor powers of $\mathcal{V}$. The associativity and other properties mean equality of two composite morphisms in tensor powers of $\mathcal{V}$.

More precisely, we use cocompletions of tensor powers of $\mathcal{V}$, where the cocompletion of an abelian $\Bbbk$-linear category $\mathcal{A}$ with finite dimensional $\Bbbk$-spaces of morphisms always means the category $\widehat{\mathcal{A}} = \operatorname{ind}-\mathcal{A}$, made of filtered inductive limits of objects of $\mathcal{A}$. If $\mathcal{A}$ is essentially small, $\widehat{\mathcal{A}}$ is equivalent to the category $\underline{\mathrm{Hom}}_{\Bbbk,\mathrm{l.e.}}(\mathcal{A}^{\mathrm{op}}, \Bbbk\text{-vect})$ of left exact functors $\mathcal{A}^{\mathrm{op}} \to \Bbbk\text{-vect}$. (See Grothendieck and Verdier [**14**].)

We assume that $\Bbbk$ is a perfect field.

Let us recall the reconstruction theorem in details following [**17**]. If $\omega : \mathcal{C} \to \Bbbk\text{-vect}$ is a faithful exact $\Bbbk$-linear functor and $\mathcal{C}$ is essentially small, then there is an equivalence F of $\mathcal{C}$ with the category of C-comodules, where C is the $\Bbbk$-coalgebra

$$C = \int^{X\in\mathcal{C}} \omega(X) \otimes_{\Bbbk} \omega(X)^*,$$

and ω is isomorphic to the composite of F and the underlying functor $\mathcal{U}$: C-comod $\to$ $\Bbbk$-vect. When $\Bbbk$-vect is replaced by an abelian $\Bbbk$-linear rigid monoidal category $\mathcal{V}$ with finite dimensional spaces $\mathrm{Hom}_{\mathcal{V}}(-,-)$ such that $\mathrm{End}\, \mathbb{I} = \Bbbk$ ($\mathbb{I}$ is the unit object) and each object has finite length, a version of the reconstruction theorem holds, although F is no longer an equivalence. It turns out that by modifying the definitions of coalgebras and comodules one can make F into an equivalence, thus recovering the original form of the theorem. Namely, instead of the coalgebra in $\widehat{\mathcal{V}}$

$$\bar{C} = \int^{X \in \mathcal{C}} \omega(X) \otimes \omega(X)^{\vee}, \tag{1}$$

one can use the squared coalgebra

$$C = \int^{X \in \mathcal{C}} \omega(X) \boxtimes \omega(X)^{\vee} \in \widehat{\mathcal{V} \boxtimes \mathcal{V}}, \tag{2}$$

where $\boxtimes : \mathcal{V} \times \mathcal{V} \to \mathcal{V} \boxtimes \mathcal{V}$ is the canonical functor of external tensor product.

To explain the definition of a squared coalgebra, we recall that the tensor product functor $\otimes : \mathcal{V} \times \mathcal{V} \to \mathcal{V}$ can be decomposed as $\mathcal{V} \times \mathcal{V} \xrightarrow{\boxtimes} \mathcal{V} \boxtimes \mathcal{V} \xrightarrow{\circledast} \mathcal{V}$ up to an isomorphism. Using the functor $\boxtimes$ and the diagonal restriction functor $\circledast$, we can construct various objects like $C_{13} \boxtimes \mathbb{I}_2 \in \widehat{\mathcal{V}^{\boxtimes 3}}$ ($\mathbb{I}$ is the unit object and the subindices indicate the tensor factors in which an object is placed), or like $C_{12'} \otimes C_{2''3} \in \widehat{\mathcal{V}^{\boxtimes 3}}$ (this is the result of applying $\circledast$ on the second and third places to $C_{12} \boxtimes C_{34} \in \widehat{\mathcal{V}^{\boxtimes 4}}$), or like $C_{1'1''} = \circledast C \in \widehat{\mathcal{V}}$ (the dash and the double dash indicate the order of multiplicands) etc. Notice that $\circledast$ applied to (2) gives (1). The system of notations is explained in details in Section 1.5. There we define generalised objects and morphisms of $\mathcal{V}^{\boxtimes I}$. Notations are illustrated in the example of coassociativity equation.

The major definitions and results of this work are illustrated in the example $\mathcal{V} = H$-comod (*finite dimensional* H-comodules) where H is a Hopf $\Bbbk$-algebra. The reader is advised to keep this example in mind throughout the whole text. In this framework $\widehat{\mathcal{V}} = H$-Comod (*all* H-comodules), $\mathcal{V}^{\boxtimes n} = H^{\otimes n}$-comod, $\widehat{\mathcal{V}^{\boxtimes n}} = H^{\otimes n}$-Comod. The functor $\boxtimes : \mathcal{V} \times \mathcal{V} \to \mathcal{V} \boxtimes \mathcal{V}$ sends a pair of H-comodules (X, Y) to the $H \otimes_{\Bbbk} H$-comodule $X \otimes_{\Bbbk} Y$ (external tensor product of comodules) and the diagonal restriction functor $\circledast : \mathcal{V} \boxtimes \mathcal{V} \to \mathcal{V}$ sends an $H \otimes_{\Bbbk} H$-comodule to an H-comodule via the homomorphism of coalgebras $m : H \otimes_{\Bbbk} H \to H$ – the multiplication. Thus, given an $H \otimes_{\Bbbk} H$-comodule C with the coaction δ : $c \mapsto c_{(1)} \otimes c_{(2)} \otimes c_{(3)} \in H \otimes_{\Bbbk} H \otimes_{\Bbbk} C$ we get the $H^{\otimes 3}$-comodule $C_{13} \boxtimes \mathbb{I}_2 = (C, \delta : c \mapsto c_{(1)} \otimes 1 \otimes c_{(2)} \otimes c_{(3)} \in H^{\otimes 3} \otimes_{\Bbbk} C)$, the $H^{\otimes 3}$-comodule $C_{12'} \otimes C_{2''3} = (C \otimes_{\Bbbk} C, c \otimes d \mapsto c_{(1)} \otimes c_{(2)} d_{(1)} \otimes d_{(2)} \otimes c_{(3)} \otimes d_{(3)} \in H^{\otimes 3} \otimes_{\Bbbk} C^{\otimes 2})$, the H-comodule $\circledast C = (C, \delta : c \mapsto c_{(1)} c_{(2)} \otimes c_{(3)} \in H \otimes_{\Bbbk} C)$ etc.

Now we may define a squared coalgebra as an object $C \in \widehat{\mathcal{V} \boxtimes \mathcal{V}}$ equipped with the comultiplication $\Delta_{123} : C_{13} \boxtimes \mathbb{I}_2 \to C_{12'} \otimes C_{2''3}$ and the counit $\varepsilon : C_{1'1''} \to \mathbb{I}_1$, which satisfy the coassociativity axiom (an equation in $\widehat{\mathcal{V}^{\boxtimes 4}}$, see (2.1.1)) and two axioms for the counit (equations (2.1.2a)-(2.1.2b) in $\widehat{\mathcal{V}^{\boxtimes 2}}$). A C-comodule is an object $X \in \mathcal{V}$ equipped with the coaction $\delta : X_1 \boxtimes \mathbb{I}_2 \to C_{12'} \otimes X_{2''} \in \widehat{\mathcal{V}^{\boxtimes 2}}$, which is coassociative (equation (2.2.1) in $\widehat{\mathcal{V}^{\boxtimes 3}}$) and counital (equation (2.2.2) in $\widehat{\mathcal{V}^{\boxtimes 2}}$). It turns out that for any object $Y \in \mathcal{V}$ the object $Y \boxtimes Y^{\vee} \in \mathcal{V} \boxtimes \mathcal{V}$ has the canonical structure of a squared coalgebra and Y is a comodule over it. We deduce that the coend (2) is a squared coalgebra as well. Moreover, the second part of the reconstruction theorem claims that any squared coalgebra is isomorphic to a coalgebra of the form (2).

Thus, the full form of the reconstruction theorem asserts equivalence of the following two categories: the category of $\Bbbk$-linear exact faithful functors from an essentially small category to $\mathcal{V}$ and the category of squared coalgebras in $\mathcal{V}$. Philosophically, categories over the category $\mathcal{V}$ are fully encoded in terms of coalgebras living in $\mathcal{V}$ (in fact, in $\widehat{\mathcal{V} \boxtimes \mathcal{V}}$) and vice versa. Comparing the category of comodules ${}^{C}\mathcal{V}$ over a squared coalgebra C and the category of comodules ${}^{\bar{C}}\mathcal{V}$ over the ordinary coalgebra (comonoid) $\bar{C} = \circledast C$ we get ${}^{\bar{C}}\mathcal{V} = {}^{C}\mathcal{V} \boxtimes \mathcal{V}$. That is expected from the description of the coend reconstructed from the functor ${}^{\bar{C}}\mathcal{V} \to \mathcal{V}$ given by Pareigis [**23**] in the case $\mathcal{V} = H$-comod for a Hopf algebra H.

The monoidal version of the reconstruction theorem also holds. Namely, the category of monoidal $\Bbbk$-linear exact faithful functors $\omega : \mathcal{C} \to \mathcal{V}$ ($\mathcal{C}$ is essentially small) and the category of squared bicoalgebras in $\mathcal{V}$ are equivalent. A *squared bicoalgebra* is defined as an object of $\widehat{\mathcal{V} \boxtimes \mathcal{V}}$ having the structure of a squared coalgebra and of an algebra in the monoidal category $\widehat{\mathcal{V} \boxtimes \mathcal{V}}$ with compatibility axioms which require that the multiplication and the unit were homomorphisms of coalgebras. (There are several monoidal structures in $\widehat{\mathcal{V} \boxtimes \mathcal{V}}$ and we chose a special one.)

The dual notion, squared bialgebras, is defined as a squared algebra structure plus a coalgebra structure in $\widehat{\mathcal{V} \boxtimes \mathcal{V}}$ with compatibility axioms. Unlike the case of vector spaces the notion of squared bicoalgebra is not self-dual, so it has to be distinguished from squared bialgebras. The choice of terminology is motivated by our primary interest in comodules rather than in modules. We shall simplify it further dropping the adjective squared and keeping the term bicoalgebra.

Notice that a braiding in $\mathcal{V}$ is not required for work with bicoalgebras. However, if $\mathcal{V}$ is braided, any bicoalgebra B generates a braided bialgebra $\circledast B$ in $\mathcal{V}$. Not every braided bialgebra comes from a bicoalgebra in that way.

To introduce Hopf algebras we require that the second dual $X^{\vee\vee}$ of an object $X \in \mathcal{V}$ be isomorphic to X via a functorial isomorphism $\zeta = \zeta_X : X \to X^{\vee\vee}$. Obviously, this condition is weaker than existence of braiding. Given a squared coalgebra C in such $\mathcal{V}$, one can define the opposite coalgebra C_{op}. A (squared) Hopf coalgebra is defined as a bicoalgebra H together with an isomorphism $\gamma : H_{\mathrm{op}} \to H \in \widehat{\mathcal{V} \boxtimes \mathcal{V}}$ – the antipode – satisfying two equations in $\widehat{\mathcal{V} \boxtimes \mathcal{V}}$. The reason for introducing Hopf coalgebras is the following: the category of comodules over a Hopf coalgebra is rigid and the rigid version of the reconstruction theorem holds: the category of monoidal $\Bbbk$-linear exact faithful functors $\omega : \mathcal{C} \to \mathcal{V}$, where $\mathcal{C}$ is rigid monoidal (and essentially small), and the category of Hopf coalgebras in $\mathcal{V}$ are equivalent. The dual notion, squared Hopf algebras, is not equivalent to the notion of Hopf coalgebras.

If, in addition, $\mathcal{V}$ is braided, we get the equivalence of the category of monoidal $\Bbbk$-linear exact faithful functors $\omega : \mathcal{C} \to \mathcal{V}$, where $\mathcal{C}$ is rigid braided, and of the category of *quasitriangular Hopf coalgebras*, appropriately defined. In particular, the category of comodules over a quasitriangular Hopf coalgebra is braided. This is not trivial, and allows us to introduce a non-obvious braiding for the bigger category of comodules over the braided Hopf algebra $\circledast H$. However, it seems impossible, in general, to introduce a braided structure of any kind for the whole category of comodules over a braided Hopf algebra not related with Hopf coalgebras. Thus the notion of a quasitriangular Hopf coalgebra is the closest to the idea of a "quantum group in a braided category".

In particular, applying the (re)construction theorem to the identity functor $\mathrm{Id} : \mathcal{V} \to \mathcal{V}$, we get a quasitriangular Hopf coalgebra structure of the coend

$$C = \int^{X \in \mathcal{V}} X \boxtimes X^{\vee},$$

and this is the most interesting case for us. Similar notions exist for ribbon categories.

A by-product of our approach is the possibility to present an abelian (monoidal) category by generators and relations relative to another abelian monoidal category.

We pay some attention also to the background on which the action develops – the higher category theory. Besides the usual definitions of monoidal, braided and symmetric categories we give a version of these definitions with n-ary tensor operations. We prove that the 2-category of such categories is 2-equivalent to its conventional counterpart. This fits into the general picture of Batanin [**3**], who shows that the definition of a weak m-category is not unique, but unique up to an equivalence.

In Appendix A we develop selected topics of 2-category theory suitable for summarising the properties of Deligne's tensor product of abelian categories. We discuss weak pasting for 2-categories. Based on that we define 3-categories. The chosen definition allows us to prove that Deligne's tensor product gives indeed a monoidal structure on a 2-category of abelian categories. An advantage of this definition of a monoidal 2-category is that minor changes transform it into a definition of a symmetric monoidal 2-category. We prove also that the 2-category of $\Bbbk$-linear abelian categories is symmetric. Notice that the usual definition of a weak symmetric monoidal 2-category as a special case of a weak 6-category does not seem unpackable at all. The semistrict case (Gray monoids) treated by Day and Street in [**6**] cannot be applied immediately to the category of abelian categories without replacing it first with a 2-equivalent category. Fortunately, the universal property of Deligne's external tensor product of abelian categories together with the result that iterated external product is also universal turns out sufficient to produce the structure of a weak symmetric monoidal 2-category.

Acknowledgements. This work was supported in part by the EPSRC research grant GR/G 42976 and by NSF grant 530666. The main results were obtained at the University of York and I use this opportunity to thank the Department of Mathematics for creating a friendly atmosphere and good work conditions. The exposition was systematized while the author was working at the National Technical University of Ukraine. I am grateful to the Department of Applied Mathematics for moral and financial support. Final touches were given to the work at the Kansas State University, and I express my gratitude to the Department of Mathematics. Some of the diagrams were composed using Paul Taylor's diagram style. I would like to thank Dr. Yu. Bespalov, Prof. C. Kassel, Prof. Yan Soibelman, Prof. A. Sudbery and Prof. D. Yetter for fruitful discussions and attention to this work.

CHAPTER 1

Tools

1.1. Tensor product of abelian categories

Everywhere in this work $\Bbbk$ will denote a perfect field, for instance, a field of characteristics 0, except Appendix A where $\Bbbk$ can be an arbitrary field.

1.1.1. Definition. We say that an abelian $\Bbbk$-linear category $\mathcal{A}$ is a *category with length* if

1. for any pair of objects $X, Y \in \mathcal{A}$ the $\Bbbk$-vector space $\mathrm{Hom}_{\mathcal{A}}(X, Y)$ is finite dimensional, and
2. any object $X \in \mathcal{A}$ has finite length.

The most impressive result from the theory of such categories is:

1.1.2. Proposition (Deligne [7] Corollary 2.17). *Let $\mathcal{A}$ be an abelian $\Bbbk$-linear category with length. Assume that there exists an object $X \in \mathcal{A}$ such that any object $Y \in \mathcal{A}$ is a subquotient of $X^n = X \oplus \cdots \oplus X$ for some n. Then the category $\mathcal{A}$ is equivalent to the category* mod-A *for some finite dimensional $\Bbbk$-algebra A.*

We recall the definition of a tensor product of abelian $\Bbbk$-linear categories with length, belonging to Deligne [7], in a modified form, using his results, valid under the assumption of perfectness of the field.

1.1.3. Definition (following Deligne [7] Definition 5.1). Let $\mathcal{A}_1, \dots, \mathcal{A}_n$ be $\Bbbk$-linear abelian categories with length. A $\Bbbk$-linear abelian category $\mathcal{A}_1 \boxtimes \cdots \boxtimes \mathcal{A}_n$, equipped with a $\Bbbk$-multilinear, exact in each variable, functor

$$\boxtimes : \mathcal{A}_1 \times \cdots \times \mathcal{A}_n \to \mathcal{A}_1 \boxtimes \cdots \boxtimes \mathcal{A}_n$$

is called a tensor product of $\mathcal{A}_1, \dots, \mathcal{A}_n$ if for each $\Bbbk$-linear abelian category $\mathcal{A}$ the induced functor

$$\mathrm{Hom}_{\mathrm{m.l.,r.e.}}(\mathcal{A}_1 \boxtimes \cdots \boxtimes \mathcal{A}_n, \mathcal{A}) \to \mathrm{Hom}_{\mathrm{m.l.,r.e.}}(\mathcal{A}_1 \times \cdots \times \mathcal{A}_n, \mathcal{A}), F \mapsto F \circ \boxtimes$$

from the category of $\Bbbk$-linear right exact functors to the category of $\Bbbk$-multilinear right exact in each variable functors is an equivalence.

1.1.4. Remark. Equivalently, one can use left exact functors in the above definition (Deligne [7, Proposition 5.13]).

1.1.5. Proposition (Deligne [7] Proposition 5.13). *The tensor product of $\Bbbk$-linear abelian categories with length exists and is a category with length. It is unique up to an equivalence. The functor similar to that of Definition 1.1.3*

$$\mathrm{Hom}_{m.l.,e.}(\mathcal{A}_1 \boxtimes \cdots \boxtimes \mathcal{A}_n, \mathcal{A}) \to \mathrm{Hom}_{m.l.,e.}(\mathcal{A}_1 \times \cdots \times \mathcal{A}_n, \mathcal{A}), F \mapsto F \circ \boxtimes$$

with the right exact functors replaced by exact functors is also an equivalence. The natural map

$$\otimes_i \mathrm{Hom}(X_i, Y_i) \to \mathrm{Hom}(\boxtimes_i X_i, \boxtimes_i Y_i)$$

is an isomorphism.

The assumption of perfectness is essential for this result which follows mainly from:

1.1.6. Proposition (Deligne [7] Proposition 5.3, Corollary 5.4). *Let $\mathcal{A}_i$ be A_i-mod, where A_i are finite dimensional $\Bbbk$-algebras, $1 \leqslant i \leqslant n$. Then $\mathcal{A} = A_1 \otimes_\Bbbk \cdots \otimes_\Bbbk A_n$-mod equipped with the external tensor product functor $\boxtimes$: $(X_1, \dots, X_n) \mapsto X_1 \otimes_\Bbbk \cdots \otimes_\Bbbk X_n$ is a tensor product of $\mathcal{A}_1, \dots, \mathcal{A}_n$.*

Notice that any category with length $\mathcal{A}$ is a filtered inductive limit of its full subcategories of the form $\langle X \rangle$ – the full subcategory formed by the objects $Y \in \mathcal{A}$, which are subquotients of X^n for some n. Here X is an object of $\mathcal{A}$. This remark together with Propositions 1.1.2, 1.1.6 were used in [7] to prove Proposition 1.1.5.

The category of finite dimensional left C-comodules $\mathcal{A} = C$-comod is filtered by full subcategories C'-comod, where $C' \subset C$ are finite dimensional subcoalgebras. If $C' \overset{i}{\hookrightarrow} C'' \subset C$ are two subcoalgebras, the corresponding inclusion C'-comod $\to$ C''-comod is simply i-comod. Since $\otimes_\Bbbk$ commutes with inductive limits, one can reinforce Proposition 1.1.6.

1.1.7. Proposition. *Let $\mathcal{A}_i$ be C_i-comod, where C_i are $\Bbbk$-coalgebras, $1 \leqslant i \leqslant n$. Then $\mathcal{A} = C_1 \otimes_\Bbbk \cdots \otimes_\Bbbk C_n$-comod equipped with the external tensor product functor $\boxtimes : (X_1, \dots, X_n) \mapsto X_1 \otimes_\Bbbk \cdots \otimes_\Bbbk X_n$ is a tensor product of $\mathcal{A}_1, \dots, \mathcal{A}_n$.*

1.1.8. Corollary. *Let $\mathcal{A}_i$ be A_i-mod, where A_i are $\Bbbk$-algebras, $1 \leqslant i \leqslant n$. Then $\mathcal{A} = A_1 \otimes_\Bbbk \cdots \otimes_\Bbbk A_n$-mod equipped with the external tensor product functor $\boxtimes : (X_1, \dots, X_n) \mapsto X_1 \otimes_\Bbbk \cdots \otimes_\Bbbk X_n$ is a tensor product of $\mathcal{A}_1, \dots, \mathcal{A}_n$.*

PROOF. For any $\Bbbk$-algebra A there is a $\Bbbk$-coalgebra $C = A^\circ$ consisting of $\Bbbk$-linear maps $f : A \to \Bbbk$, for which there exists a two-sided ideal I of A such that $\dim_\Bbbk A/I < \infty$ and $f(I) = 0$. The category A-mod is identified with comod-C – the category of finite dimensional right C-comodules [1]. The isomorphism $A_1^\circ \otimes_\Bbbk \cdots \otimes_\Bbbk A_n^\circ \simeq (A_1 \otimes_\Bbbk \cdots \otimes_\Bbbk A_n)^\circ$ proves the corollary. □

1.1.9. Proposition (Deligne [7] Lemma 5.9). *Let $\mathcal{A}_1, \dots, \mathcal{A}_n$ be $\Bbbk$-linear abelian categories with length.*

(i) *If objects $S_i \in \mathcal{A}_i$ are simple, then $\boxtimes_i S_i \in \mathcal{A}_1 \boxtimes \cdots \boxtimes \mathcal{A}_n$ is semisimple.*
(ii) *Each simple object of $\mathcal{A}_1 \boxtimes \cdots \boxtimes \mathcal{A}_n$ is a direct summand of $\boxtimes_i S_i$ for some simple $S_i \in \mathcal{A}_i$.*

Let $\mathcal{A}_1, \dots, \mathcal{A}_n, \mathcal{B}_1, \dots, \mathcal{B}_n$ be $\Bbbk$-linear abelian categories with length and let $T_i : \mathcal{A}_i \to \mathcal{B}_i$ be $\Bbbk$-linear left (resp. right) exact functors. By definition, there exists a $\Bbbk$-linear left (resp. right) exact functor

$$T = \boxtimes_i T_i : \boxtimes_i \mathcal{A}_i \to \boxtimes_i \mathcal{B}_i, \qquad T(\boxtimes_i X_i) = \boxtimes_i T(X_i).$$

1.1.10. Proposition (Deligne [7] Proposition 5.14). *If T_i are exact (resp. exact and faithful, resp. equivalence of $\mathcal{A}_i$ with a full subcategory of $\mathcal{B}_i$ stable with respect to subquotients), then T has the same property.*

1.1.11. Proposition. *Let S be a simple object of an abelian category $\mathcal{A}$. Assume that the field $\Bbbk = \operatorname{End} S$ is commutative. Then the subcategory $\langle S \rangle \subset \mathcal{A}$ is equivalent to $\Bbbk$-vect via the functor $\Bbbk$-vect $\to \mathcal{A}$, $V \mapsto V \otimes_\Bbbk S$ ($\Bbbk^n \mapsto S^n$).*

1.2. Monoidal categories

We shall recall first the traditional definition.

1.2.1. Definition. A *monoidal category* $(\mathcal{C}, \otimes, a, \mathbb{I}, l, r)$ is a category $\mathcal{C}$, a functor $\otimes : \mathcal{C} \times \mathcal{C} \to \mathcal{C}$ – the tensor product, a functorial isomorphism $a : X \otimes (Y \otimes Z) \to (X \otimes Y) \otimes Z$ – the associativity constraint, a unit object $\mathbb{I}$ and two functorial isomorphisms $l : \mathbb{I} \otimes X \to X$, $r : X \otimes \mathbb{I} \to X$ such that

$$\begin{array}{ccccc} X \otimes (Y \otimes (Z \otimes W)) & \xrightarrow{a} & (X \otimes Y) \otimes (Z \otimes W) & \xrightarrow{a} & ((X \otimes Y) \otimes Z) \otimes W \\ \downarrow X \otimes a & & & & \uparrow a \otimes W \\ X \otimes ((Y \otimes Z) \otimes W) & & \xrightarrow{a} & & (X \otimes (Y \otimes Z)) \otimes W \end{array}$$

commutes (the pentagon equation) and

$$a_{X,\mathbb{I},Y} = \big(X \otimes (\mathbb{I} \otimes Y) \xrightarrow{X \otimes l_Y} X \otimes Y \xrightarrow{r_X^{-1} \otimes Y} (X \otimes \mathbb{I}) \otimes Y\big). \tag{1.2.1}$$

Notice that in a monoidal category the equations

$$a_{\mathbb{I},X,Y} = \big(\mathbb{I} \otimes (X \otimes Y) \xrightarrow{l} X \otimes Y \xrightarrow{l_X^{-1} \otimes Y} (\mathbb{I} \otimes X) \otimes Y\big),$$
$$a_{X,Y,\mathbb{I}} = \big(X \otimes (Y \otimes \mathbb{I}) \xrightarrow{X \otimes r_Y} X \otimes Y \xrightarrow{r^{-1}} (X \otimes Y) \otimes \mathbb{I}\big)$$

also hold.

We want to modify this definition to make it closer to the definition of monoidal 2-category given in Appendix A. We follow the ideas of Deligne and Milne [**9**].

Denote by $\mathcal{O}$ a category, whose objects are linearly ordered sets $\mathbf{n} = \{1 < 2 < \cdots < n\}$ including the empty set $\mathbf{0}$. Thus, $\operatorname{Ob} \mathcal{O} \simeq \mathbb{Z}_{\geqslant 0}$, morphisms are monotonic maps. For a map $f : I \to J$ of $\mathcal{O}$ and an element $j \in J$, denote by $f^{-1}(j)$ the object of $\mathcal{O}$ which is the pull-back in $\mathcal{O}$ of f and $j : \mathbf{1} \to J$, $j(1) = j$,

$$\begin{array}{ccc} f^{-1}(j) & \xhookrightarrow{e_j} & I \\ \downarrow & \lrcorner & \downarrow f \\ \mathbf{1} & \xhookrightarrow{j} & J \end{array} \quad . \tag{1.2.2}$$

This pull-back is unique. Morphisms $f : \mathbf{k} \to \mathbf{n}$ of $\mathcal{O}$ are in bijection with partitions $(k_1, \dots, k_n)$ of k, $k = k_1 + \cdots + k_n$, $k_j \geqslant 0$, $k_j = \# f^{-1}(j)$.

For an object I of $\mathcal{O}$, denote by $\mathcal{C}^I$ the category of functions on I with values in $\mathcal{C}$:

$$\operatorname{Ob} \mathcal{C}^I = \{ \text{ maps } I \to \operatorname{Ob} \mathcal{C}\}, \qquad \operatorname{Mor} \mathcal{C}^I = \{ \text{ maps } I \to \operatorname{Mor} \mathcal{C}\}.$$

$\mathcal{C}^{\mathbf{n}}$ is the usual $\mathcal{C}^n$, $\mathcal{C}^{\mathbf{0}}$ is the category with one object and one morphism.

To distinguish the traditional and the new version of monoidal categories we capitalize the latter. It will be shown later that these definitions are equivalent.

1.2.2. Definition. A *Monoidal category*[1] $(\mathcal{C}, \otimes^I, \lambda_f)_{f \in \mathcal{O}}$ consists of

1. A category $\mathcal{C}$.

[1] I am indebted to Yan Soibelman for this definition.

2. A functor $\otimes^I : \mathcal{C}^I \to \mathcal{C}$ for every object I of $\mathcal{O}$, such that $\otimes^{\mathbf{1}} = \mathrm{Id}_{\mathcal{C}}$.

   For a map $f : I \to J$ of $\mathcal{O}$ introduce a functor $\otimes^f_* : \mathcal{C}^I \to \mathcal{C}^J$, which to a function $X : I \to \mathcal{C}$, $i \mapsto X_i$ assigns the function

$$j \longmapsto \otimes^{f^{-1}(j)}(f^{-1}(j) \xrightarrow{e_j} I \xrightarrow{X} \mathcal{C}) = \otimes^{f^{-1}(j)}(X_{i_1}, \dots, X_{i_m}),$$

   where $\{i_1, \dots, i_m\} = \{i \in I \mid f(i) = j\}$, $i_1 < \dots < i_m$. $\mathcal{C}$ above stands for $\operatorname{Ob}\mathcal{C}$ or $\operatorname{Mor}\mathcal{C}$.

3. An isomorphism of functors λ_f for every map $f : I \to J$ of $\mathcal{O}$

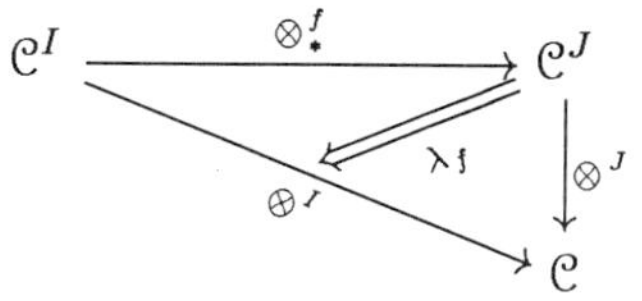

such that

$$\lambda_{\mathbf{0}\to\mathbf{0}} = \mathrm{id}, \qquad \lambda_{I\to\mathbf{1}} = \mathrm{id}, \qquad \lambda_{\mathrm{id}_I} = \mathrm{id},$$

and for any pair of composable maps $I \xrightarrow{f} J \xrightarrow{g} K$ from $\mathcal{O}$ an equation holds:

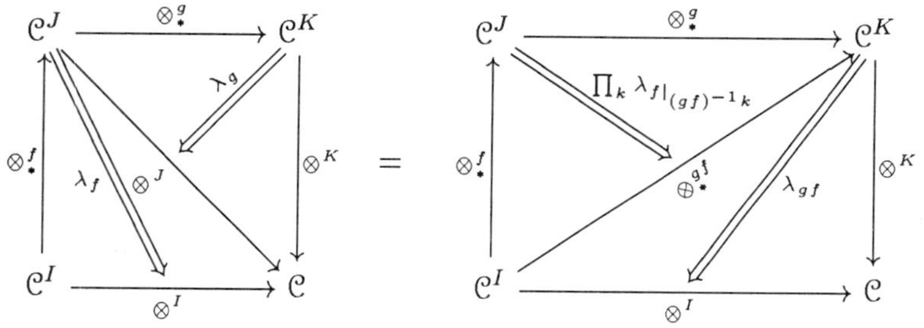

(1.2.3)

Note that for a monotonic map $f : I \to J$ we have $\otimes^f_* = \prod_{j\in J} \otimes^{f^{-1}(j)} : \mathcal{C}^I \to \mathcal{C}^J$. There exists a version of the above definition, in which I runs over all finite linearly ordered sets and $f^{-1}(j)$ retains its usual meaning.

To establish the relationship between these definitions, remark first that each Monoidal structure on a category $\mathcal{C}$ produces a monoidal structure as follows:

$$\otimes = \otimes^{\mathbf{2}} : \mathcal{C}^2 \to \mathcal{C};$$

$$\mathbb{I} = \otimes^{\mathbf{0}}(\text{object}), \qquad \text{where } \otimes^{\mathbf{0}} : \mathbf{1} \to \mathcal{C};$$

$a =$

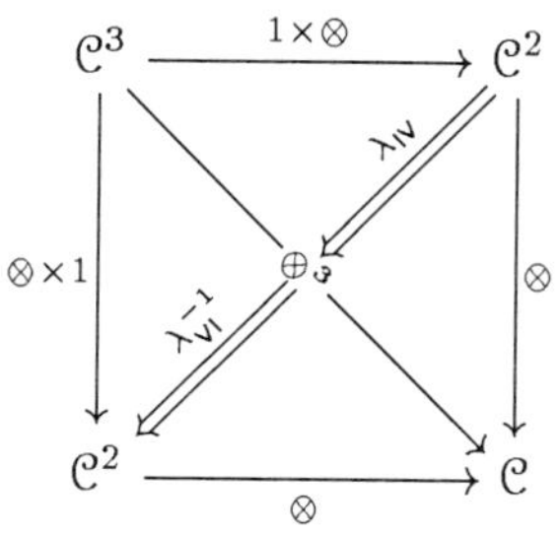

$l =$

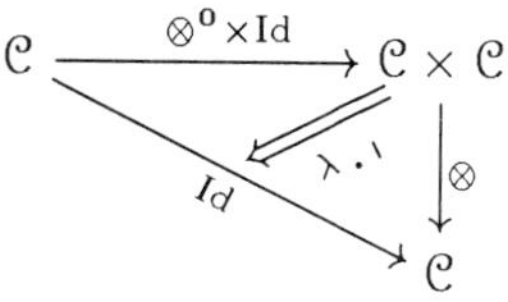

, $r =$

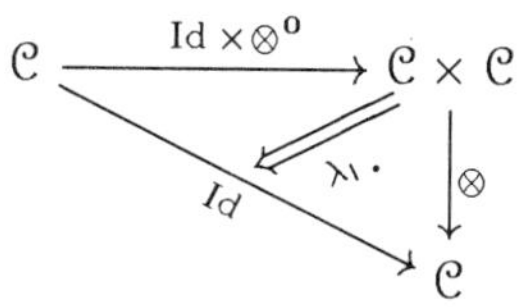

where the maps are

$$\begin{aligned}
\mathsf{IV} = f : \mathbf{3} \to \mathbf{2}, &\qquad f(1) = 1, \quad f(2) = f(3) = 2,\\
\mathsf{VI} = g : \mathbf{3} \to \mathbf{2}, &\qquad f(1) = f(2) = 1, \quad f(3) = 2,\\
\mathsf{I}. = e_1 : \mathbf{1} \to \mathbf{2}, &\qquad e_1(1) = 1,\\
.\mathsf{I} = e_2 : \mathbf{1} \to \mathbf{2}, &\qquad e_2(1) = 2.
\end{aligned}$$

The pentagon for associativity is proven below.

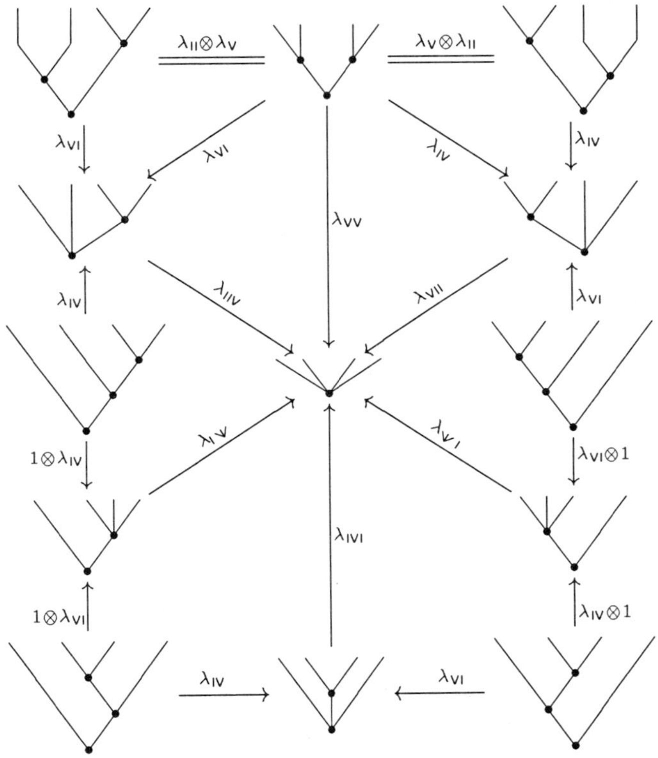

Each tree denotes a repeated tensor product. A vertex with n incoming branches denotes $\otimes^{\mathbf{n}}$. Squares and triangles commute due to axiom (1.2.3).

Condition (1.2.1) holds as the following diagram shows:

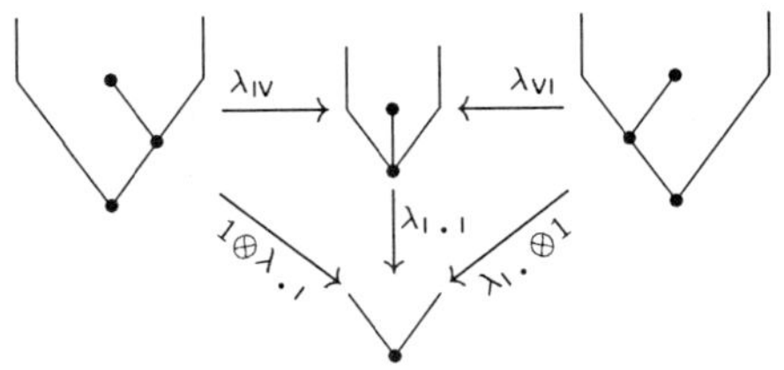

1.2.3. Definition. A *monoidal functor* $(F, \phi, \boldsymbol{f}) : (\mathcal{C}, \otimes) \to (\mathcal{D}, \otimes)$ is a functor $F : \mathcal{C} \to \mathcal{D}$, a functorial isomorphism $\phi = \phi_{X,Y} : F(X) \otimes F(Y) \to F(X \otimes Y) \in \mathcal{D}$

and an isomorphism $\boldsymbol{f} : \mathbb{1} \to F\mathbb{1} \in \mathcal{D}$ such that

$$\begin{array}{ccccc}
FX \otimes (FY \otimes FZ) & \xrightarrow{1\otimes\phi} & FX \otimes F(Y \otimes Z) & \xrightarrow{\phi} & F(X \otimes (Y \otimes Z)) \\
a \downarrow & & & & \downarrow Fa \\
(FX \otimes FY) \otimes FZ & \xrightarrow{\phi\otimes 1} & F(X \otimes Y) \otimes FZ & \xrightarrow{\phi} & F((X \otimes Y) \otimes Z)
\end{array} \quad , \quad (1.2.4)$$

$$\begin{array}{ccc ccc}
F\mathbb{1} \otimes FX & \xrightarrow{\phi} & F(\mathbb{1} \otimes X) & FX \otimes F\mathbb{1} & \xrightarrow{\phi} & F(X \otimes \mathbb{1}) \\
\boldsymbol{f}\otimes 1 \uparrow & & \downarrow Fl & 1\otimes\boldsymbol{f} \uparrow & & \downarrow Fr \\
\mathbb{1} \otimes FX & \xrightarrow{l} & FX & FX \otimes \mathbb{1} & \xrightarrow{r} & FX
\end{array} \quad , \quad (1.2.5)$$

commute. A *morphism of monoidal functors* $\lambda : (F, \phi, \boldsymbol{f}) \to (G, \psi, \boldsymbol{g})$ is a functorial morphism $\lambda : F \to G$ such that

$$\begin{array}{ccc}
FX \otimes FY & \xrightarrow{\phi} & F(X \otimes Y) \\
\lambda \otimes \lambda \downarrow & & \downarrow \lambda \\
GX \otimes GY & \xrightarrow{\psi} & G(X \otimes Y)
\end{array} \quad ,$$

$$\boldsymbol{g} = (\mathbb{1} \xrightarrow{\boldsymbol{f}} F\mathbb{1} \xrightarrow{\lambda} G\mathbb{1}).$$

The $\boldsymbol{f}$ datum of a monoidal functor $(F, f, \boldsymbol{f})$ is uniquely determined by the (F, f) data, so we often denote a monoidal functor as (F, f).

1.2.4. Definition. A *Monoidal functor*

$$(F, \phi^{\mathbf{n}}) : (\mathcal{C}, \otimes^{\mathbf{n}}, \lambda_f) \to (\mathcal{D}, \otimes^{\mathbf{n}}, \mu_f)$$

consists of

i) a functor $F : \mathcal{C} \to \mathcal{D}$,
ii) a functorial isomorphism for each $n \geq 0$

$$\begin{array}{ccc}
\mathcal{C}^n & \xrightarrow{F^n} & \mathcal{D}^n \\
\otimes^n \downarrow & \phi^n \swarrow & \downarrow \otimes^n \\
\mathcal{C} & \xrightarrow{F} & \mathcal{D}
\end{array}$$

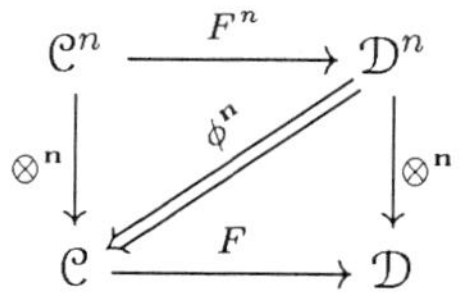

such that $\phi^{\mathbf{1}} = \mathrm{id}_F$ and for every map $f : I \to J$ of $\mathcal{O}$ the following equation holds:

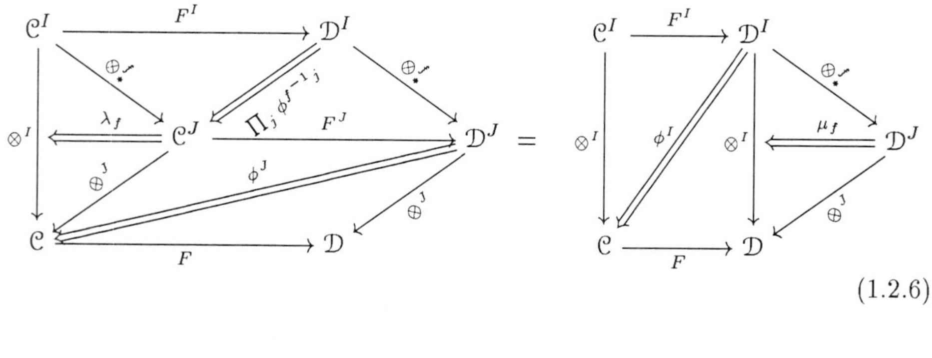

$$(1.2.6)$$

A Monoidal functor $(F, \phi^{\mathbf{n}})$ produces a monoidal functor $(F, \phi^{\mathbf{2}}, \phi^{\mathbf{0}})$, $\phi^{\mathbf{2}} : FX \otimes FY \to F(X \otimes Y)$, $\phi^{\mathbf{0}} : \mathbb{1} \to F\mathbb{1}$. Indeed, (1.2.4) follows from

$$\begin{array}{ccccc}
FX \otimes (FY \otimes FZ) & \xrightarrow{1 \otimes \phi^{\mathbf{2}}} & FX \otimes F(Y \otimes Z) & \xrightarrow{\phi^{\mathbf{2}}} & F(X \otimes (Y \otimes Z)) \\
\downarrow{\scriptstyle \mu_{\mathsf{IV}}} & & & & \downarrow{\scriptstyle F\lambda_{\mathsf{IV}}} \\
FX \otimes FY \otimes FZ & & \xrightarrow{\qquad\qquad \phi^{\mathbf{3}} \qquad\qquad} & & F(X \otimes Y \otimes Z) \\
\uparrow{\scriptstyle \mu_{\mathsf{VI}}} & & & & \uparrow{\scriptstyle F\lambda_{\mathsf{VI}}} \\
(FX \otimes FY) \otimes FZ & \xrightarrow{\phi^{\mathbf{2}} \otimes 1} & F(X \otimes Y) \otimes FZ & \xrightarrow{\phi^{\mathbf{2}}} & F((X \otimes Y) \otimes Z)
\end{array}$$

Particular cases of (1.2.6) for $f = \,.\mathsf{I}$ and $f = \mathsf{I}.$ give equations (1.2.5).

1.2.5. Definition. A *morphism of Monoidal functors*

$$t : (F, \phi^{\mathbf{n}}) \to (G, \psi^{\mathbf{n}}) : (\mathcal{C}, \otimes^{\mathbf{n}}, \lambda_f) \to (\mathcal{D}, \otimes^{\mathbf{n}}, \mu_f)$$

is a natural transformation $t : F \to G$ such that for every $n \geqslant 0$

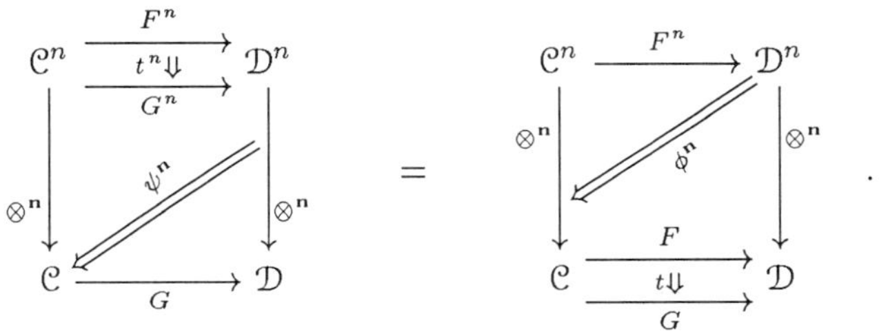

A Monoidal transformation $t : (F, \phi^{\mathbf{n}}) \to (G, \psi^{\mathbf{n}})$ defined above is, in particular, a monoidal transformation $t : (F, \phi^{\mathbf{2}}, \phi^{\mathbf{0}}) \to (G, \psi^{\mathbf{2}}, \psi^{\mathbf{0}})$.

1.2.6. Comparison of monoidal and Monoidal categories. It is clear how to define the composition of Monoidal functors. The class of Monoidal categories, Monoidal functors and their morphisms is a 2-category [**4, 12**] denoted Moncat. Summing up, we have a 2-functor from Moncat to the 2-category of monoidal categories, moncat. This 2-functor is a 2-equivalence as the following observation shows.

Each monoidal category is equivalent to a strict monoidal category (a, l, r are identities) [**21**]. We shall see that each Monoidal category is equivalent to a strict Monoidal category (λ_f are identities). Notice that strict Monoidal categories are in bijection with strict monoidal categories and Monoidal functors between strict monoidal functors are in bijection with monoidal functors between strict monoidal categories. Therefore, the 2-functor strict Moncat $\to$ strict moncat is an isomorphism. Hence, the 2-functor Moncat $\to$ moncat is a 2-equivalence. One of the quasi-inverse 2-functors is provided by the system of left parenthesized tensor products.

1.2.7. Theorem. *A Monoidal category* $(\mathcal{C}, \otimes^{\mathbf{n}}, \lambda_f)$ *is equivalent to a strict Monoidal category* $(\tilde{\mathcal{C}}, \tilde{\otimes}^{\mathbf{n}}, \mathrm{id})$, *where*

(a) *Objects of* $\tilde{\mathcal{C}}$ *are associative words in the alphabet* $\operatorname{Ob}\mathcal{C}$.

(b) *Morphisms of* $\tilde{\mathcal{C}}$ *are*

$$\tilde{\mathcal{C}}(X_1 \dots X_k, Y_1 \dots Y_m) = \mathcal{C}(X_1 \otimes \cdots \otimes X_k, Y_1 \otimes \cdots \otimes Y_m).$$

We denote $\otimes^{\mathbf{k}}(X_1 \dots X_k)$ *by* $X_1 \otimes \cdots \otimes X_k$.

(c) *The functor* $\tilde{\otimes}^{\mathbf{n}} : \tilde{\mathcal{C}}^n \to \tilde{\mathcal{C}}$ *is defined on objects as the concatenation of words.*

(d) *Given morphisms* $a^1 : X_1^1 \dots X_{k_1}^1 \to Y_1^1 \dots Y_{m_1}^1, \dots, a^n : X_1^n \dots X_{k_n}^n \to Y_1^n \dots Y_{m_n}^n$ *we define the morphism*

$$\tilde{\otimes}^{\mathbf{n}}(a^1, \dots, a^n) : X_1^1 \dots X_{k_1}^1 \dots X_1^n \dots X_{k_n}^n \to Y_1^1 \dots Y_{m_1}^1 \dots Y_1^n \dots Y_{m_n}^n \in \tilde{\mathcal{C}}$$

via the diagram

$$\begin{array}{ccc}
(X_1^1 \otimes \cdots \otimes X_{k_1}^1) \otimes \cdots \otimes (X_1^n \otimes \cdots \otimes X_{k_n}^n) & \xrightarrow{\otimes^{\mathbf{n}}(a^1, \dots, a^n)} & (Y_1^1 \otimes \cdots \otimes Y_{m_1}^1) \otimes \cdots \otimes (Y_1^n \otimes \cdots \otimes Y_{m_n}^n) \\
\lambda_f \Big\downarrow \wr & = & \wr \Big\downarrow \lambda_g \\
X_1^1 \otimes \cdots \otimes X_{k_1}^1 \otimes \cdots \otimes X_1^n \otimes \cdots \otimes X_{k_n}^n & \xrightarrow{\tilde{\otimes}^{\mathbf{n}}(a^1, \dots, a^n)} & Y_1^1 \otimes \cdots \otimes Y_{m_1}^1 \otimes \cdots \otimes Y_1^n \otimes \cdots \otimes Y_{m_n}^n
\end{array}$$

where f *(resp.* g*) corresponds to the partition* $(k_1, \dots, k_n)$ *of* $k_1 + \dots + k_n$ *(resp. partition* $(m_1, \dots, m_n)$ *of* $m_1 + \dots + m_n$*).*

PROOF. The essential part is to show that, indeed, $\tilde{\mathcal{C}}$ is a Monoidal category. The tensor multiplication $\tilde{\otimes}^I$ is strictly associative on objects. Let us show that it is strictly associative on morphisms.

Consider a map $h : J \to L$ from $\mathcal{O}$, $J = \mathbf{n}$, $L = \mathbf{p}$, corresponding to the partition $(n_1, \dots, n_p)$ of n. We have to show that

$$(\tilde{\mathcal{C}}^J \xrightarrow{\tilde{\otimes}^h_\bullet} \tilde{\mathcal{C}}^L \xrightarrow{\tilde{\otimes}^L} \tilde{\mathcal{C}}) = (\tilde{\mathcal{C}}^J \xrightarrow{\tilde{\otimes}^J} \tilde{\mathcal{C}}). \tag{1.2.7}$$

Apply these functors to n-tuple of morphisms of $\tilde{\mathcal{C}}$

$$a^1 : X_1^1 \dots X_{k_1}^1 \to Y_1^1 \dots Y_{m_1}^1, \dots, a^n : X_1^n \dots X_{k_n}^n \to Y_1^n \dots Y_{m_n}^n.$$

Denote $K \in \operatorname{Ob} \mathcal{O}$ (resp. $M \in \operatorname{Ob} \mathcal{O}$) the set of $k_1 + \dots + k_n$ (resp. $m_1 + \dots + m_n$) elements. Introduce a map $f : K \to J \in \mathcal{O}$ (resp. $g : M \to J \in \mathcal{O}$) corresponding to the partition $(k_1, \dots, k_n)$ (resp. $(m_1, \dots, m_n)$).

The left hand side of (1.2.7) gives a morphism $\tilde{\otimes}^{l \in L}(\tilde{\otimes}^{j \in h^{-1}l}(a^j))$ of $\tilde{\mathcal{C}}$ determined from the following diagram:

$$\begin{array}{ccc}
\otimes^{l \in L}(\otimes^{j \in h^{-1}l}(\otimes^{k \in f^{-1}j}(X_k^j))) & \xrightarrow{\otimes^{l \in L}(\otimes^{j \in h^{-1}l}(a^j))} & \otimes^{l \in L}(\otimes^{j \in h^{-1}l}(\otimes^{m \in g^{-1}j}(Y_m^j))) \\
\otimes^{l \in L}(\lambda_{f|_{f^{-1}h^{-1}l}}) \Big\downarrow & = & \Big\downarrow \otimes^{l \in L}(\lambda_{g|_{g^{-1}h^{-1}l}}) \\
\otimes^{l \in L}(\otimes^{j \in h^{-1}l, k \in f^{-1}j}(X_k^j)) & \xrightarrow{\otimes^{l \in L}(\tilde{\otimes}^{j \in h^{-1}l}(a^j))} & \otimes^{l \in L}(\otimes^{j \in h^{-1}l, m \in g^{-1}j}(Y_m^j)) \\
\lambda_{hf} \Big\downarrow & = & \Big\downarrow \lambda_{hg} \\
X_1^1 \otimes \cdots \otimes X_{k_1}^1 \otimes \cdots \otimes X_1^n \otimes \cdots \otimes X_{k_n}^n & \xrightarrow[\tilde{\otimes}^{l \in L}(\tilde{\otimes}^{j \in h^{-1}l}(a^j))]{} & Y_1^1 \otimes \cdots \otimes Y_{m_1}^1 \otimes \cdots \otimes Y_1^n \otimes \cdots \otimes Y_{m_n}^n
\end{array}$$

By axiom (1.2.3) the columns can be replaced with other columns:

$$\begin{array}{ccc}
\otimes^{l\in L}(\otimes^{j\in h^{-1}l}(\otimes^{k\in f^{-1}j}(X^j_k))) & \xrightarrow{\otimes^{l\in L}(\otimes^{j\in h^{-1}l}(a^j))} & \otimes^{l\in L}(\otimes^{j\in h^{-1}l}(\otimes^{m\in g^{-1}j}(Y^j_m))) \\
{\scriptstyle\lambda_h}\downarrow & = & \downarrow{\scriptstyle\lambda_h} \\
(X^1_1\otimes\cdots\otimes X^1_{k_1})\otimes\cdots\otimes(X^n_1\otimes\cdots\otimes X^n_{k_n}) & \xrightarrow{a^1\otimes\cdots\otimes a^n} & (Y^1_1\otimes\cdots\otimes Y^1_{m_1})\otimes\cdots\otimes(Y^n_1\otimes\cdots\otimes Y^n_{m_n}) \\
{\scriptstyle\lambda_f}\downarrow & = & \downarrow{\scriptstyle\lambda_g} \\
X^1_1\otimes\cdots\otimes X^1_{k_1}\otimes\cdots\otimes X^n_1\otimes\cdots\otimes X^n_{k_n} & \xrightarrow[\tilde{\otimes}^{l\in L}(\tilde{\otimes}^{j\in h^{-1}l}(a^j))]{\tilde{\otimes}^J(a^1,\dots,a^n)} & Y^1_1\otimes\cdots\otimes Y^1_{m_1}\otimes\cdots\otimes Y^n_1\otimes\cdots\otimes Y^n_{m_n}
\end{array}$$

The exterior of the diagram with the same upper and lower horizontal arrows still commutes. However, by naturality of λ_h it commutes also if we replace the lowest horizontal arrow with $\tilde{\otimes}^J(a^1,\dots,a^n)$ – the right hand side of (1.2.7). Hence,

$$\tilde{\otimes}^{l\in L}(\tilde{\otimes}^{j\in h^{-1}l}(a^j)) = \tilde{\otimes}^J(a^1,\dots,a^n),$$

and $\tilde{\mathcal{C}}$ is a strict Monoidal category.

Consider the functor $E:\mathcal{C}\to\tilde{\mathcal{C}}$, which to an object X of $\mathcal{C}$ assigns a 1-letter word X, and which is "identity" on morphisms. Introduce a natural isomorphism $\phi^{\mathbf{n}}$, which comes from $\mathrm{id}_{X_1\otimes\cdots\otimes X_n}\in\operatorname{Mor}\mathcal{C}$

$$\phi^{\mathbf{n}}:\tilde{\otimes}^{\mathbf{n}}\circ E^n\to E\circ\otimes^{\mathbf{n}}:\mathcal{C}^n\to\tilde{\mathcal{C}},$$

$$\phi^{\mathbf{n}}:X_1\dots X_n\to(X_1\otimes\cdots\otimes X_n)\in\operatorname{Mor}\tilde{\mathcal{C}}$$

with the 1-letter target word. Equation (1.2.6) for $\phi^{\mathbf{n}}$ reduces to $\lambda_f\circ\lambda_f^{-1}=1$. Therefore, $(E,\phi^{\mathbf{n}}):(\mathcal{C},\otimes^{\mathbf{n}},\lambda_f)\to(\tilde{\mathcal{C}},\tilde{\otimes}^{\mathbf{n}},\mathrm{id})$ is a Monoidal functor.

Consider the functor $P:\tilde{\mathcal{C}}\to\mathcal{C}$, $P(X_1\dots X_n)=X_1\otimes\cdots\otimes X_n$, "identity" on morphisms. Define a natural transformation

$$\psi^{\mathbf{n}}:\otimes^{\mathbf{n}}\circ P^n\to P\circ\tilde{\otimes}^{\mathbf{n}}:\tilde{\mathcal{C}}^n\to\mathcal{C}$$

on object $(X^1_1\dots X^1_{k_1},\dots,X^n_1\dots X^n_{k_n})$ as

$$\begin{aligned}\psi^{\mathbf{n}}=\lambda_h:(X^1_1\otimes\cdots\otimes X^1_{k_1})\otimes\cdots\otimes(X^n_1\otimes\cdots\otimes X^n_{k_n})&\\ \to X^1_1\otimes\cdots\otimes X^1_{k_1}\otimes\cdots\otimes X^n_1\otimes\cdots\otimes X^n_{k_n}&,\end{aligned}$$

where h is associated with the partition $(k_1,\dots,k_n)$. Equation (1.2.6) for ψ on the object $(X^1_1\dots X^1_{k_1},\dots,X^n_1\dots X^n_{k_n})$ coincides with equation (1.2.3) for λ written for maps $K\xrightarrow{h}I\xrightarrow{f}J$, where h is associated with the partition $(k_1,\dots,k_n)$. Therefore, $(P,\psi^{\mathbf{n}}):(\tilde{\mathcal{C}},\tilde{\otimes}^{\mathbf{n}},\mathrm{id})\to(\mathcal{C},\otimes^{\mathbf{n}},\lambda_f)$ is a Monoidal functor.

The composition of these Monoidal functors in one order is the identity functor:

$$(\mathcal{C}\xrightarrow{(E,\phi^{\mathbf{n}})}\tilde{\mathcal{C}}\xrightarrow{(P,\psi^{\mathbf{n}})}\mathcal{C})=(\mathrm{Id},\mathrm{id}).$$

Composition in the other order is

$$\tilde{\mathcal{C}}\xrightarrow{P}\mathcal{C}\xrightarrow{E}\tilde{\mathcal{C}},X_1\dots X_n\mapsto(X_1\otimes\cdots\otimes X_n)\text{ – 1-letter word.}$$

The isomorphism of functors

$$t:\mathrm{Id}\xrightarrow{\sim}E\circ P:\tilde{\mathcal{C}}\to\tilde{\mathcal{C}},$$

$$t_{X_1\dots X_n}:X_1\dots X_n\to(X_1\otimes\cdots\otimes X_n),$$

coming from $\mathrm{id}_{X_1\otimes\cdots\otimes X_n}$ is, in fact, an isomorphism of Monoidal functors. Therefore, E and P are Monoidal equivalences quasi-inverse to each other. □

1.2.8. Definition. A *rigid category* $\mathcal{C}$ is a monoidal category in which for every object $X \in \mathcal{C}$ its dual objects $X^\vee$, ${}^\vee X \in \mathcal{C}$ and morphisms of evaluation and coevaluation

$\mathrm{ev} : X \otimes X^\vee \to \mathbb{I} =$  , $\quad$ $\mathrm{ev} \;: {}^\vee X \otimes X \to \mathbb{I} =$  ,

$\mathrm{coev} : \mathbb{I} \to X^\vee \otimes X =$  , $\quad$ $\mathrm{coev}\colon \mathbb{I} \to X \otimes {}^\vee X =$  .

are chosen so that the composites

$$X \xrightarrow{r^{-1}} X \otimes \mathbb{I} \xrightarrow{1\otimes\mathrm{coev}} X \otimes (X^\vee \otimes X) \xrightarrow{a} (X \otimes X^\vee) \otimes X \xrightarrow{\mathrm{ev}\otimes 1} \mathbb{I} \otimes X \xrightarrow{l} X,$$

$$X \xrightarrow{l^{-1}} \mathbb{I} \otimes X \xrightarrow{\mathrm{coev}\otimes 1} (X \otimes {}^\vee X) \otimes X \xrightarrow{a^{-1}} X \otimes ({}^\vee X \otimes X) \xrightarrow{1\otimes\mathrm{ev}} X \otimes \mathbb{I} \xrightarrow{r} X,$$

$$X^\vee \xrightarrow{l^{-1}} \mathbb{I} \otimes X^\vee \xrightarrow{\mathrm{coev}\otimes 1} (X^\vee \otimes X) \otimes X^\vee \to \\ \xrightarrow{a^{-1}} X^\vee \otimes (X \otimes X^\vee) \xrightarrow{1\otimes\mathrm{ev}} X^\vee \otimes \mathbb{I} \xrightarrow{r} X^\vee,$$

$${}^\vee X \xrightarrow{r^{-1}} {}^\vee X \otimes \mathbb{I} \xrightarrow{1\otimes\mathrm{coev}} {}^\vee X \otimes (X \otimes {}^\vee X) \to \\ \xrightarrow{a} ({}^\vee X \otimes X) \otimes {}^\vee X \xrightarrow{\mathrm{ev}\otimes 1} \mathbb{I} \otimes {}^\vee X \xrightarrow{l} {}^\vee X$$

are all identity morphisms.

By the *opposite tensor product* we mean $\otimes_{\mathrm{op}} = \otimes \circ P : \mathcal{C} \times \mathcal{C} \to \mathcal{C}$, where $P : \mathcal{C} \times \mathcal{C} \to \mathcal{C} \times \mathcal{C}$, $(X, Y) \mapsto (Y, X)$ is the permutation functor.

In a rigid monoidal category $\mathcal{C}$ there is a pairing

$$(X \otimes Y) \otimes (Y^\vee \otimes X^\vee) \xrightarrow{\sim} (X \otimes (Y \otimes Y^\vee)) \otimes X^\vee \to \\ \xrightarrow{X\otimes\mathrm{ev}\otimes X^\vee} (X \otimes \mathbb{I}) \otimes X^\vee \xrightarrow{r\otimes X^\vee} X \otimes X^\vee \xrightarrow{\mathrm{ev}} \mathbb{I},$$

which induces an isomorphism $j_{+X,Y} : Y^\vee \otimes X^\vee \to (X \otimes Y)^\vee$, such that the above pairing coincides with

$$(X \otimes Y) \otimes (Y^\vee \otimes X^\vee) \xrightarrow{1\otimes j_+} (X \otimes Y) \otimes (X \otimes Y)^\vee \xrightarrow{\mathrm{ev}} \mathbb{I}.$$

The equation

$$\mathrm{coev}_{X\otimes Y} = (\mathbb{I} \xrightarrow{\mathrm{coev}_Y} Y^\vee \otimes Y \simeq Y^\vee \otimes \mathbb{I} \otimes Y \xrightarrow{1\otimes\mathrm{coev}_X\otimes 1} \\ Y^\vee \otimes X^\vee \otimes X \otimes Y \xrightarrow{j_+\otimes 1} (X \otimes Y)^\vee \otimes (X \otimes Y))$$

also holds. Similarly, there is an isomorphism $j_{-X,Y} : {}^\vee Y \otimes {}^\vee X \to {}^\vee(X \otimes Y)$.

There are unique isomorphisms $d : \mathbb{I} \to \mathbb{I}^\vee$, $d_- : \mathbb{I} \to {}^\vee\mathbb{I}$ such that

$$r_{\mathbb{I}} = (\mathbb{I} \otimes \mathbb{I} \xrightarrow{\mathbb{I}\otimes d} \mathbb{I} \otimes \mathbb{I}^\vee \xrightarrow{\mathrm{ev}} \mathbb{I}),$$

$$r_{\mathbb{I}} = (\mathbb{I} \otimes \mathbb{I} \xrightarrow{d_-\otimes\mathbb{I}} {}^\vee\mathbb{I} \otimes \mathbb{I} \xrightarrow{\mathrm{ev}} \mathbb{I}).$$

One can easily prove that

$$(\text{-}^\vee, j_+, d) : (\mathcal{C}^{\mathrm{op}}, \otimes, \mathbb{I}) \to (\mathcal{C}, \otimes_{\mathrm{op}}, \mathbb{I}), \quad X \mapsto X^\vee, \quad f \mapsto f^t,$$
$$({}^\vee\text{-}, j_-, d_-) : (\mathcal{C}^{\mathrm{op}}, \otimes, \mathbb{I}) \to (\mathcal{C}, \otimes_{\mathrm{op}}, \mathbb{I}), \quad X \mapsto {}^\vee X, \quad f \mapsto {}^t f.$$

are monoidal equivalences of categories. The "square" of the first one

$$(\text{-}^{\vee\vee}, j_{+2}, d_2) : (\mathcal{C}, \otimes, \mathbb{I}) \to (\mathcal{C}, \otimes, \mathbb{I}), \ X \mapsto X^{\vee\vee}, \ f \mapsto f^{tt},$$

$$j_{+2X,Y} = \left(X^{\vee\vee} \otimes Y^{\vee\vee} \xrightarrow{j_+} (Y^\vee \otimes X^\vee)^\vee \xrightarrow{j_+^{-1t}} (X \otimes Y)^{\vee\vee}\right),$$
$$d_2 = \left(\mathbb{I} \xrightarrow{d} \mathbb{I}^\vee \xrightarrow{d^{-1t}} \mathbb{I}^{\vee\vee}\right),$$

is a monoidal self-equivalence of $\mathcal{C}$.

There are canonical isomorphisms

$$X \to {}^\vee(X^\vee), \qquad X \to ({}^\vee X)^\vee.$$

To simplify notations we assume the functors $\text{-}^\vee$ and ${}^\vee\text{-}$ inverse to each other (we can always achieve this replacing the category by an equivalent one). We shall denote the iterated duals by $X^{(n\vee)} = X^{\vee\cdots\vee}$ (n times) and $X^{(-n\vee)} = {}^{\vee\cdots\vee}X$ (n times) for $n \geqslant 0$.

1.2.9. Definition. A *braided category* $(\mathcal{C}, c)$ is a monoidal category $\mathcal{C}$ equipped with a functorial isomorphism $c = c_{X,Y} : X \otimes Y \to Y \otimes X$ – the braiding, or the commutativity constraint – such that the two hexagons commute

$$\begin{array}{ccccc} X \otimes (Y \otimes Z) & \xrightarrow{1 \otimes c^{\pm 1}} & X \otimes (Z \otimes Y) & \xrightarrow{a} & (X \otimes Z) \otimes Y \\ {\scriptstyle a}\downarrow & & & & \downarrow{\scriptstyle c^{\pm 1} \otimes 1} \\ (X \otimes Y) \otimes Z & \xrightarrow{c^{\pm 1}} & Z \otimes (X \otimes Y) & \xrightarrow{a} & (Z \otimes X) \otimes Y \end{array} \tag{1.2.8}$$

(one for c and one for c^{-1}).

The graphical notation for the braiding and its inverse is

$c = (c_{X,Y} : X \otimes Y \to Y \otimes X) =$ 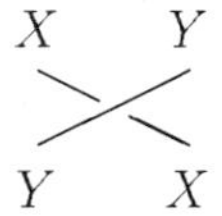

, $c^{-1} =$ 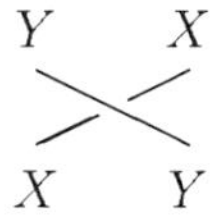

.

In a braided category the following equations hold [27]

$$c_{X,\mathbb{I}} = \left(X \otimes \mathbb{I} \xrightarrow{r} X \xrightarrow{l^{-1}} \mathbb{I} \otimes X\right),$$
$$c_{\mathbb{I},X} = \left(\mathbb{I} \otimes X \xrightarrow{l} X \xrightarrow{r^{-1}} X \otimes \mathbb{I}\right).$$

In a rigid monoidal braided category there are functorial isomorphisms

$u_1^2 =$ 

, $u_{-1}^2 =$ 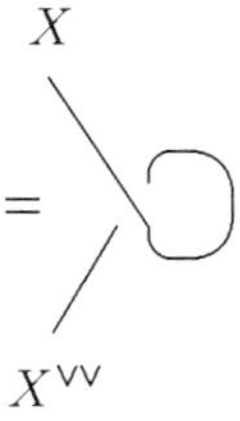

, $u_1^{-2} =$ 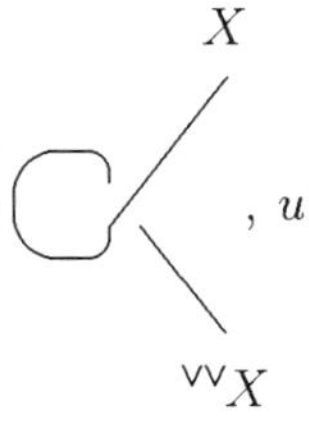

, $u_{-1}^{-2} =$ 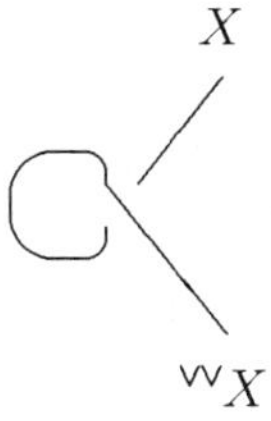

1.2.10. Definition. A *ribbon category* is a rigid braided category equipped with a functorial endomorphism $\nu = \nu_X : X \to X$, the ribbon twist, which is self-adjoint, that is, $\nu_{X^\vee} = \nu_X^t : X^\vee \to X^\vee$ and satisfies $\nu_{\mathbb{I}} = \mathrm{id}_{\mathbb{I}}$ and

$$\nu_{X\otimes Y} = \bigl(X \otimes Y \xrightarrow{c^2} X \otimes Y \xrightarrow{\nu_X \otimes \nu_Y} X \otimes Y\bigr).$$

1.2.11. Proposition. $\nu^2 = u_1^2 \circ u_{\bar{1}}^{-2}$.

PROOF. Since $\nu_{\mathbb{I}} = \mathrm{id}_{\mathbb{I}}$, the commutative diagram

$$\begin{array}{ccccc} X \otimes X^\vee & \xrightarrow{c^{-2}} & X \otimes X^\vee & \xrightarrow{\mathrm{ev}} & \mathbb{I} \\ \Big\| & & \Big\downarrow{\scriptstyle \nu_{X\otimes X^\vee}} & & \Big\downarrow{\scriptstyle \nu_{\mathbb{I}}} \\ X \otimes X^\vee & \xrightarrow{\nu_X \otimes \nu_{X^\vee}} & X \otimes X^\vee & \xrightarrow{\mathrm{ev}} & \mathbb{I} \end{array}$$

implies that

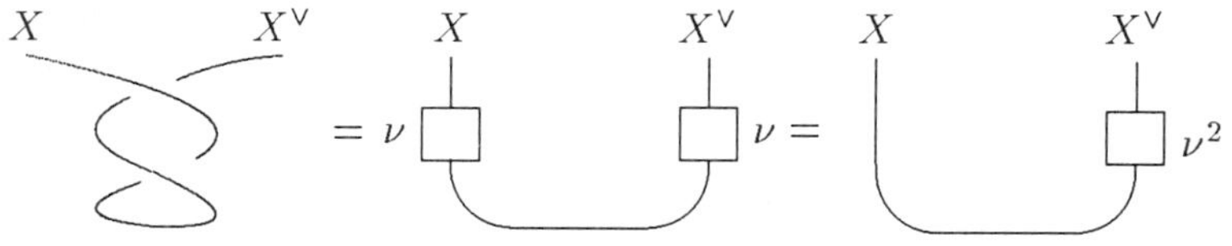

Hence, $\nu_{X^\vee}^2 = u_1^2 \circ u_{\bar{1}}^{-2} : X^\vee \to X^\vee$. □

1.2.12. Corollary. *The ribbon twist* ν *is invertible.*

We consider also the functorial isomorphisms

$$u_0^2 = u_1^2 \circ \nu^{-1} = u_{-1}^2 \circ \nu : X \to X^{\vee\vee}, \quad u_{\bar{0}}^{-2} = u_{\bar{1}}^{-2} \circ \nu^{-1} = u_{-\bar{1}}^{-2} \circ \nu : X \to {}^{\vee\vee}X.$$

1.2.13. Definition. A *symmetric category* is a braided category $(\mathcal{C}, c)$ for which the braiding satisfies the equation

$$\bigl(X \otimes Y \xrightarrow{c_{X,Y}} Y \otimes X \xrightarrow{c_{Y,X}} X \otimes Y\bigr) = \mathrm{id}_{X\otimes Y}$$

for all objects X, Y of $\mathcal{C}$.

Deligne and Milne [9] define symmetric categories differently. Their definition practically coincides with the following.

Let $\mathcal{S}$ be a category of finite sets, $\mathrm{Ob}\,\mathcal{S} = \mathrm{Ob}\,\mathcal{O} = \{\mathbf{n}\}_{n\geqslant 0}$,

$$\mathrm{Mor}(\mathbf{n}, \mathbf{m}) = \{\text{ all maps } \mathbf{n} \to \mathbf{m} \text{ (ignoring the ordering) }\}.$$

The notation $f^{-1}(j)$ for a map $f : I \to J$ of $\mathcal{S}$ has the same meaning as for $\mathcal{O}$. In particular, $e_j : f^{-1}(j) \hookrightarrow I$ is monotonic. With this extra condition pull-back (1.2.2) is still unique.

1.2.14. Definition. A *symmetric Monoidal category* $(\mathcal{C}, \otimes^I, \lambda_f)_{f\in\mathcal{S}}$ is defined exactly in the same way as Monoidal category (see Definition 1.2.2) with the only difference that $\mathcal{O}$ is replaced with $\mathcal{S}$.

For symmetric Monoidal categories the family of isomorphisms λ_f is larger than for Monoidal categories. The family of equations which need to be satisfied is also larger. The obvious inclusion $\mathcal{O} \to \mathcal{S}$ implies that each symmetric Monoidal category is, in particular, Monoidal.

A symmetric Monoidal category $(\mathcal{C}, \otimes^I, \lambda_f)_{f \in \mathcal{S}}$ produces a symmetric monoidal category $(\mathcal{C}, \otimes, \mathbb{1}, a, r, l, c)$, where

$$c = \lambda_{\mathsf{X}}^{-1}, \qquad \lambda_{\mathsf{X}} : \otimes \circ P \to \otimes : \mathcal{C}^2 \to \mathcal{C},$$
$$\mathsf{X} : \mathbf{2} \to \mathbf{2}, \qquad \mathsf{X}(1) = 2, \qquad \mathsf{X}(2) = 1.$$

Note that $\otimes_*^{\mathsf{X}} = P : \mathcal{C}^2 \to \mathcal{C}^2$ is the permutation functor. Let us check the axioms for the braiding. Equation (1.2.3) for the pair of maps $\mathbf{2} \xrightarrow{\mathsf{X}} \mathbf{2} \xrightarrow{\mathsf{X}} \mathbf{2}$ implies

$$\bigl(A \otimes B \xrightarrow{(\lambda_{\mathsf{X}})_{B,A}} B \otimes A \xrightarrow{(\lambda_{\mathsf{X}})_{A,B}} A \otimes B\bigr) = \mathrm{id}_{A \otimes B}$$

for objects A, B of $\mathcal{C}$. Proof of the hexagon equation (1.2.8) is given in diagram (1.2.10).

Definition of a *symmetric Monoidal functor* coincides formally with definition of a Monoidal functor 1.2.4 with the only difference that $f : I \to J$ is taken from $\mathcal{S}$, and not only from $\mathcal{O}$. Definition of morphisms of Monoidal functors 1.2.5 is used also for morphisms of symmetric Monoidal functors.

1.2.15. Proposition. *The 2-categories of symmetric strict Monoidal categories and of symmetric strict monoidal categories are isomorphic.*

PROOF. First we prove that all structure morphisms λ_f of a symmetric strict Monoidal category are expressed through the inverse braiding λ_{X}. Indeed, any map $f : I \to J$ from $\mathcal{S}$ can be represented uniquely as a composition

$$f = \bigl(I \xrightarrow{\sigma(f)} I \xrightarrow{\pi(f)} J\bigr), \tag{1.2.9}$$

where $\pi(f)$ is monotonic, $\sigma(f)$ is bijective and $\sigma(f)|_{f^{-1}j} : f^{-1}j \to I$ is monotonic for all $j \in J$. The map $\pi(f)$ corresponds to the partition $(\#f^{-1}(1), \dots, \#f^{-1}(m))$ of $\#I$, where $J = \mathbf{m}$. Axiom (1.2.3) applied to (1.2.9)

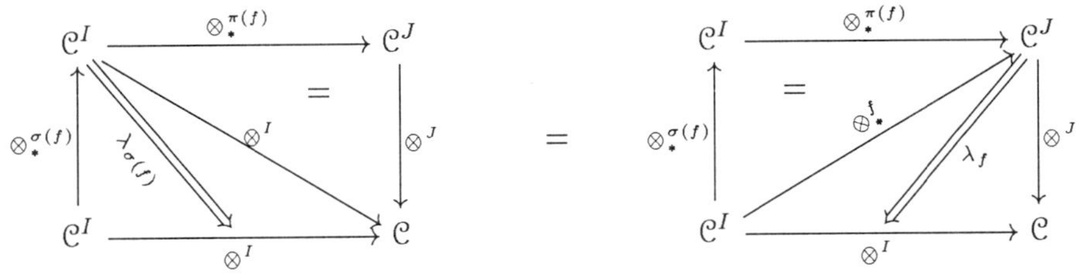

shows that $\lambda_f = \lambda_{\sigma(f)}$.

We look now at permutations $f, g : I \to I$ and corresponding permutation functors $\otimes_*^f, \otimes_*^g : \mathcal{C}^I \to \mathcal{C}^I$. For them axiom (1.2.3) implies

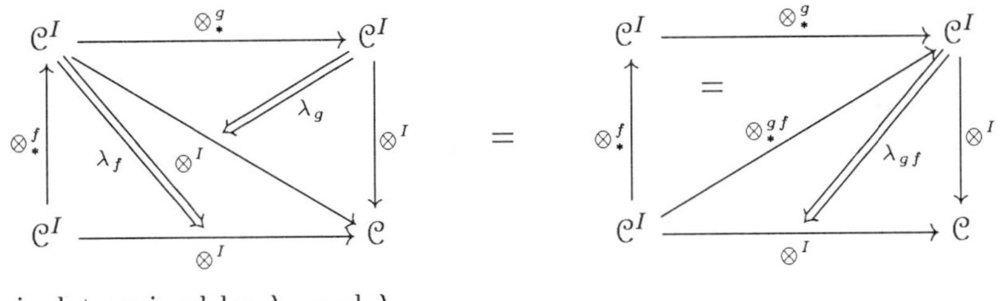

so λ_{gf} is determined by λ_g and λ_f.

Suppose that intervals $I = \sqcup_{k=1}^p I_k$, $J = \sqcup_{k=1}^p J_k \in \operatorname{Ob} \mathcal{S}$ are disjoint unions of intervals I_k, J_k, where $I_k < I_{k+1}$, $J_k < J_{k+1}$. Consider arbitrary maps $f_k : I_k \to$

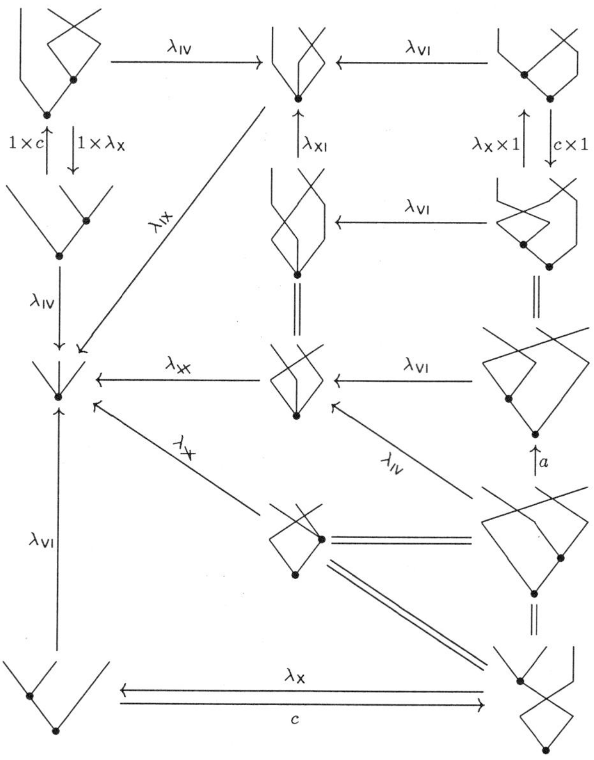

(1.2.10)

$J_k \in \mathcal{S}$ and their union $f = \sqcup_{k=1}^{p} f_k : I \to J$. Axiom (1.2.3) applied to f and $g : J \to \mathbf{p}$, $g(J_k) = k$ gives

$$\lambda_f = \otimes^{\mathbf{P}}(\lambda_{f_1}, \dots, \lambda_{f_p}) : \otimes^J \circ \otimes_*^f \to \otimes^I, \tag{1.2.11}$$

since $gf = g$ is monotonic.

Summing up, any λ_h is a product of some $1 \otimes \lambda_{\mathsf{X}} \otimes 1$, and the 2-functor Ψ : symmetric strict Moncat $\to$ symmetric strict moncat is injective on objects.

Now we extend a given symmetric strict monoidal structure on $\mathcal{C}$ to a symmetric strict Monoidal structure. We already have the strict Monoidal structure, in particular, $\otimes_*^f = \otimes_*^{\pi(f)} \circ \otimes_*^{\sigma(f)}$ for each $f \in \operatorname{Mor} \mathcal{S}$. Using generators $1 \otimes \lambda_{\mathsf{X}} \otimes 1 = 1 \otimes c^{-1} \otimes 1$ of the symmetric group we construct the isomorphisms λ_s for permutations $s : I \to I$, and we set $\lambda_f = \lambda_{\sigma(f)}$ for arbitrary $f \in \operatorname{Mor} \mathcal{S}$. It remains to prove equation (1.2.3). We shall do it first in particular cases.

A) Equation (1.2.3) holds for bijective f and g.

B) Assume that $f : J \to J$ is bijective and $g : J \to K$ is monotonic. Then there exists a permutation $r = \sigma(gf) : J \to J$ such that

$$(J \xrightarrow{f} J \xrightarrow{g} K) = (J \xrightarrow{r} J \xrightarrow{g} K)$$

and $r|_{f^{-1}g^{-1}k} : f^{-1}g^{-1}k \to J$ is monotonic for all $k \in K$. Denote $u = f \circ r^{-1} : J \to J$. Then $u(g^{-1}k) = g^{-1}k$ for all $k \in K$. The right hand side of (1.2.3) is

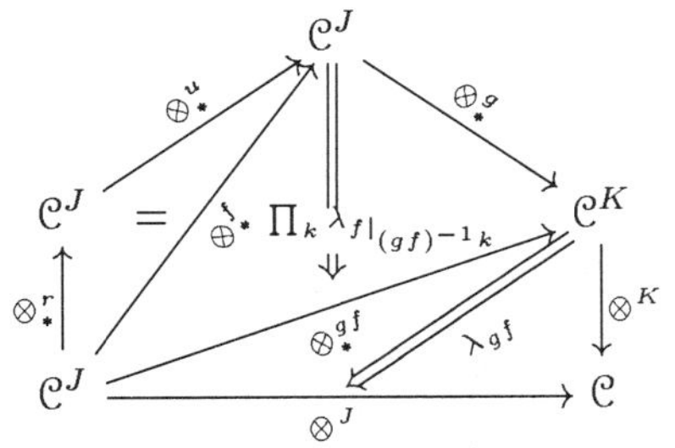

$\stackrel{A)}{=}$

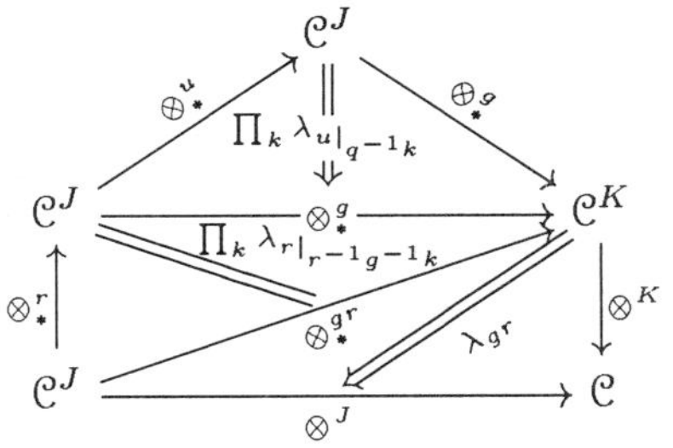

$\stackrel{\text{def}}{=}$

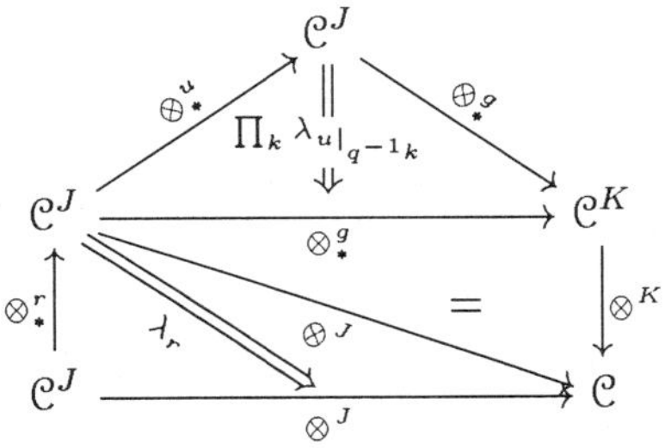

$=$

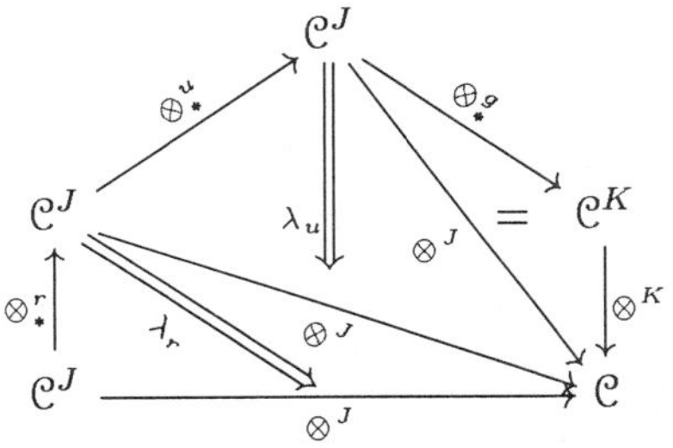

by property (1.2.11). By A) the last isomorphism equals to the left hand side of (1.2.3).

C) Assume that $f : J \to J$ is bijective, $g : J \to K$ is arbitrary. Decompose g into $t = \sigma(g) : J \to J$ and $q = \pi(g) : J \to K$. Denote $h = t \circ f : J \to J$, then $q \circ h = g \circ f$. Applying B) to h and q we get

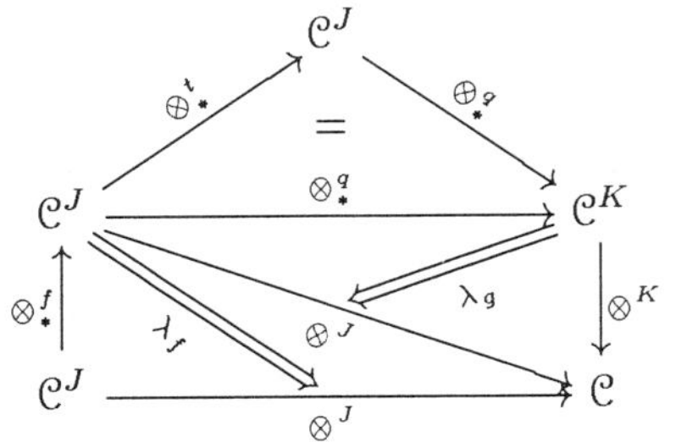

$\stackrel{\text{def}}{=}$

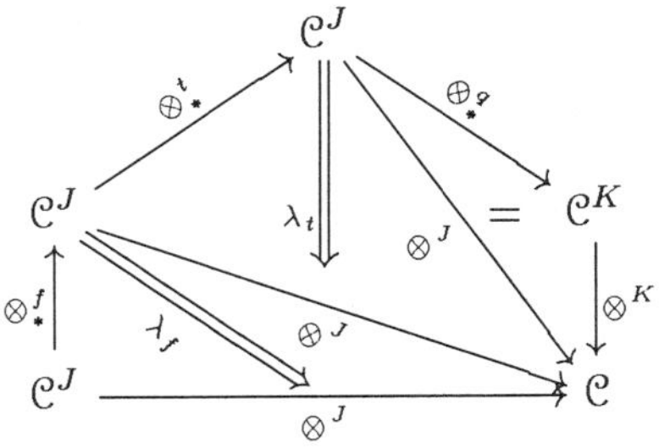

$\stackrel{A)}{=}$

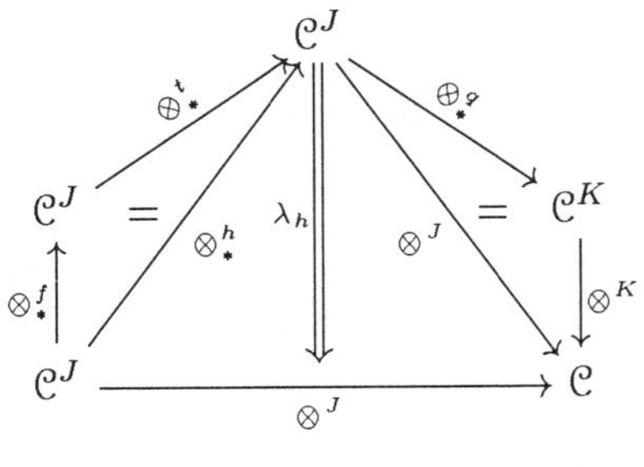

$\stackrel{B)}{=}$

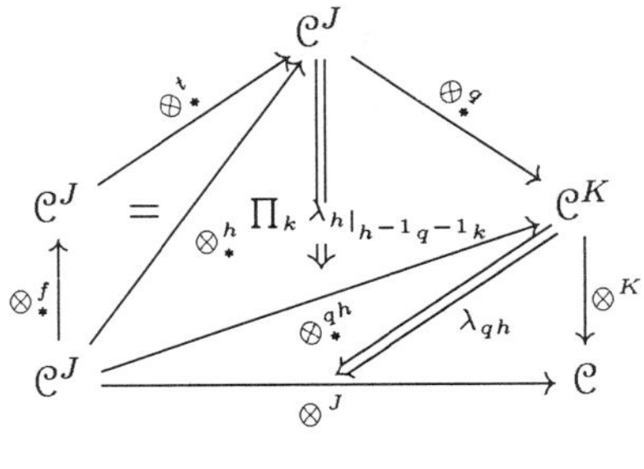

$\overset{\text{def}}{=}$

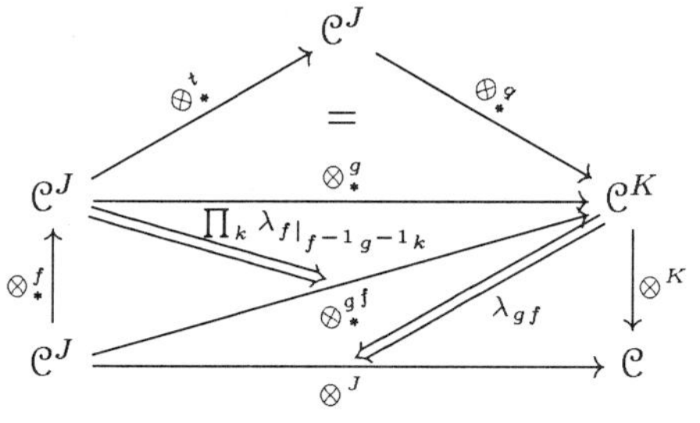

since in the decomposition

$$h|_{h^{-1}q^{-1}k} : f^{-1}t^{-1}q^{-1}k \xrightarrow[\sim]{f|} t^{-1}q^{-1}k \xrightarrow[\sim]{t|} q^{-1}k$$

the map $t|_{t^{-1}q^{-1}k}$ is a monotonic bijection, that is an identity in our conventions.

D) Assume that $f : I \to J$ is monotonic, $g : J \to K$ is arbitrary. Set $t = \sigma(g) : J \to J$, $q = \pi(g) : J \to K$. Consider also $u = \sigma(t \circ f) : I \to I$, $r = \pi(t \circ f) : I \to J$. Then by definition

$$(I \xrightarrow{f} J \xrightarrow[\sim]{t} J) = (I \xrightarrow[\sim]{u} I \xrightarrow{r} J),$$

and $u|_{u^{-1}r^{-1}j} : u^{-1}r^{-1}j \to I$ is monotonic for all $j \in J$.

Consider $j', j'' \in q^{-1}k$, $j' < j''$. Then $t^{-1}j' < t^{-1}j''$, since $t|_{q^{-1}k}$ is monotonic. Therefore, $f^{-1}t^{-1}j' < f^{-1}t^{-1}j''$. Also $u(f^{-1}t^{-1}j') = r^{-1}j' < r^{-1}j'' = u(f^{-1}t^{-1}j'')$. Thus u maps an interval $f^{-1}t^{-1}j$ to an interval $r^{-1}j$ preserving the order of intervals, when j runs over $q^{-1}k$. Since the restriction of u to $f^{-1}t^{-1}j$ is monotonic, we deduce that $u|_{f^{-1}t^{-1}q^{-1}k}$ is monotonic. Hence, $u = \sigma(g \circ f) : I \to I$, $q \circ r = \pi(g \circ f) : I \to K$.

The left hand side of (1.2.3) is

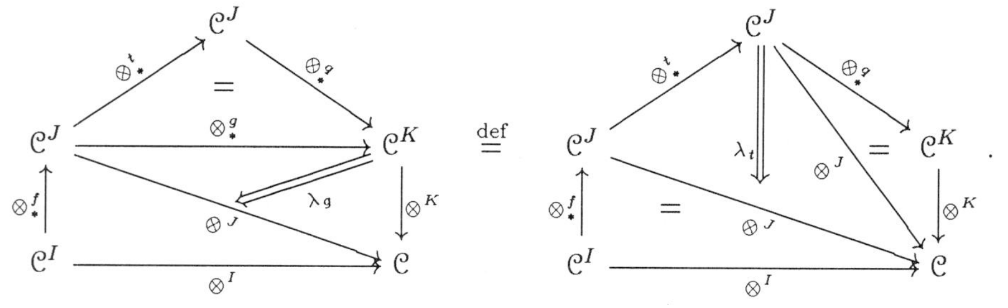

The right hand side of (1.2.3) is

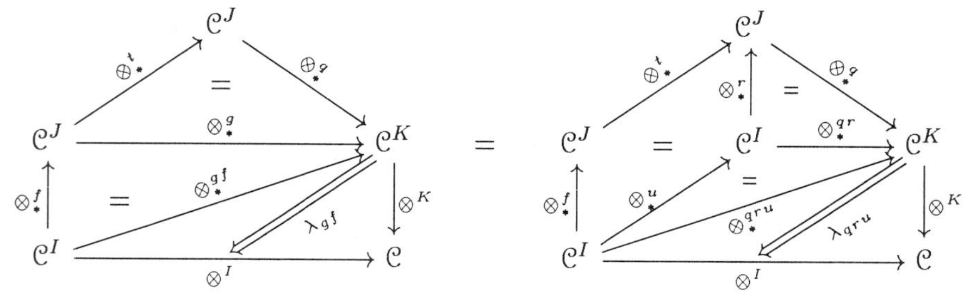

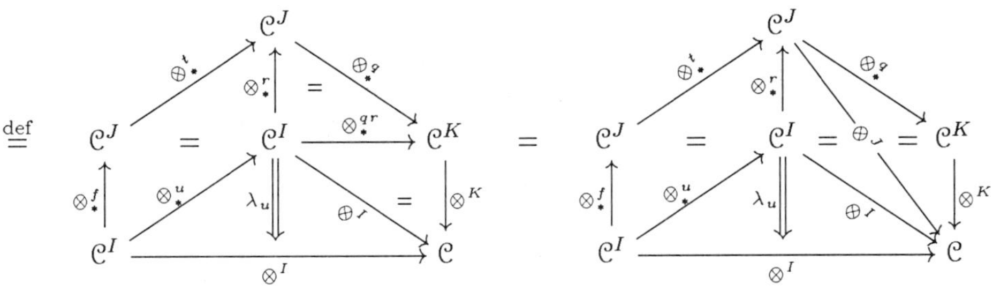

Therefore, equation (1.2.3) follows from the equation

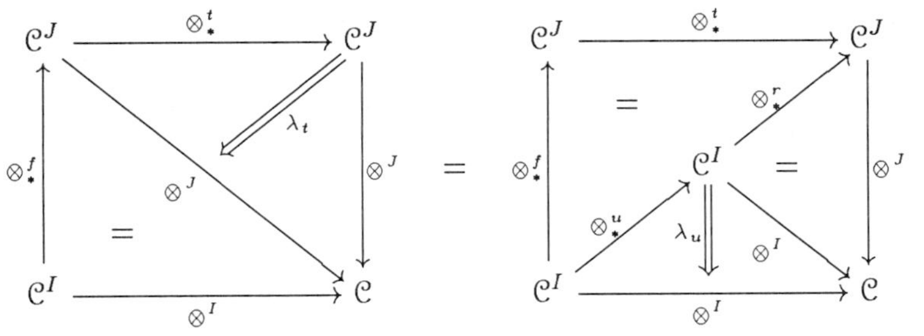

(1.2.12)

in which q and K do not appear. This equation is proved directly in a symmetric strict monoidal category. This is a formalization of the fact that braiding of tensor products can be performed consecutively. It follows from hexagon axiom (1.2.8).

E) Consider arbitrary $f : I \to J$, $g : J \to K$ from $\mathcal{S}$. Denote $s = \sigma(f) : I \to I$, $p = \pi(f) : I \to J$. The right hand side of (1.2.3) is

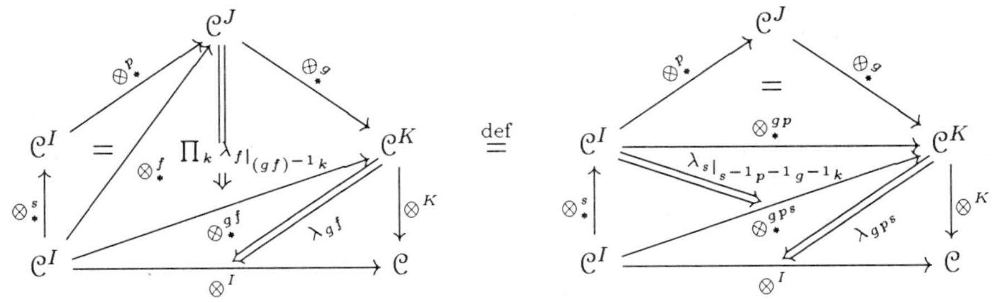

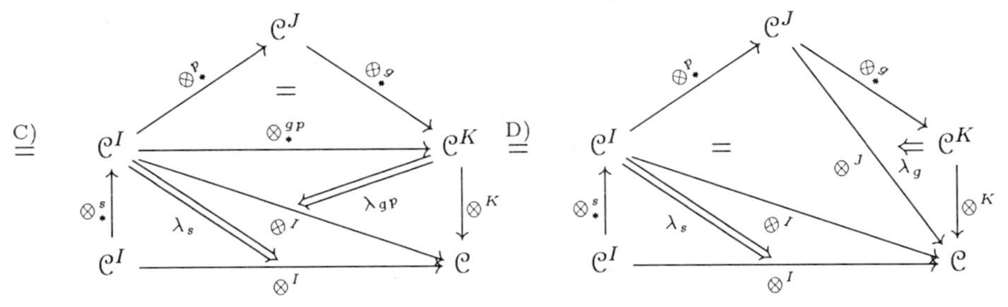

$\stackrel{\text{def}}{=}$

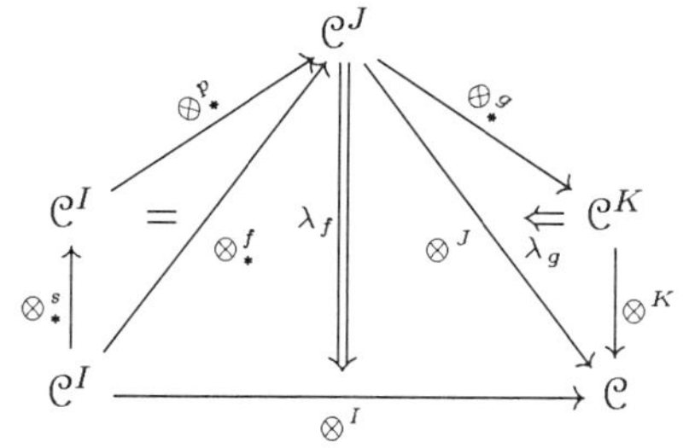

which is the left hand side of (1.2.3).

Therefore, a symmetric strict monoidal structure extends to a symmetric strict Monoidal one, and the 2-functor Ψ is bijective on objects. $\square$

The origin of the following definition of braided Monoidal categories is in the work [**8**] by Deligne. He uses a presentation of the braid group B_n via generators $\sigma(w)$, $w \in \mathfrak{S}_n$, which obey relations $\sigma(w'w'') = \sigma(w')\sigma(w'')$ if length of $w'w''$ is the sum of lengths of w' and of w''.

1.2.16. Definition. A *braided Monoidal category* $(\mathcal{C}, \otimes^I, \lambda_f)_{f \in \mathcal{S}}$ is defined similarly to a Monoidal category (see Definition 1.2.2), where $\mathcal{O}$ is replaced with $\mathcal{S}$ and equation (1.2.3) is imposed on λ for any pair of composable maps $f : I \to J$, $g : J \to K$ from $\mathcal{S}$ satisfying the following property †:

Property † for any pair of elements $a, b \in I$ inequalities $a < b$ and $f(a) > f(b)$ imply $gf(a) \geqslant gf(b)$.

A braided Monoidal category $(\mathcal{C}, \otimes^I, \lambda_f)_{f \in \mathcal{S}}$ produces a braided monoidal category $(\mathcal{C}, \otimes, \mathbb{1}, a, l, r, c)$, where $c = \lambda_{\mathsf{X}}^{-1}$. Diagram (1.2.10) proves one of the hexagons, its mirror image reversing left and right sides proves the other hexagon.

1.2.17. Proposition. *The 2-categories of braided strict Monoidal categories and of braided strict monoidal categories are isomorphic.*

Proof. Repeat the proof of Proposition 1.2.15 replacing "symmetric" with "braided". We have to check additionally condition †, when appropriate. The paragraphs A)–E) below describe statements, which need to be added to parts A)–E) of the proof of Proposition 1.2.15.

A) Note that for bijective $f, g : \mathbf{n} \to \mathbf{n}$ condition † is equivalent to the condition: length of gf is the sum of lengths of f and of g. The transformation λ_f^{-1} is defined through an arbitrary reduced expression $s_{i_1} \dots s_{i_m} = f$, where elementary transpositions s_i are replaced with $1^{\otimes i-1} \otimes \lambda_{\mathsf{X}}^{-1} \otimes 1^{\otimes n-i-1}$. This makes (1.2.3) obvious.

B) We have to check property † for the pair (r, u). Indeed, $a < b$, $r(a) > r(b)$ for $a, b \in J$ imply $gur(a) > gur(b)$, hence, $ur(a) > ur(b)$.

C) The property † for a pair (f, g) implies property † for the pair (f, t). Indeed, $a < b$, $f(a) > f(b)$ for $a, b \in J$ imply $gf(a) \geqslant gf(b)$, that is, $qtf(a) \geqslant qtf(b)$. If $qtf(a) > qtf(b)$, then $tf(a) > tf(b)$. If $qtf(a) = qtf(b) = k$, then $f(a), f(b) \in t^{-1}q^{-1}k$. Since $t|_{t^{-1}q^{-1}k}$ is monotonic, $f(a) > f(b)$ implies $tf(a) > tf(b)$.

D) Equation (1.2.12) holds in braided strict monoidal categories.

E) The property † for a pair (f, g) implies property † for the pair (s, pg). Indeed, $a < b$, $s(a) > s(b)$ for $a, b \in I$ imply $f(a) \geqslant f(b)$. If $f(a) = f(b)$, then $gf(a) = gf(b)$. If $f(a) > f(b)$, then $gf(a) \geqslant gf(b)$ by †. $\square$

1.3. Monoidal abelian categories

Let $\mathcal{V} = (\mathcal{V}, \otimes, a, \mathbb{I}, r, l)$ be a monoidal abelian category. Then it has a canonical $\Bbbk$-linear structure, where $\Bbbk = \operatorname{End}_{\mathcal{V}} \mathbb{I}$ is a commutative ring, $\mathbb{I}$ is the unit object. For $\lambda \in \operatorname{End}_{\mathcal{V}} \mathbb{I}$, $f \in \operatorname{Hom}_{\mathcal{V}}(X, Y)$ the morphism λf is defined as $(X \xrightarrow{l^{-1}} \mathbb{I} \otimes X \xrightarrow{\lambda \otimes f} \mathbb{I} \otimes Y \xrightarrow{l} Y)$. Assume that objects of $\mathcal{V}$ have finite length and $\mathcal{V}$ is rigid.

1.3.1. Proposition (cf. Deligne and Milne [9]). *Let $\mathcal{V}$ be an abelian rigid monoidal category whose objects have finite length. Then the unit object is decomposed as $\mathbb{I} = S_1 \oplus S_2 \oplus \cdots \oplus S_r$, where S_i are pairwise non-isomorphic simple objects, such that*

$$S_i^\vee \simeq S_i \simeq {}^\vee S_i, \quad S_i \otimes S_i \simeq S_i, \quad S_i \otimes S_j = 0 \quad \text{for } i \neq j.$$

1.3.2. Corollary. *In hypotheses of Proposition 1.3.1*

$$\Bbbk = \operatorname{End} \mathbb{I} \simeq \prod_{i=1}^{r} \operatorname{End} S_i = \prod_{i=1}^{r} \Bbbk_i$$

is a product of fields.

Therefore, if $\Bbbk$ is a field, then $\mathbb{I}$ is simple.

We assume more: $\mathcal{V}$ is a rigid monoidal abelian category with length and $\Bbbk$ is a perfect field.

The tensor product functor $X \otimes$ - (resp. - $\otimes X$) has a right adjoint $X^\vee \otimes$ - (resp. - $\otimes {}^\vee X$) and a left adjoint ${}^\vee X \otimes$ - (resp. - $\otimes X^\vee$). Therefore, the functor $\otimes : \mathcal{V} \times \mathcal{V} \to \mathcal{V}$ is exact in each variable and it is $\Bbbk$-bilinear by the choice of $\Bbbk$-linear structure. By Proposition 1.1.5 there exists a $\Bbbk$-linear exact functor $\circledast : \mathcal{V} \boxtimes \mathcal{V} \to \mathcal{V}$ called the diagonal restriction functor, such that $\otimes$ is isomorphic to the composite

$$\mathcal{V} \times \mathcal{V} \xrightarrow{\boxtimes} \mathcal{V} \boxtimes \mathcal{V} \xrightarrow{\circledast} \mathcal{V}.$$

The functors $X \mapsto X^\vee$, $X \mapsto {}^\vee X$, quasiinverse to each other, are also exact.

1.3.3. Proposition. *The functor $\circledast : \mathcal{V} \boxtimes \mathcal{V} \to \mathcal{V}$ is faithful.*

PROOF. Assume the contrary: for some non-zero morphism $f \in \mathcal{V} \boxtimes \mathcal{V}$ we have $\circledast f = 0$. Since $\circledast$ is exact, $\operatorname{Im} f \neq 0$, but $\circledast \operatorname{Im} f = 0$. So there is a simple object $S \in \mathcal{V} \boxtimes \mathcal{V}$ such that $\circledast S = 0$. There exist simple objects $S_1, S_2 \in \mathcal{V}$ such that $S_1 \boxtimes S_2 \simeq S \oplus K$ for some $K \in \mathcal{V} \boxtimes \mathcal{V}$ by Proposition 1.1.9. Therefore, there is a nontrivial idempotent in the algebra $\operatorname{End}(S_1 \boxtimes S_2)$, whose image under $\circledast$ vanishes. However, the composite

$$\phi = \big(\operatorname{End} S_1 \otimes_{\Bbbk} \operatorname{End} S_2 \overset{\boxtimes}{\simeq} \operatorname{End}(S_1 \boxtimes S_2) \xrightarrow{\circledast} \operatorname{End}(S_1 \otimes S_2)\big)$$

is simply the tensor product of morphisms. All we have to do is to show that ϕ is injective.

Using the isomorphisms $\operatorname{End} S_1 \simeq \operatorname{Hom}(\mathbb{I}, S_1^\vee \otimes S_1)$, $\operatorname{End} S_2 \simeq \operatorname{Hom}(S_2 \otimes S_2^\vee, \mathbb{I})$, $\operatorname{End}(S_1 \otimes S_2) \simeq \operatorname{Hom}(S_2 \otimes S_2^\vee, S_1^\vee \otimes S_1)$ we transform ϕ to the isomorphic map – the composition:

$$\circ : \operatorname{Hom}(\mathbb{I}, S_1^\vee \otimes S_1) \otimes_{\Bbbk} \operatorname{Hom}(S_2 \otimes S_2^\vee, \mathbb{I}) \to \operatorname{Hom}(S_2 \otimes S_2^\vee, S_1^\vee \otimes S_1)$$

Denote $X = S_2 \otimes S_2^\vee$, $Y = S_1^\vee \otimes S_1$. It suffices to restrict the category $\mathcal{V}$ to $\langle X \oplus Y \oplus \mathbb{I} \rangle$. Therefore, we may assume that $X, Y, \mathbb{I} \in$ mod-A for some finite dimensional $\Bbbk$-algebra A, and $\mathbb{I}$ is a simple module such that $\operatorname{End}_A \mathbb{I} = \Bbbk$. We have to prove that

$$\circ : \operatorname{Hom}_A(\mathbb{I}, Y) \otimes_\Bbbk \operatorname{Hom}_A(X, \mathbb{I}) \to \operatorname{Hom}_A(X, Y) \tag{1.3.1}$$

is injective. This map decomposes as

$$\operatorname{Hom}(\mathbb{I}, \operatorname{soc} Y) \otimes_\Bbbk \operatorname{Hom}(X/R_X, \mathbb{I}) \xrightarrow{\circ} \operatorname{Hom}(X/R_X, \operatorname{soc} Y) \hookrightarrow \operatorname{Hom}(X, Y),$$

where R_X is the radical of X, and $\operatorname{soc} Y$ is the socle of Y. So we have to consider only the case of semisimple X, Y. Moreover, we have to look only at the isotypic component of the type $\mathbb{I}$. In the remaining case, $X = \mathbb{I}^n$, $Y = \mathbb{I}^m$, we have $\operatorname{Hom}(\mathbb{I}, Y) = \Bbbk^m$, $\operatorname{Hom}(X, \mathbb{I}) = \Bbbk^n$, $\operatorname{Hom}(X, Y) \simeq \Bbbk^{mn}$ and the map (1.3.1) is a bijection. □

1.4. Symmetric monoidal 2-category of abelian categories

This subject is fully developed in Appendix A. Here we shall only name the main properties of objects we deal with. The 2-category $\mathfrak{Ab}^l_\Bbbk$ (resp. $\mathfrak{Ab}^r_\Bbbk$) is formed by

0-morphisms, or objects: essentially small $\Bbbk$-linear abelian categories $\mathcal{A}$ with length,
1-morphisms: $\Bbbk$-linear left (resp. right) exact functors $F : \mathcal{A} \to \mathcal{B}$,
2-morphisms: natural transformations (morphisms of functors).

The reader is referred to Appendix A.1 for a definition of a 2-category or to Bénabou [**4**] for a definition of a bicategory.

Monoidal 2-categories are defined and studied by Kapranov and Voevodsky [**18**]. More general notion of a monoidal bicategory is proposed by Gordon, Power and Street [**11**]. The relationship between those theories is clarified by Baez and Neuchl [**2**].

The monoidal structure of $\mathfrak{Ab}^l_\Bbbk$ is given by Deligne's tensor product of categories $\boxtimes : (\mathcal{A}, \mathcal{B}) \mapsto \mathcal{A} \boxtimes \mathcal{B}$. The tensor product of 1-morphisms $F : \mathcal{A} \to \mathcal{C}$ and $G : \mathcal{B} \to \mathcal{D}$ is the lowest arrow in the diagram

$$\begin{array}{ccc} \mathcal{A} \times \mathcal{B} & \xrightarrow{F \times G} & \mathcal{C} \times \mathcal{D} \\ {\scriptstyle\boxtimes}\downarrow & & \downarrow{\scriptstyle\boxtimes} \\ \mathcal{A} \boxtimes \mathcal{B} & \xrightarrow{\exists F \boxtimes G} & \mathcal{C} \boxtimes \mathcal{D} \end{array}$$

constructed by definition as a left exact functor. Of course, this determines $F \boxtimes G$ only up to an isomorphism, so we obtain not one, but a family of monoidal structures. It is shown in Appendix A that each choice gives one symmetric monoidal structure on the category $\mathfrak{Ab}^l_\Bbbk$. The unit object is the category $\mathcal{I} = \Bbbk\text{-vect}$.

1.5. System of notations

1.5.1. Contractible groupoids. Before explaining some special notation, we need some technical definitions. The following terminology is compatible with Quillen's work [**25**].

1.5.2. Definition. A *contractible groupoid* G is a category equivalent to **1**, a category with one object and one morphism. That is, $\operatorname{Ob} G$ is not empty, and for every pair X, Y of objects of G, there is exactly one morphism $\alpha_{X,Y}$ from X to Y.

To specify a contractible groupoid G amounts to specify the class of objects $\operatorname{Ob} G$. Let $\mathfrak{G}$ be a non-empty set of groupoids and let $\mathcal{C}$ be a category. A *generalised object* of $\mathcal{C}$ is a functor $A : G \to \mathcal{C}$, where $G \in \mathfrak{G}$. When $A : \operatorname{Ob} G \to \operatorname{Ob} \mathcal{C}$ is injective, it is customary to say that A is an object of $\mathcal{C}$ defined up to a canonical isomorphism. We shall extend this way of thinking also on non-injective $A : \operatorname{Ob} G \to \operatorname{Ob} \mathcal{C}$.

A *generalised morphism* h from $A : G \to \mathcal{C}$ to $B : G' \to \mathcal{C}$ is a collection of morphisms $h_{X,Y} : AX \to BY$ for objects X of G and objects Y of G', such that

$$\begin{array}{ccc} AX & \xrightarrow{h_{X,Y}} & BY \\ {\scriptstyle A(\alpha_{X,\tilde{X}})}\downarrow & = & \downarrow{\scriptstyle B(\alpha_{Y,\tilde{Y}})} \\ A\tilde{X} & \xrightarrow{h_{\tilde{X},\tilde{Y}}} & B\tilde{Y} \end{array} \tag{1.5.1}$$

for all objects X, $\tilde{X}$ of G and all objects Y, $\tilde{Y}$ of G'. Clearly, to give a generalised morphism $h : A \to B$ it suffices to choose one object X of G and one object Y of G' and to specify $h_{X,Y} : AX \to BY$. More generally, it suffices to specify a collection $(h_{X,Y})_{X \in S, Y \in S'}$ for non-empty subclasses $S \subset \operatorname{Ob} G$, $S' \subset \operatorname{Ob} G'$, which satisfies (1.5.1).

The composition f of generalised morphisms $h : A \to B$ and $g : B \to C$, $C : G'' \to \mathcal{C}$ is defined as follows. Given $X \in \operatorname{Ob} G$ and $Z \in \operatorname{Ob} G''$, choose an arbitrary $Y \in \operatorname{Ob} G'$ and set

$$f_{X,Z} = \big(AX \xrightarrow{h_{X,Y}} BY \xrightarrow{g_{Y,Z}} CZ\big).$$

Diagrams (1.5.1) imply that $f_{X,Z}$ does not depend on the choice of Y, hence, $f = (f_{X,Z})$ is a well-defined generalised morphism $A \to C$. Thus, generalised objects and morphisms form a category $\underline{\mathcal{C}}$. If $\mathfrak{G}$ is not empty, then there is an equivalence $\underline{\mathcal{C}} \to \mathcal{C}$.

1.5.3. Notation. The above trivial considerations turn out to be useful for the following notation. Assume that $\mathcal{V}$ is a $\Bbbk$-linear abelian Monoidal category with length. Given $A^{(1)} \in \operatorname{Ob} \mathcal{V}^{\boxtimes n_1}, \dots, A^{(k)} \in \operatorname{Ob} \mathcal{V}^{\boxtimes n_k}$, we want to define

$$A^{(1)}_{i^1_1 \dots i^1_{n_1}} \boxtimes \dots \boxtimes A^{(k)}_{i^k_1 \dots i^k_{n_k}}$$

as a generalised object of the category $\mathcal{V}^{\boxtimes l}$, where the sequence $(i^1_1, \dots, i^1_{n_1}, \dots, i^k_1, \dots, i^k_{n_k})$ is a permutation of the sequence $(1', 1'', 1''', \dots, 1^{m_1}, \dots, l', l'', \dots, l^{m_l})$. Discrete parameters associated with this notation are:

a sequence $(n_1, \dots, n_k)$, $n_i \in \mathbb{Z}_{\geqslant 0}$;
a sequence $(m_1, \dots, m_l)$, $m_j \in \mathbb{Z}_{\geqslant 0}$ such that

$$n = n_1 + \dots + n_k = m_1 + \dots + m_l;$$

a permutation $\sigma \in \mathfrak{S}_n$.

To these data we assign a contractible groupoid $G = G(n_1, \dots, n_k; m_1, \dots, m_l; \sigma)$, whose objects are pairs (F, ϕ), where F is a $\Bbbk$-multilinear right exact in each variable functor, and ϕ is an isomorphism of functors in the diagram

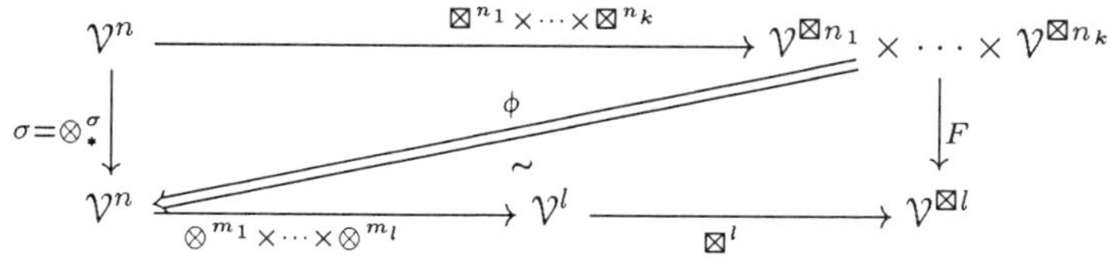

A morphism $\alpha : (F,\phi) \to (\tilde F,\tilde\phi)$ is a morphism of functors $\alpha : F \to \tilde F$, such that the following equation holds

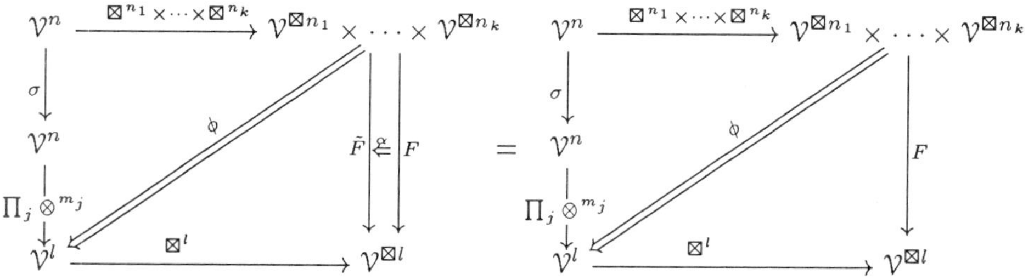

Proposition A.5.4 implies existence and uniqueness of such α, which will be denoted $\alpha_{F,\phi,\tilde F,\tilde\phi}$. Thus, $G = G(n_1,\dots,n_k;m_1,\dots,m_l;\sigma)$ is a contractible groupoid. Fix l and vary $n_1,\dots,n_k;m_1,\dots,m_l;\sigma$. The set G_l of corresponding groupoids G is used to define the class of objects of $\underline{\mathcal{V}^{\boxtimes l}}$.

Now we define a generalised object

$$A^{(1)}_{i^1_1\dots i^1_{n_1}} \boxtimes \dots \boxtimes A^{(k)}_{i^k_1\dots i^k_{n_k}} : G(n_1,\dots,n_k;m_1,\dots,m_l;\sigma) \to \mathcal{V}^{\boxtimes l}$$

as a functor which assigns to (F,ϕ) the object $F(A^{(1)},\dots,A^{(k)})$, and to $\alpha_{F,\phi,\tilde F,\tilde\phi}$ the morphism

$$\alpha_{F,\phi,\tilde F,\tilde\phi;A^{(1)},\dots,A^{(k)}} : F(A^{(1)},\dots,A^{(k)}) \to \tilde F(A^{(1)},\dots,A^{(k)}).$$

Among the groupoids from G_l there is $G(l;1,\dots,1;1)$, where $k=1$. It contains $(\mathrm{Id},\mathrm{id})$ as an object. Therefore, $\underline{\mathcal{V}^{\boxtimes l}}$ is equivalent to $\mathcal{V}^{\boxtimes l}$. The above construction is functorial in $A^{(1)},\dots,A^{(k)}$, so we get a functor

$$-_{i^1_1\dots i^1_{n_1}} \boxtimes \dots \boxtimes -_{i^k_1\dots i^k_{n_k}} : \mathcal{V}^{\boxtimes n_1} \times \dots \times \mathcal{V}^{\boxtimes n_k} \to \underline{\mathcal{V}^{\boxtimes l}}. \tag{1.5.2}$$

For purely aesthetic reasons we will sometimes replace some of the separating symbols $\boxtimes$ with $\otimes$. This does not mean any change in the above definition of this functor.

1.5.4. Substitution. Now we have to understand how to use such functors as arguments for another such functor. That is, given $(l_1,\dots,l_r)$, $(t_1,\dots,t_s)$, $\sigma \in \mathfrak{S}_l$, such that $l_1+\dots+l_r = l = t_1+\dots+t_s$, which parametrise the functor

$$-_{i^1_1\dots i^1_{l_1}} \boxtimes \dots \boxtimes -_{i^r_1\dots i^r_{l_r}} : \mathcal{V}^{\boxtimes l_1} \times \dots \times \mathcal{V}^{\boxtimes l_r} \to \underline{\mathcal{V}^{\boxtimes l}}$$

we want to define also a functor

$$-_{i^1_1\dots i^1_{l_1}} \boxtimes \dots \boxtimes -_{i^r_1\dots i^r_{l_r}} : \underline{\mathcal{V}^{\boxtimes l_1}} \times \dots \times \underline{\mathcal{V}^{\boxtimes l_r}} \to \underline{\mathcal{V}^{\boxtimes l}}. \tag{1.5.3}$$

To define it on object, $(A^j_{1;\dots} \boxtimes \dots \boxtimes A^j_{k_j;\dots})_{1\leqslant j\leqslant r}$, assume given sequences $(n^j_1,\dots,n^j_{k_j})$, $(m^j_1,\dots,m^j_{l_j})$ and permutations $\sigma^j \in \mathfrak{S}_{n^j}$ for $1\leqslant j\leqslant r$, such that $n^j = \sum_{p=1}^{k_j} n^j_p =$

$\sum_{q=1}^{l_j} m_q^j$, and assume given objects A_p^j of $\mathcal{V}^{\boxtimes n_p^j}$. Set $n = \sum_{j=1}^r n^j$, $l = \sum_{j=1}^r l_j$. The following diagram suggests to assign to the above object the generalised object

$$A^1_{1;\dots} \boxtimes \dots \boxtimes A^1_{k_1;\dots} \boxtimes \dots \boxtimes A^r_{1;\dots} \boxtimes \dots \boxtimes A^r_{k_r;\dots} :$$
$$G(n_1^1, \dots, n_{k_1}^1, \dots, n_1^r, \dots, n_{k_r}^r; y_1, \dots, y_s; \tilde{\sigma} \circ \prod_{j=1}^r \sigma_j) \to \mathcal{V}^{\boxtimes s},$$

where $\tilde{\sigma} \in \mathfrak{S}_n$ is the permutation which permutes blocks of size m_q^j the same way that σ permutes elements; $(m_{q(z)}^{j(z)})_{z=1}^l$ is the sequence $(m_1^1, \dots, m_{l_1}^1, \dots, m_1^r, \dots, m_{l_r}^r)$ permuted by σ; monotonic maps $g : \mathbf{n} \to \mathbf{l}$, $f : \mathbf{l} \to \mathbf{s}$ are associated with the partition $(m_{q(z)}^{j(z)})_{z=1}^l$ of n and the partition $(t_x)_{x=1}^s$ of l, respectively; the isomorphisms λ are part of the Monoidal structure. The following pasting is denoted ψ.

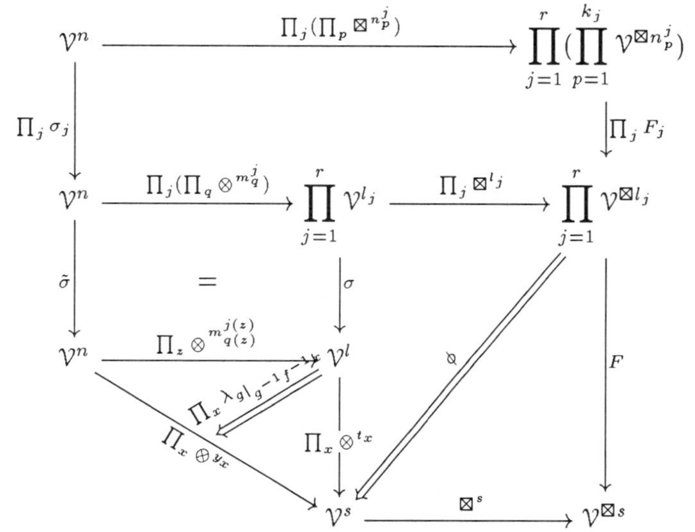

This diagram also defines a functor

$$((F_1, \phi_1), \dots, (F_r, \phi_r), (F, \phi)) \mapsto (F \circ \prod_{j=1}^r F_j, \psi), \tag{1.5.4}$$

presented at the upper row of the following diagram

$$\begin{array}{ccc}
\prod_{j=1}^r G((n_p^j)_{p=1}^{k_j}; (m_q^j)_{q=1}^{l_j}; \sigma_j) \times G((l_j)_{j=1}^r; (t_x)_{x=1}^s; \sigma) & \longrightarrow & G((n_p^j); (y_x)_{x=1}^s; \tilde{\sigma} \circ \prod \sigma_j) \\
\downarrow & = & \downarrow \\
\prod_{j=1}^r \operatorname{Hom}(\prod_{p=1}^{k_j} \mathcal{V}^{\boxtimes n_p^j}, \mathcal{V}^{\boxtimes l_j}) \times \operatorname{Hom}(\prod_{j=1}^r \mathcal{V}^{\boxtimes l_j}, \mathcal{V}^{\boxtimes s}) & \longrightarrow & \operatorname{Hom}(\prod_{j=1}^r \prod_{p=1}^{k_j} \mathcal{V}^{\boxtimes n^j}, \mathcal{V}^{\boxtimes s})
\end{array}$$

The vertical functors forget the second component of a pair (F, ϕ). The lower row is the substitution functor. Uniqueness of the morphism $\alpha_{-,-}$ implies that for objects

(F_j, ϕ_j, F, ϕ), $\tilde{F}$, $\tilde{\phi}_j$, $\tilde{F}$, $\tilde{\phi}$ of top left category

$$\Big(\prod_{j=1}^{r} \alpha^j_{F_j,\phi_j,\tilde{F}_j,\tilde{\phi}_j}\Big) \centerdot \alpha_{F,\phi,\tilde{F},\tilde{\phi}} = \alpha_{F\circ\prod F_j,\psi,\tilde{F}\circ\prod \tilde{F}_j,\tilde{\psi}}. \tag{1.5.5}$$

Therefore, the above diagram of functors strictly commutes. The product of 2-cells in the left hand side has two interpretations, one of which is

$$F(F_1,\dots,F_r) \xrightarrow{F(\prod_j \alpha^j_{F_j,\phi_j,\tilde{F}_j,\tilde{\phi}_j})} F(\tilde{F}_1,\dots,\tilde{F}_r) \xrightarrow{\alpha_{F,\phi,\tilde{F},\tilde{\phi}}} \tilde{F}(\tilde{F}_1,\dots,\tilde{F}_r).$$

Now we shall define functor (1.5.3) on morphisms. Suppose we are given generalised morphisms

$$h^j : A^j_{1;\dots} \boxtimes \cdots \boxtimes A^j_{k_j;\dots} \to B^j_{1;\dots} \boxtimes \cdots \boxtimes B^j_{k'_j;\dots} \in \operatorname{Mor} \underline{\mathcal{V}^{\boxtimes l_j}}$$

for $1 \leqslant j \leqslant r$. They are systems of morphisms

$$h^j_{F_j,\phi_j,F'_j,\phi'_j} : F_j(A^j_1,\dots,A^j_{k_j}) \to F'_j(B^j_1,\dots,B^j_{k'_j}),$$

depending on objects (F_j,ϕ_j) of $G((n^j_p);(m^j_q);\sigma_j)$ and (F'_j,ϕ'_j) of $G((n'^j_p);(m'^j_q);\sigma'_j)$ in such a way that

$$\begin{array}{ccc}
F_j(A^j_1,\dots,A^j_{k_j}) & \xrightarrow{h^j_{F_j,\phi_j,F'_j,\phi'_j}} & F'_j(B^j_1,\dots,B^j_{k'_j}) \\
{\scriptstyle \alpha_{F_j,\phi_j,\tilde{F}_j\cdot\tilde{\phi}_j}}\downarrow & = & \downarrow{\scriptstyle \alpha_{F'_j,\phi'_j,\tilde{F}'_j\cdot\tilde{\phi}'_j}} \\
\tilde{F}_j(A^j_1,\dots,A^j_{k_j}) & \xrightarrow{h^j_{\tilde{F}_j,\tilde{\phi}_j,\tilde{F}'_j,\tilde{\phi}'_j}} & \tilde{F}'_j(B^j_1,\dots,B^j_{k'_j})
\end{array}$$

Consider also objects (F,ϕ), (F',ϕ'), $(\tilde{F},\tilde{\phi})$, $(\tilde{F}',\tilde{\phi}')$ of $G((l_j)_j;(t_x)_x;\sigma)$ and a system of maps

$$\begin{aligned}
F(F_1((A^1_p)_p),\dots,F_r((A^r_p)_p)) &\xrightarrow{F(\prod_j h^j_{F_j,\phi_j,F'_j,\phi'_j})} F(F'_1((B^1_b)_b),\dots,F'_r((B^r_b)_b)) \\
&\xrightarrow{\alpha_{F,\phi,F',\phi'}} F'(F'_1((B^1_b)_b),\dots,F'_r((B^r_b)_b))
\end{aligned} \tag{1.5.6}$$

depending on F, ϕ, F_j, ϕ_j, F', ϕ', F'_j, ϕ'_j. In fact, the above morphism depends only on the object $(F\circ\prod F_j,\psi)$ of $G((n^j_p);(y_x);\tilde{\sigma}\circ\prod\sigma_j)$ and the object $(F'\circ\prod F'_j,\psi')$ of $G((n'^j_p);(y'_x);\tilde{\sigma}'\circ\prod\sigma'_j)$ (see (1.5.4)). Indeed, the following diagram commutes

$$\begin{array}{ccccc}
F(F_1(A^1_\bullet),\dots,F_r(A^r_\bullet)) & \xrightarrow{F(\prod_j h^j_{F_j,\phi_j,F'_j,\phi'_j})} & F(F'_1(B^1_\bullet),\dots,F'_r(B^r_\bullet)) & \xrightarrow{\alpha_{F,\phi,F',\phi'}} & F'(F'_1(B^1_\bullet),\dots,F'_r(B^r_\bullet)) \\
{\scriptstyle F(\prod_j \alpha_{F_j,\phi_j,\tilde{F}_j\cdot\tilde{\phi}_j})}\downarrow & & {\scriptstyle F(\prod_j \alpha_{F'_j,\phi'_j,\tilde{F}'_j\cdot\tilde{\phi}'_j})}\downarrow & & \downarrow{\scriptstyle F'(\prod_j \alpha_{F'_j,\phi'_j,\tilde{F}'_j\cdot\tilde{\phi}'_j})} \\
F(\tilde{F}_1(A^1_\bullet),\dots,\tilde{F}_r(A^r_\bullet)) & \xrightarrow{F(\prod_j h^j_{\tilde{F}_j,\tilde{\phi}_j,\tilde{F}'_j,\tilde{\phi}'_j})} & F(\tilde{F}'_1(B^1_\bullet),\dots,\tilde{F}'_r(B^r_\bullet)) & \xrightarrow{\alpha_{F,\phi,\tilde{F}',\tilde{\phi}'}} & F'(\tilde{F}'_1(B^1_\bullet),\dots,\tilde{F}'_r(B^r_\bullet)) \\
{\scriptstyle \alpha_{F,\phi,\tilde{F},\tilde{\phi}}}\downarrow & & {\scriptstyle \alpha_{F,\phi,\tilde{F}',\tilde{\phi}'}}\downarrow & & \downarrow{\scriptstyle \alpha_{F',\phi',\tilde{F}',\tilde{\phi}'}} \\
\tilde{F}(\tilde{F}_1(A^1_\bullet),\dots,\tilde{F}_r(A^r_\bullet)) & \xrightarrow{\tilde{F}(\prod_j h^j_{\tilde{F}_j,\tilde{\phi}_j,\tilde{F}'_j,\tilde{\phi}'_j})} & \tilde{F}(\tilde{F}'_1(B^1_\bullet),\dots,\tilde{F}'_r(B^r_\bullet)) & \xrightarrow{\alpha_{\tilde{F},\tilde{\phi},\tilde{F}',\tilde{\phi}'}} & \tilde{F}'(\tilde{F}'_1(B^1_\bullet),\dots,\tilde{F}'_r(B^r_\bullet))
\end{array}$$

The left and right columns are

$$\alpha_{F\circ\prod F_j,\psi,\tilde{F}\circ\prod\tilde{F}_j,\tilde{\psi}} \text{ and } \alpha_{F'\circ\prod F'_j,\psi',\tilde{F}'\circ\prod\tilde{F}'_j,\tilde{\psi}'}, \tag{1.5.7}$$

respectively, by (1.5.5). If $(F \circ \prod F_j, \psi) = (\tilde{F} \circ \prod \tilde{F}_j, \tilde{\psi})$ and $(F' \circ \prod F'_j, \psi') = (\tilde{F}' \circ \prod \tilde{F}'_j, \tilde{\psi}')$, then morphisms (1.5.7) are identities. Therefore, the top row equals the bottom row. Now we can denote the composite morphism (1.5.6) by

$$h_{F\circ\prod F_j,\psi,F'\circ\prod F'_j,\psi'} : F(F_1(A^1_\bullet), \dots, F_r(A^r_\bullet)) \to F'(F'_1(B^1_\bullet), \dots, F'_r(B^r_\bullet)).$$

Commutativity of the exterior of the above diagram implies that so defined system of morphisms determines a generalised morphism, denoted

$$h = h^1_{i^1_1 \dots i^1_{k_1}} \boxtimes \dots \boxtimes h^r_{i^r_1 \dots i^r_{k_r}} \in \operatorname{Mor} \underline{\mathcal{V}^{\boxtimes l}},$$

where $(i^1_1 \dots i^1_{k_1}; \dots; i^r_1 \dots i^r_{k_r})$ corresponds to (l_j), (t_x), σ. So constructed correspondence (1.5.3) preserves the composition, therefore, it is a functor.

1.5.5. Permutation isomorphisms. Let us discuss the isomorphism of generalised objects arising from a permutation of factors. Given a permutation $\sigma \in \mathfrak{S}_k$, $\tau = \sigma^{-1}$, we define an isomorphism of generalised objects

$$\overline{\sigma} : A^{(1)}_{i^1_1 \dots i^1_{n_1}} \boxtimes \dots \boxtimes A^{(k)}_{i^k_1 \dots i^k_{n_k}} \xrightarrow{\sim} A^{(\tau 1)}_{i^{\tau 1}_1 \dots i^{\tau 1}_{n_{\tau 1}}} \boxtimes \dots \boxtimes A^{(\tau k)}_{i^{\tau k}_1 \dots i^{\tau k}_{n_{\tau k}}}.$$

Assume that the source is associated with partitions $(n_1, \dots, n_k)$, $(m_1, \dots, m_l)$, $\sum n_i = n = \sum m_j$, and a permutation $\sigma_1 \in \mathfrak{S}_n$. Then the target is associated with the sequences $(n_{\tau 1}, \dots, n_{\tau k})$, $(m_1, \dots, m_l)$ and a permutation $\sigma_2 = \sigma_1 \circ \tilde{\sigma}^{-1}$, where $\tilde{\sigma} \in \mathfrak{S}_n$ permutes blocks of size n_1, ... ,n_k the same way that σ permutes elements of k. The isomorphism $\overline{\sigma}$ is defined as the system of isomorphisms

$$\overline{\sigma}_{F_1,\phi_1,F_2,\phi_2} : F_1(A^{(1)}, \dots, A^{(k)}) \to F_2(A^{(\tau 1)}, \dots, A^{(\tau k)})$$

for which the following equation holds:

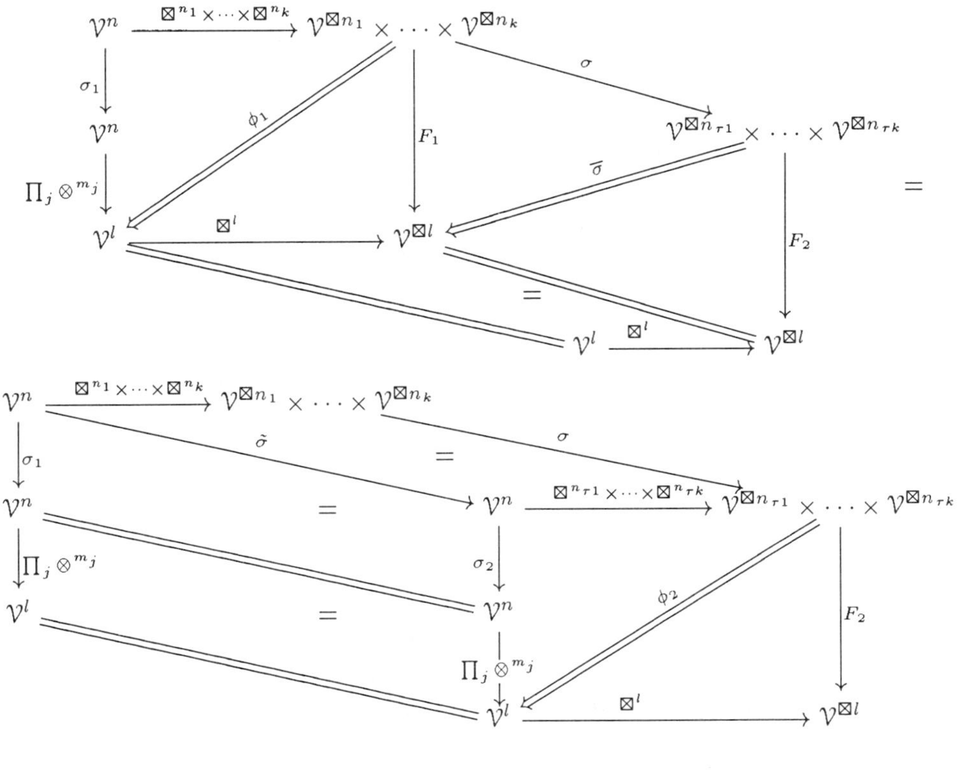

Clearly, condition (1.5.1) is satisfied and $\overline{\sigma}$ is indeed an isomorphism of functors

$$\overline{\sigma}: -_{i^1_1 \dots i^1_{n_1}} \boxtimes \dots \boxtimes -_{i^k_1 \dots i^k_{n_k}} \to -_{i^{\tau 1}_1 \dots i^{\tau 1}_{n_{\tau 1}}} \boxtimes \dots \boxtimes -_{i^{\tau k}_1 \dots i^{\tau k}_{n_{\tau k}}} \circ \sigma .$$

These isomorphisms obey the rule $\overline{\sigma} \circ \overline{\theta} = \overline{\sigma \circ \theta}$, hence, the collection of permuted functors (1.5.2) and the isomorphisms $\overline{\sigma}$, $\sigma \in \mathfrak{S}_n$, also forms a contractible groupoid.

1.5.6. Practical example – coassociativity. Let us consider in minor details a practical example. Let $\mathcal{V}$ be a $\Bbbk$-linear abelian rigid Monoidal category with length. The unit object is denoted $\mathbb{1}$. Let C be an object of $\mathcal{V} \boxtimes \mathcal{V}$. Then $C_{13} \boxtimes \mathbb{1}_2$ is the functor

$$G(2,1;1,1,1;(23)) \to \mathcal{V}^{\boxtimes 3}, (F, \phi) \mapsto F(C, \mathbb{1}),$$

where ϕ is an isomorphism

$$\begin{array}{ccc} \mathcal{V}^3 & \xrightarrow{\boxtimes^2 \times 1} & \mathcal{V}^{\boxtimes 2} \times \mathcal{V} \\ {\scriptstyle (23)} \downarrow & \overset{\phi}{\Leftarrow} & \downarrow {\scriptstyle F} \\ \mathcal{V}^3 & \xrightarrow{\boxtimes^3} & \mathcal{V}^{\boxtimes 3} \end{array} .$$

$C_{12'} \boxtimes C_{2''3}$ denotes the functor

$$G(2,2;1,2,1;1) \to \mathcal{V}^{\boxtimes 3}, (F', \phi') \mapsto F'(C, C),$$

where ϕ' is an isomorphism

$$\begin{array}{ccc} \mathcal{V}^4 & \xrightarrow{\boxtimes^2 \times \boxtimes^2} & \mathcal{V}^{\boxtimes 2} \times \mathcal{V}^{\boxtimes 2} \\ {\scriptstyle 1 \times \otimes \times 1} \downarrow & \overset{\phi'}{\Leftarrow} & \downarrow {\scriptstyle F'} \\ \mathcal{V}^3 & \xrightarrow{\boxtimes^3} & \mathcal{V}^{\boxtimes 3} \end{array} .$$

Consider now a generalised morphism of $\underline{\mathcal{V}^{\boxtimes 3}}$

$$\Delta : C_{13} \boxtimes \mathbb{1}_2 \to C_{12'} \boxtimes C_{2''3}.$$

It will be called the comultiplication. This is a system of morphisms

$$\Delta_{F, \phi, F', \phi'} : F(C, \mathbb{1}) \to F'(C, C),$$

satisfying the equation

$$\begin{array}{ccc} F(C, \mathbb{1}) & \xrightarrow{\Delta_{F, \phi, F', \phi'}} & F'(C, C) \\ {\scriptstyle \alpha_{F, \phi, \tilde{F}, \tilde{\phi}}} \downarrow & = & \downarrow {\scriptstyle \alpha_{F, \phi, \tilde{F}', \tilde{\phi}'}} \\ \tilde{F}(C, \mathbb{1}) & \xrightarrow{\Delta_{\tilde{F}, \tilde{\phi}, \tilde{F}', \tilde{\phi}'}} & \tilde{F}'(C, C) \end{array}$$

We can apply various functors to Δ and obtain the morphisms of $\underline{\mathcal{V}^4}$ present in the diagram

$$\begin{array}{ccc} C_{14} \boxtimes \mathbb{1}_2 \boxtimes \mathbb{1}_3 & \xrightarrow{\Delta_{124} \boxtimes \mathbb{1}_3} & C_{12'} \otimes C_{2''4} \boxtimes \mathbb{1}_3 \\ {\scriptstyle \Delta_{134} \boxtimes \mathbb{1}_2} \downarrow & & \downarrow {\scriptstyle C_{12'} \otimes \Delta_{2''34}} \\ C_{13'} \boxtimes \mathbb{1}_2 \boxtimes C_{3''4} & \xrightarrow{\Delta_{123'} \otimes C_{3''4}} & C_{12'} \otimes C_{2''3'} \otimes C_{3''4} \end{array} .$$

This diagram is called the coassociativity equation. Actually, one of the morphisms in this diagram is not defined, since permutation isomorphisms are deliberately omitted. The above diagram contains the essential morphisms of the following diagram; it is a shorthand for

$$\begin{array}{ccccc}
C_{14}\boxtimes \mathbb{I}_2\boxtimes \mathbb{I}_3 & \xrightarrow{\Delta_{124}\boxtimes \mathbb{I}_3} & C_{12'}\otimes C_{2''4}\boxtimes \mathbb{I}_3 & \xrightarrow{C_{12'}\otimes\Delta_{2''34}} & C_{12'}\otimes C_{2''3'}\otimes C_{3''4} \\
\downarrow{\scriptstyle \overline{(34)}} & & & & \uparrow{\scriptstyle \Delta_{123'}\otimes C_{3''4}} \\
C_{14}\boxtimes \mathbb{I}_3\boxtimes \mathbb{I}_2 & \xrightarrow{\Delta_{134}\boxtimes \mathbb{I}_2} & C_{13'}\otimes C_{3''4}\boxtimes \mathbb{I}_2 & \xrightarrow{\overline{(345)}} & C_{13'}\boxtimes \mathbb{I}_2\boxtimes C_{3''4}
\end{array}$$

The above 6 morphisms are well-defined in $\underline{\mathcal{V}}^4$, and it makes sense to speak of commutativity of the above diagram in $\underline{\mathcal{V}}^4$.

When one uses the definitions of these morphisms and represents each of them as a system of morphisms in $\mathcal{V}^4$, one finds that the target of the horizontal left top arrow is not equal but is canonically isomorphic to the source of horizontal right top arrow. The isomorphism is that of $G(2,2,1;1,2,1,1;(45))$. Similarly in the top right corner a canonical isomorphism of $G(2,2,2;1,2,2,1;1)$ is inserted. Altogether, 8 morphisms appear in the unpacked coassociativity equation – twice the original amount. The upper path of the equation is realised as follows.

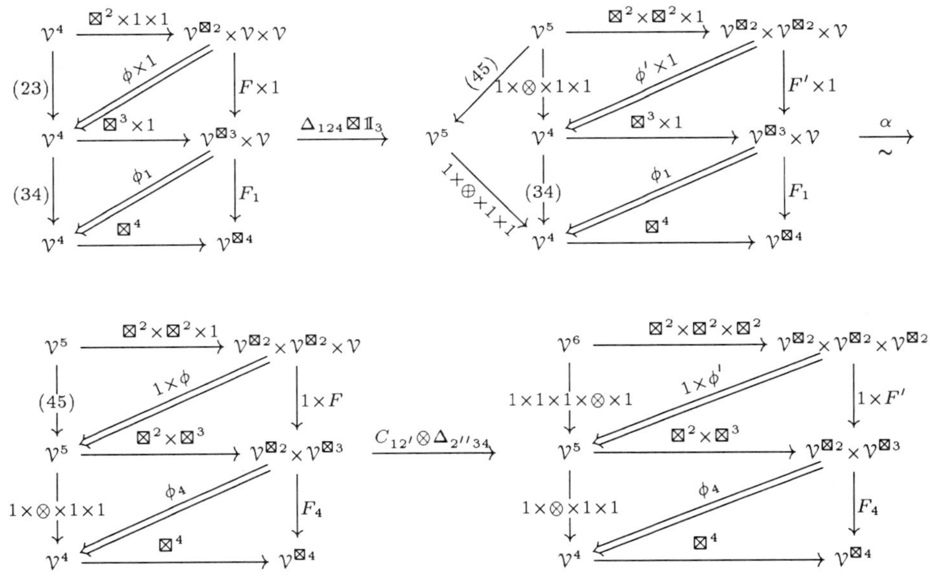

The beginning and the end of this path are related by isomorphisms $\overline{(34)}$ and the canonical isomorphism α', respectively, with the beginning and the end of the lower path:

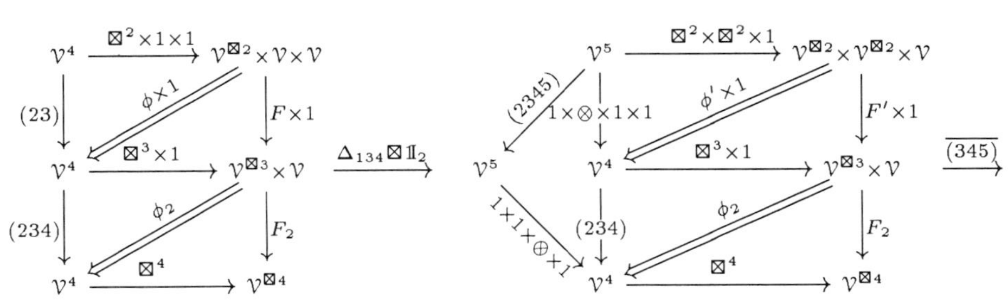

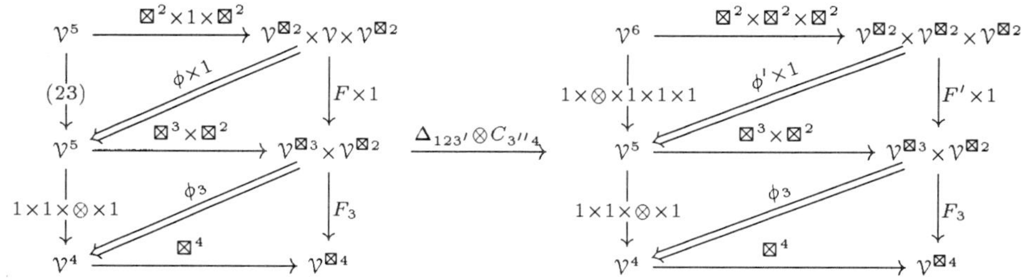

The features of this example will be encountered in many other equations between generalised morphisms. We will not mention explicitly the omissions of permutation isomorphisms in these equations.

Examples. If $X, Y \in \mathcal{V}$, $C \in \mathcal{V} \boxtimes \mathcal{V}$ then $X_{1'} \otimes Y_{1''}$ denotes the usual tensor product of X and Y, $X_{1''} \otimes Y_{1'}$ is $Y \otimes X$. Similarly, $C_{1'1''}$ is $\circledast C$ and $C_{1''1'}$ is $\circledast PC$, where $P : \mathcal{V} \boxtimes \mathcal{V} \to \mathcal{V} \boxtimes \mathcal{V}$ is the permutation functor, $P(X \boxtimes Y) = Y \boxtimes X$. The three objects $X_1 \boxtimes C_{23}$, $X_2 \boxtimes C_{13}$, $C_{12} \boxtimes X_3$ of $\mathcal{V}^{\boxtimes 3}$ differ by the place where X goes. Applying $\circledast \boxtimes \mathrm{Id}_{\mathcal{V}} : \mathcal{V}^{\boxtimes 3} \to \mathcal{V}^{\boxtimes 2}$ to $X_1 \boxtimes C_{23}$ we get $X_{1'} \otimes C_{1''2}$. Applying $\mathrm{Id}_{\mathcal{V}} \boxtimes \circledast : \mathcal{V}^{\boxtimes 3} \to \mathcal{V}^{\boxtimes 2}$ to $C_{12} \boxtimes X_3$ we get $C_{12'} \otimes X_{2''}$.

1.6. Monoidal structures on $\mathcal{V} \boxtimes \mathcal{V}$

Let $\mathcal{V}$ be a $\Bbbk$-linear abelian monoidal category with length. Then $\mathcal{V} \boxtimes \mathcal{V}$ has a monoidal structure as well. In fact, there are four monoidal structures, since we may choose between $(\mathcal{V}, \otimes)$ and $(\mathcal{V}, \otimes_{\mathrm{op}})$ in both factors. The main monoidal structure, which we fix from now on, is:

$$\bar{\otimes} : (\mathcal{V} \boxtimes \mathcal{V}) \times (\mathcal{V} \boxtimes \mathcal{V}) \longrightarrow \mathcal{V} \boxtimes \mathcal{V}$$
$$(A_{12}, B_{12}) \longmapsto A_{1'2''} \otimes B_{1''2'}$$
$$(X \boxtimes Y, V \boxtimes W) \longmapsto (X \otimes V) \boxtimes (W \otimes Y).$$

1.6.1. Theorem. *If $(\mathcal{V}, \otimes)$ is rigid, then $(\mathcal{V} \boxtimes \mathcal{V}, \bar{\otimes})$ is rigid as well.*

PROOF. By Proposition 1.1.10 there are functors $\mathcal{V} \boxtimes \mathcal{V} \to (\mathcal{V} \boxtimes \mathcal{V})^{\mathrm{op}}$, $M \mapsto M^{\vee}$, $X \boxtimes Y \mapsto X^{\vee} \boxtimes {}^{\vee}Y$ and $M \mapsto {}^{\vee}M$, $X \boxtimes Y \mapsto {}^{\vee}X \boxtimes Y^{\vee}$. They are exact equivalences, quasiinverse to each other.

Let M be an object of $\mathcal{V} \boxtimes \mathcal{V}$. By Proposition 1.1.2 M can be included into an exact sequence

$$X \boxtimes Y \xrightarrow{\sum h_j \boxtimes k_j} V \boxtimes W \xrightarrow{p} M \to 0 \qquad (1.6.1)$$

for some $h_j, k_j \in \mathcal{V}$. In the diagram

$$\begin{array}{c}
\mathbb{1} \boxtimes \mathbb{1} \xrightarrow{\mathrm{coev} \boxtimes \mathrm{coev}} (X^{\vee} \boxtimes {}^{\vee}Y) \bar{\otimes} (X \boxtimes Y) \\
{\scriptstyle \mathrm{coev} \boxtimes \mathrm{coev}} \searrow \qquad \qquad \downarrow {\scriptstyle 1 \bar{\otimes} (\sum h_j \boxtimes k_j)} \\
\exists \mathrm{coev} \downarrow \quad (V^{\vee} \boxtimes {}^{\vee}W) \bar{\otimes} (V \boxtimes W) \xrightarrow{(\sum h_j^t \boxtimes {}^t k_j) \bar{\otimes} 1} (X^{\vee} \boxtimes {}^{\vee}Y) \bar{\otimes} (V \boxtimes W) \\
\qquad \qquad \downarrow {\scriptstyle 1 \bar{\otimes} p} \qquad \qquad \qquad \downarrow {\scriptstyle 1 \bar{\otimes} p} \\
0 \to M^{\vee} \bar{\otimes} M \xrightarrow{p^t \bar{\otimes} 1} (V^{\vee} \boxtimes {}^{\vee}W) \bar{\otimes} M \xrightarrow{(\sum h_j^t \boxtimes {}^t k_j) \bar{\otimes} 1} (X^{\vee} \boxtimes {}^{\vee}Y) \bar{\otimes} M
\end{array}$$

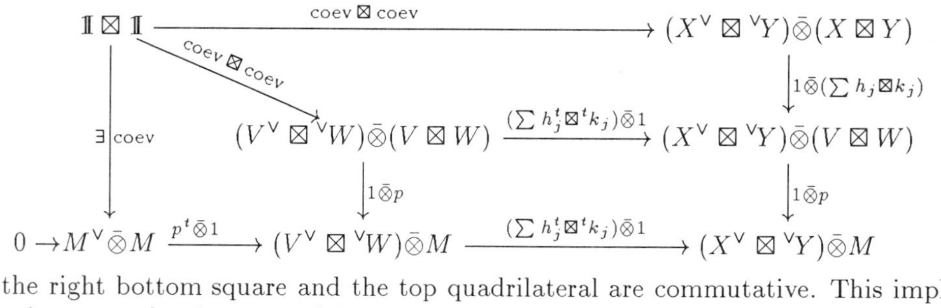

the right bottom square and the top quadrilateral are commutative. This implies existence and uniqueness of the vertical leftmost arrow $\mathrm{coev} : \mathbb{1} \boxtimes \mathbb{1} \to M^{\vee} \bar{\otimes} M$, such that the left quadrilateral commutes. Reversing arrows and using the resolution

$$0 \to M \xrightarrow{i} A\boxtimes B \xrightarrow{\sum f_l\boxtimes g_l} C\boxtimes D \tag{1.6.2}$$

we construct the morphism $\mathrm{ev} : M\bar{\otimes}M^\vee \to \mathbb{1}\boxtimes\mathbb{1}$, such that

$$\begin{array}{ccc} M\bar{\otimes}(A^\vee\boxtimes{}^\vee B) & \xrightarrow{1\bar{\otimes}i^t} & M\bar{\otimes}M^\vee \\ {\scriptstyle i\bar{\otimes}1}\downarrow & & \downarrow{\scriptstyle \mathrm{ev}} \\ (A\boxtimes B)\bar{\otimes}(A^\vee\boxtimes{}^\vee B) & \xrightarrow{\mathrm{ev}} & \mathbb{1}\boxtimes\mathbb{1} \end{array}$$

commutes. Similarly for $\mathrm{ev} : {}^\vee M\bar{\otimes}M \to \mathbb{1}\boxtimes\mathbb{1}$ and $\mathrm{coev} : \mathbb{1}\boxtimes\mathbb{1} \to M\bar{\otimes}{}^\vee M$.

To prove the required relations between ev and coev we use the projection p, the embedding i, and their composite

$$u = \sum r_n\boxtimes s_n = \left(V\boxtimes W \xrightarrow{p} M \xrightarrow{i} A\boxtimes B\right).$$

First of all, the diagram

$$\begin{array}{ccc} \mathbb{1}\boxtimes\mathbb{1} & \xrightarrow{\mathrm{coev}} & M^\vee\bar{\otimes}M \\ {\scriptstyle \mathrm{coev}\boxtimes\mathrm{coev}}\downarrow & & \downarrow{\scriptstyle 1\bar{\otimes}i} \\ (A^\vee\boxtimes{}^\vee B)\bar{\otimes}(A\boxtimes B) & \xrightarrow{i^t\bar{\otimes}1} & M^\vee\bar{\otimes}(A\boxtimes B) \end{array}$$

is commutative. This follows from the diagram

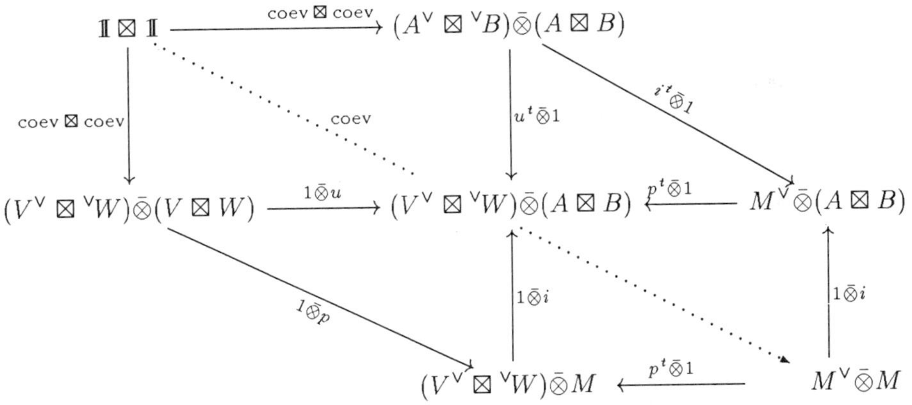

Now we prove that the composite

$$M \simeq M\bar{\otimes}(\mathbb{1}\boxtimes\mathbb{1}) \xrightarrow{1\bar{\otimes}\,\mathrm{coev}} M\bar{\otimes}M^\vee\bar{\otimes}M \xrightarrow{\mathrm{ev}\,\bar{\otimes}1} (\mathbb{1}\boxtimes\mathbb{1})\bar{\otimes}M \simeq M$$

is the identity. Indeed,

$$\begin{aligned}
&\left(M \xrightarrow{i} A\boxtimes B\right) \\
&= \Big(M \xrightarrow{i} A\boxtimes B \simeq (A\boxtimes B)\bar{\otimes}(\mathbb{1}\boxtimes\mathbb{1}) \xrightarrow{1\bar{\otimes}\,\mathrm{coev}} \\
&\qquad (A\boxtimes B)\bar{\otimes}(A^\vee\boxtimes{}^\vee B)\bar{\otimes}(A\boxtimes B) \xrightarrow{\mathrm{ev}\,\bar{\otimes}1} (\mathbb{1}\boxtimes\mathbb{1})\bar{\otimes}(A\boxtimes B) \simeq A\boxtimes B\Big) \\
&= \Big(M \simeq M\bar{\otimes}(\mathbb{1}\boxtimes\mathbb{1}) \xrightarrow{1\bar{\otimes}\,\mathrm{coev}} M\bar{\otimes}(A^\vee\boxtimes{}^\vee B)\bar{\otimes}(A\boxtimes B) \xrightarrow{1\bar{\otimes}i^t\bar{\otimes}1} \\
&\qquad M\bar{\otimes}M^\vee\bar{\otimes}(A\boxtimes B) \xrightarrow{\mathrm{ev}\,\bar{\otimes}1} (\mathbb{1}\boxtimes\mathbb{1})\bar{\otimes}(A\boxtimes B) \simeq A\boxtimes B\Big) \\
&= \Big(M \simeq M\bar{\otimes}(\mathbb{1}\boxtimes\mathbb{1}) \xrightarrow{1\bar{\otimes}\,\mathrm{coev}} M\bar{\otimes}M^\vee\bar{\otimes}M \xrightarrow{\mathrm{ev}\,\bar{\otimes}1} (\mathbb{1}\boxtimes\mathbb{1})\bar{\otimes}M \simeq M \xrightarrow{i} A\boxtimes B\Big)
\end{aligned}$$

and i is a monomorphism.

The composite

$$M^\vee \simeq (\mathbb{I}\boxtimes\mathbb{I})\bar{\otimes}M^\vee \xrightarrow{\operatorname{coev}\bar{\otimes}1} M^\vee\bar{\otimes}M\bar{\otimes}M^\vee \xrightarrow{1\bar{\otimes}\operatorname{ev}} M^\vee\bar{\otimes}(\mathbb{I}\boxtimes\mathbb{I}) \simeq M^\vee$$

equals identity. Indeed,

$$\begin{aligned}
&\big(A^\vee \boxtimes {}^\vee B \xrightarrow{i^t} M^\vee \simeq (\mathbb{I}\boxtimes\mathbb{I})\bar{\otimes}M^\vee \xrightarrow{\operatorname{coev}\bar{\otimes}1} \\
&\qquad M^\vee\bar{\otimes}M\bar{\otimes}M^\vee \xrightarrow{1\bar{\otimes}\operatorname{ev}} M^\vee\bar{\otimes}(\mathbb{I}\boxtimes\mathbb{I}) \simeq M^\vee\big)\\
&= \big(A^\vee\boxtimes{}^\vee B \simeq (\mathbb{I}\boxtimes\mathbb{I})\bar{\otimes}(A^\vee\boxtimes{}^\vee B) \xrightarrow{\operatorname{coev}\bar{\otimes}1} M^\vee\bar{\otimes}M\bar{\otimes}(A^\vee\boxtimes{}^\vee B) \xrightarrow{1\bar{\otimes}i\bar{\otimes}1} \\
&\qquad M^\vee\bar{\otimes}(A\boxtimes B)\bar{\otimes}(A^\vee\boxtimes{}^\vee B) \xrightarrow{1\bar{\otimes}\operatorname{ev}} M^\vee\bar{\otimes}(\mathbb{I}\boxtimes\mathbb{I}) \simeq M^\vee\big)\\
&= \big(A^\vee\boxtimes{}^\vee B \simeq (\mathbb{I}\boxtimes\mathbb{I})\bar{\otimes}(A^\vee\boxtimes{}^\vee B) \xrightarrow{\operatorname{coev}\bar{\otimes}1} \\
&\qquad (A^\vee\boxtimes{}^\vee B)\bar{\otimes}(A\boxtimes B)\bar{\otimes}(A^\vee\boxtimes{}^\vee B) \xrightarrow{i^t\bar{\otimes}\operatorname{ev}} M^\vee\bar{\otimes}(\mathbb{I}\boxtimes\mathbb{I}) \simeq M^\vee\big)\\
&= \big(A^\vee\boxtimes{}^\vee B \xrightarrow{i^t} M^\vee\big)
\end{aligned}$$

and i^t is an epimorphism. Therefore, $M^\vee$ is the right dual to M.

Since the evaluation for M is defined through resolution (1.6.2) and the coevaluation is defined through resolution (1.6.1), we conclude also that ev and coev do not depend on the choice of resolutions. Similarly for ${}^\vee M$. □

1.7. Ind-objects

Following Grothendieck and Verdier [**14**] we consider the category of ind-objects of a given $\Bbbk$-linear abelian category $\mathcal{A}$. Recall that an ind-object of $\mathcal{A}$ is a functor $X : I \to \mathcal{A}$ from a filtered partially ordered set I, in particular, an arrow x_{ij} : $X_i \to X_j \in \mathcal{A}$ is given for $i < j$, $i, j \in I$. The set of morphisms from $X : I \to \mathcal{A}$ to $Y : J \to \mathcal{A}$ is

$$\varprojlim_I \varinjlim_J \operatorname{Hom}(X_i, Y_j).$$

We denote the category of ind-objects of $\mathcal{A}$ by $\widehat{\mathcal{A}}$ as a synonym of standard notation $\operatorname{Ind}(\mathcal{A})$. The category $\widehat{\mathcal{A}}$ is a $\Bbbk$-linear abelian category [**14**, Exercise 8.9.9]. Small projective and inductive limits in $\widehat{\mathcal{A}}$ are representable in $\widehat{\mathcal{A}}$ (Grothendieck and Verdier [**14**, Propositions 8.9.1 and 8.9.5]).

1.7.1. Theorem (Grothendieck and Verdier [14] Theorem 8.3.3). *Let $\mathcal{A}$ be essentially small, then the functor*

$$\begin{aligned}
\widehat{\mathcal{A}} &\longrightarrow \operatorname{Hom}_{\Bbbk,l.e.}(\mathcal{A}^{\mathrm{op}}, \Bbbk\text{-Vect}),\\
(X : I \to \mathcal{A}) &\longmapsto (Y \mapsto \varinjlim_I \operatorname{Hom}(Y, X_i))
\end{aligned}$$

with values in the category of $\Bbbk$-linear left exact functors is an equivalence of categories.

If $\mathcal{A}, \mathcal{B}$ are $\Bbbk$-linear abelian essentially small categories, then any functor F : $\mathcal{A} \to \mathcal{B}$ extends to a functor $\hat{F} : \widehat{\mathcal{A}} \to \widehat{\mathcal{B}}$, $X \mapsto F \circ X$. If F is $\Bbbk$-linear (resp. right exact, resp. exact), so is $\hat{F}$ (Grothendieck and Verdier [**14**, Corollary 8.9.8]). The functor $\hat{F} : \widehat{\mathcal{A}} \to \widehat{\mathcal{B}}$ commutes with filtered inductive limits [**14**, Proposition 8.6.3]. If F is right exact, $\hat{F}$ commutes with arbitrary inductive limits.

1.7.2. Proposition. *Let any object of $\mathcal{A}$ have finite length. Then the category $\widehat{\mathcal{A}}$ is equivalent to its full subcategory consisting of functors $X : I \to \mathcal{A}$ such that $x_{ij} : X_i \to X_j$ is a monomorphism for any pair $i < j$.*

PROOF. Let us consider an object $X : I \to \mathcal{A} \in \widehat{\mathcal{A}}$. Let us fix an element $i \in I$ and consider the inductive system $(\operatorname{Ker}(x_{ij} : X_i \to X_j))_{j>i}$ of subobjects of X_i. By hypothesis, there exists an element $i' \geqslant i$ such that for any $i'' \geqslant i'$ the morphism

$$\operatorname{Ker}(x_{ii''} : X_i \to X_{i''}) \to Y_i \stackrel{\text{def}}{=} \varinjlim_{j>i} \operatorname{Ker}(x_{ij} : X_i \to X_j)$$

is an isomorphism. Define $\bar{X}_i = X_i/Y_i$.

For any ordered pair $i < j$ choose an element $k \in I$ such that $k \geqslant i', j'$. Then $\operatorname{Ker}(x_{ik} : X_i \to X_k) \simeq Y_i$, $\operatorname{Ker}(x_{jk} : X_j \to X_k) \simeq Y_j$. Considering the composite $x_{ik} = \left(X_i \xrightarrow{x_{ij}} X_j \xrightarrow{x_{jk}} X_k\right)$ we check that in obvious conventions $\overset{-1}{x_{ij}}(Y_j) = Y_i$. Therefore, x_{ij} induces a factor-morphism $\bar{x}_{ij} : \bar{X}_i \to \bar{X}_j$, which is a monomorphism. This defines an object $\bar{X} : I \to \mathcal{A}$ of $\widehat{\mathcal{A}}$. Clearly, the canonical projections $p_i : X_i \to X_i/Y_i = \bar{X}_i$ yield a morphism $X \to \bar{X}$ in $\widehat{\mathcal{A}}$. The inverse morphism is given by the uniquely determined family $t_i : \bar{X}_i \to X_{i'}$, such that $x_{ii'} = \left(X_i \xrightarrow{p_i} \bar{X}_i \xrightarrow{t_i} X_{i'}\right)$. $\square$

1.7.3. Remark. Let $X : I \to \mathcal{A}$ be in $\widehat{\mathcal{A}}$ and let $J \subset I$ be a cofinal set. Then the ind-object $X' = X|_J : J \to \mathcal{A}$ is isomorphic to X in $\widehat{\mathcal{A}}$.

Let $X : I \to \mathcal{A}$ be an ind-object such that $x_{ij} : X_i \to X_j$ are monomorphisms, $i < j$. We say that X is *closed under intersections* if

(a) for any subset $J \subset I$ there is an element $i = \cap J \in I$ such that $i \leqslant j$ for any $j \in J$ and i is the biggest element with this property;
(b) for any subset $J \subset I$ there is a finite subset $J' \subset J$ such that for any finite K, $J' \subset K \subset J$, and any $r \geqslant K$ the subobject X_i is the intersection of subobjects X_k, $k \in K$, in X_r, that is, the canonical morphism

$$X_i \to \varprojlim_{k \in K} (X_k \xrightarrow{x_{kr}} X_r)$$

is an isomorphism.

1.7.4. Proposition. *Any* ind-*object is isomorphic in $\widehat{\mathcal{A}}$ to an* ind-*object closed under intersections.*

PROOF. We may assume that $x_{ij} : X_i \to X_j$ in a given $X : I \to \mathcal{A}$ are monomorphisms. Define $\tilde{I}$ as the set of finite subsets of I. For $\tilde{\imath} \in \tilde{I}$ let $r \in I$ be such that $r \geqslant \tilde{\imath}$ and let

$$Y_{\tilde{\imath}} = \varprojlim_{k \in \tilde{\imath}} (X_k \xrightarrow{x_{kr}} X_r) = \text{“} \cap_{k \in \tilde{\imath}} X_k \text{”}$$

(this does not depend on r). Denote $z_{\tilde{\imath}r} : Y_{\tilde{\imath}} \to X_r$ the canonical embedding. Introduce in $\tilde{I}$ the natural preorder of subobjects, namely $\tilde{\jmath} \succcurlyeq \tilde{\imath}$ if for r such that $r \geqslant \tilde{\imath}$, $r \geqslant \tilde{\jmath}$ there exists a morphism $y_{\tilde{\imath}\tilde{\jmath}} : Y_{\tilde{\imath}} \to Y_{\tilde{\jmath}}$ such that $z_{\tilde{\imath}r} = (Y_{\tilde{\imath}} \xrightarrow{y_{\tilde{\imath}\tilde{\jmath}}} Y_{\tilde{\jmath}} \xrightarrow{z_{\tilde{\jmath}r}} X_r)$. Then $y_{\tilde{\imath}\tilde{\jmath}}$ is necessarily unique and monic. Let $\bar{I}$ be a set of representatives of equivalence classes with respect to the preorder $\succcurlyeq$. Clearly, $Y : \bar{I} \to \mathcal{A}$ is an ind-object, which is isomorphic to X.

If J is a finite subset of $\bar{I}$, the element $\tilde{\imath} = \cap J \in \bar{I}$ is defined as (the representative of the equivalence class of) $\cup_{\tilde{\jmath} \in J} \tilde{\jmath} \subset I$. It is easy to check (a) and (b) for finite J. Since objects in $\mathcal{A}$ have finite length, it follows that Y is closed under intersections. □

1.7.5. Proposition. *Let $X : I \to \mathcal{A}$, $Y : J \to \mathcal{A}$ be ind-objects. Assume that Y is closed under intersections. Then any morphism $f : X \to Y \in \widehat{\mathcal{A}}$ can be represented by a monotonic map $m : I \to J$ and a family of morphisms $f_i : X_i \to Y_{m(i)}$.*

PROOF. Indeed, for any $i \in I$ we have a morphism $\tilde{f}_i : X_i \to Y_j$ for some $j \in J$. Let $K \subset J$ be the set of $k \in J$ for which there exists $g_{ik} : X_i \to Y_k$ such that

$$(X_i \xrightarrow{g_{ik}} Y_k \xrightarrow{y_{kr}} Y_r) = (X_i \xrightarrow{\tilde{f}_i} Y_j \xrightarrow{y_{jr}} Y_r)$$

for any $r \geqslant k, j$. Set $m(i) = \cap K$. The morphism $f_i : X_i \to Y_{m(i)}$ is determined uniquely by the property (b). □

Let $\mathcal{A}, \mathcal{B}$ be $\Bbbk$-linear abelian categories. Then so is $\mathcal{A} \times \mathcal{B}$.

1.7.6. Proposition. *The categories $\widehat{\mathcal{A}} \times \widehat{\mathcal{B}}$ and $\widehat{\mathcal{A} \times \mathcal{B}}$ are equivalent.*

PROOF. There is a functor $F : \widehat{\mathcal{A} \times \mathcal{B}} \to \widehat{\mathcal{A}} \times \widehat{\mathcal{B}}$, $(X_i \times Y_i)_{i \in I} \mapsto (X_i)_{i \in I} \times (Y_i)_{i \in I}$. There is a functor $G : \widehat{\mathcal{A}} \times \widehat{\mathcal{B}} \to \widehat{\mathcal{A} \times \mathcal{B}}$, $(X_i)_{i \in I} \times (Y_j)_{j \in J} \mapsto (X_i \times Y_j)_{(i,j) \in I \times J}$. The composite $G \circ F$ is isomorphic to $\mathrm{Id}_{\widehat{\mathcal{A} \times \mathcal{B}}}$, that is, $(X_i \times Y_i)_{i \in I} \simeq (X_i \times Y_j)_{i,j \in I}$ since the diagonal is a cofinal set in $I \times I$. The composite $F \circ G$ is isomorphic to $\mathrm{Id}_{\widehat{\mathcal{A}} \times \widehat{\mathcal{B}}}$, that is, $(X_i)_{i \in I} \times (Y_j)_{j \in J} \simeq (X_i)_{i \in I, j \in J} \times (Y_j)_{i \in I, j \in J}$ since the ind-objects $I \to \mathcal{A}$, $i \mapsto X_i$ and $I \times J \to \mathcal{A}$, $(i,j) \mapsto X_i$ are isomorphic. □

Assume now that $\mathcal{V}$ is a $\Bbbk$-linear abelian rigid monoidal category. The functor $\otimes : \mathcal{V} \times \mathcal{V} \to \mathcal{V}$ extends to a functor $\widehat{\otimes} : \widehat{\mathcal{V} \times \mathcal{V}} \to \widehat{\mathcal{V}}$. Using Proposition 1.7.6 we can introduce the functor

$$\otimes : \widehat{\mathcal{V}} \times \widehat{\mathcal{V}} \xrightarrow{G} \widehat{\mathcal{V} \times \mathcal{V}} \xrightarrow{\widehat{\otimes}} \widehat{\mathcal{V}}, \quad (X_i)_{i \in I} \times (Y_j)_{j \in J} \mapsto (X_i \otimes Y_j)_{i \in I, j \in J}$$

which turn $\widehat{\mathcal{V}}$ into a monoidal category. In general, $\widehat{\mathcal{V}}$ is not rigid.

1.7.7. Proposition ([20]). *If $\mathcal{V}$ is essentially small, then $\widehat{\mathcal{V}}$ is a closed monoidal category.*

PROOF. For given $X, Y \in \mathrm{Ob}\,\widehat{\mathcal{V}}$ introduce a category $\mathcal{D} = \mathcal{D}_{X,Y}$, whose objects are morphisms $Z \otimes X \to Y \in \widehat{\mathcal{V}}$, where $Z \in \mathrm{Ob}\,\mathcal{V}$. By definition,

$$\mathrm{Hom}_{\mathcal{D}}(\phi : Z \otimes X \to Y, \psi : W \otimes X \to Y) = \{f : Z \to W \mid \psi \circ (f \otimes X) = \phi\}.$$

$\mathcal{D}$ is essentially small. There is a functor $\mathcal{D} \to \widehat{\mathcal{V}}$, $(Z \otimes X \to Y) \mapsto Z$. The left inner homomorphism object is given by the colimit

$$\underline{\mathrm{Hom}}^l(X, Y) = \varinjlim(\mathcal{D} \to \widehat{\mathcal{V}}).$$

Replacing $\otimes$ by $\otimes_{\mathrm{op}}$ we get $\underline{\mathrm{Hom}}^r(X, Y)$. □

1.8. Rigid-abelian categories

We define rigid bicategories in the particular case of braided monoidal bicategories. In the general case we would ask for left and right duals.

1.8.1. Definition. An object A of a braided monoidal bicategory $\mathfrak{B}$ is *rigid* if it has a dual object $\mathcal{A}^\vee \in \mathfrak{B}$ and a pair of 1-morphisms $\mathrm{ev} : \mathcal{A}^\vee \boxtimes \mathcal{A} \to \mathcal{I}$, $\mathrm{coev} : \mathcal{I} \to \mathcal{A} \boxtimes \mathcal{A}^\vee$, where $\mathcal{I}$ is the unit object, such that the composite

$$\mathcal{A} \simeq \mathcal{I} \boxtimes \mathcal{A} \xrightarrow{\mathrm{coev} \boxtimes \mathcal{A}} (\mathcal{A} \boxtimes \mathcal{A}^\vee) \boxtimes \mathcal{A} \simeq \mathcal{A} \boxtimes (\mathcal{A}^\vee \boxtimes \mathcal{A}) \xrightarrow{\mathcal{A} \boxtimes \mathrm{ev}} \mathcal{A} \boxtimes \mathcal{I} \simeq \mathcal{A} \tag{1.8.1}$$

is isomorphic to $\mathrm{Id}_{\mathcal{A}}$, and the composite

$$\mathcal{A}^\vee \simeq \mathcal{A}^\vee \boxtimes \mathcal{I} \xrightarrow{\mathcal{A}^\vee \boxtimes \mathrm{coev}} \mathcal{A}^\vee \boxtimes (\mathcal{A} \boxtimes \mathcal{A}^\vee)$$
$$\simeq (\mathcal{A}^\vee \boxtimes \mathcal{A}) \boxtimes \mathcal{A}^\vee \xrightarrow{\mathrm{ev} \boxtimes \mathcal{A}^\vee} \mathcal{I} \boxtimes \mathcal{A}^\vee \simeq \mathcal{A}^\vee \tag{1.8.2}$$

is isomorphic to $\mathrm{Id}_{\mathcal{A}^\vee}$. A bicategory is rigid if any object is rigid.

Let $\mathcal{A} = A$-mod for a finite dimensional associative algebra A. Then $\mathcal{A}^{\mathrm{op}} \to$ mod-A, $X \mapsto X^* = \mathrm{Hom}_{\Bbbk}(X, \Bbbk)$, $f \mapsto f^t$ is an equivalence. Hence, $\mathcal{A} \boxtimes \mathcal{A}^{\mathrm{op}}$ can be viewed as A-bimod.

1.8.2. Lemma. *The A-bimodule $\int^{X \in \mathcal{A}} X \boxtimes X^*$, corresponding to $\int^{X \in \mathcal{A}} X \boxtimes X \in \widehat{\mathcal{A} \boxtimes \mathcal{A}^{\mathrm{op}}}$, is isomorphic to $({}_A A_A)^*$.*

Proof. The category A-mod can be reconstructed from the functor $p : \mathcal{P} \to \Bbbk$-vect, where $\mathrm{Ob}\,\mathcal{P}$ consists of one object A, $pA = A$ and $\mathrm{Mor}\,\mathcal{P} \simeq A$ is projected by p to $\{R_a : A \to A \mid a \in A\}$, $R_a b = ba$. By Theorem 2.8.3 $\int^{X \in \mathcal{P}} X \boxtimes X^* \simeq \int^{X \in \mathcal{A}} X \boxtimes X^*$ as an A-bimodule. By definition

$$\oplus_{a \in A} A \otimes A^* \xrightarrow{1 \otimes R_a^t - R_a \otimes 1} A \otimes A^* \to \int^{X \in \mathcal{P}} X \otimes X^* \to 0$$

is exact. Therefore,

$$\int^{X \in \mathcal{P}} X \otimes X^* = {}_A A_A \otimes_A {}_A(A^*)_A = {}_A(A^*)_A = ({}_A A_A)^*. \quad \square$$

1.8.3. Proposition. *Let $\pi : B \twoheadrightarrow A$ be a $\Bbbk$-algebra epimorphism, $\dim_{\Bbbk} B < \infty$. The full embedding of categories $\mathcal{A} = A$-mod $\hookrightarrow B$-mod $= \mathcal{B}$ induces a morphism of coends $\int^{X \in \mathcal{A}} X \boxtimes X \to \int^{X \in \mathcal{B}} X \boxtimes X \in \mathcal{B} \boxtimes \mathcal{B}^{\mathrm{op}}$, which identifies with $\pi^* : A^* \to B^*$ in B-bimod.*

Proof. We use the same functor $p : \mathcal{P} \to \Bbbk$-vect as in Lemma 1.8.2 viewing the object A as a left B-module, and A^* as a right B-module. The exact sequence $0 \to \mathrm{Ker}\,\pi \to B \xrightarrow{\pi} A \to 0$ identifies $\mathrm{Im}(i_A : A \otimes A^* \to \int^{X \in \mathcal{B}} X \otimes X^*)$ with a subquotient of $\mathrm{Im}(i_B : B \otimes B^* \to \int^{X \in \mathcal{B}} X \otimes X^*)$ by Step 6 on page 64. Namely, $a \otimes \alpha \sim b \otimes \pi^*(\alpha)$, where $\pi(b) = a \in A$, $\alpha \in A^*$. The induced map $A \otimes_A A^* \to B \otimes_B B^*$, $1 \otimes \alpha \mapsto 1 \otimes \pi^*(\alpha)$ is precisely $\pi^* : A^* \to B^*$. $\square$

1.8.4. Corollary. *If $\mathcal{C} = C$-comod, then $\int^{X \in \mathcal{C}} X \boxtimes X^* \simeq C$ as a C-bicomodule.*

1.8.5. Proposition. *Let $\mathcal{A} \in \mathrm{Ob}\,\mathfrak{Ab}_{\Bbbk}^{l}$. The coend $\int^{X \in \mathcal{A}} X \boxtimes X$ exists in $\mathcal{A} \boxtimes \mathcal{A}^{\mathrm{op}}$ iff $\mathcal{A}$ is equivalent to A-mod, $\dim_{\Bbbk} A < \infty$.*

PROOF. If $\mathcal{A}$ is equivalent to A-mod, the existence of $\int^{X\in\mathcal{A}} X\boxtimes X$ in $\mathcal{A}\boxtimes\mathcal{A}^{\mathrm{op}}$ follows from Lemma 1.8.2.

Let $\mathcal{A}\in\mathfrak{Ab}_{\Bbbk}$. Choose an additively closed family $\{X_i\}_{i\in I}$ of objects of $\mathcal{A}$ so that $\mathcal{A}_i=\langle X_i\rangle$ form an inductive system with the colimit $\mathcal{A}$. Then

$$\int^{X\in\mathcal{A}} X\boxtimes X\simeq\varinjlim_i\int^{X\in\mathcal{A}_i} X\boxtimes X.$$

Indeed, any $X\in\mathcal{A}$ belongs to one of $\mathcal{A}_i$, say $\mathcal{A}_n$, and the corresponding canonical morphism is

$$i_X: X\boxtimes X\to\int^{X\in\mathcal{A}_n} X\boxtimes X\to\varinjlim_i\int^{X\in\mathcal{A}_i} X\boxtimes X.$$

The compatibility and universality check is easy.

Denote $C_i=\int^{X\in\mathcal{A}_i} X\boxtimes X\in\mathcal{A}\boxtimes\mathcal{A}^{\mathrm{op}}$. If $i\prec j$, the category $\mathcal{A}_i$ is a full subcategory of $\mathcal{A}_j$ stable with respect to subquotients. Hence, the functor $\mathcal{A}_i\hookrightarrow\mathcal{A}_j$ is equivalent to the functor A_i-mod $\to A_j$-mod for some algebra epimorphism $\pi: A_j \twoheadrightarrow A_i$, or coalgebra monomorphism $\pi^*: A_i^*\hookrightarrow A_j^*$. If $\mathcal{A}_i\hookrightarrow\mathcal{A}_j$ is not an equivalence, then $C_i\hookrightarrow C_j$ is not an isomorphism by Proposition 1.8.3. Hence, $\mathrm{length}_{\mathcal{A}}\, C_i<\mathrm{length}_{\mathcal{A}}\, C_j$.

If $\mathrm{colim}_i\, C_i\in\mathcal{A}\boxtimes\mathcal{A}^{\mathrm{op}}$, then its length is finite. Thus, a chain $\mathcal{A}_{i_1}\subset\mathcal{A}_{i_2}\subset\dots\subset\mathcal{A}_{i_k}\subset\dots$, such that $X_{i_k}=X_{i_{k-1}}\oplus Y_k$, $Y_k\notin\mathrm{Ob}\,\mathcal{A}_{i_{k-1}}$, can not be infinite. Therefore, $\mathcal{A}$ is equivalent to $\mathcal{A}_j$ for some $j\in I$. □

1.8.6. Theorem. *An object $\mathcal{A}$ of $\mathfrak{Ab}^l_{\Bbbk}$ is rigid iff $\mathcal{A}$ is equivalent to a category A-mod for some finite dimensional $\Bbbk$-algebra A.*

PROOF. Assume that $\mathcal{A}$ is equivalent to A-mod, $\dim_{\Bbbk} A<\infty$. Then $\mathcal{A}$ has enough injectives.

Set $\mathcal{A}^\vee=\mathcal{A}^{\mathrm{op}}$, define $\mathrm{ev}:\mathcal{A}^{\mathrm{op}}\boxtimes\mathcal{A}\to\Bbbk$-vect as a left exact $\Bbbk$-linear functor such that

$$\mathrm{Hom}\simeq\big(\mathcal{A}^{\mathrm{op}}\times\mathcal{A}\xrightarrow{\boxtimes}\mathcal{A}^{\mathrm{op}}\boxtimes\mathcal{A}\xrightarrow{\mathrm{ev}}\Bbbk\text{-vect}\big), \tag{1.8.3}$$

and define

$$\mathrm{coev}:\Bbbk\text{-vect}\to\mathcal{A}\boxtimes\mathcal{A}^{\mathrm{op}},\qquad V\mapsto V\otimes\int^{X\in\mathcal{A}} X\boxtimes X.$$

We have to check that (1.8.1) and (1.8.2) are isomorphic to identity functors.

For (1.8.1) this amounts to a functorial isomorphism

$$(1_1\boxtimes\mathrm{ev}_{23})\int^{X\in\mathcal{A}} X\boxtimes X\boxtimes W\simeq W, \tag{1.8.4}$$

$W\in\mathcal{A}$. Recall that for any object $G\in\mathcal{A}\boxtimes\mathcal{A}^{\mathrm{op}}$ there is an exact sequence

$$0\to G\to B\boxtimes C\xrightarrow{\sum f_l\boxtimes g_l} D\boxtimes E\in\mathcal{A}\boxtimes\mathcal{A}^{\mathrm{op}}$$

and $1\boxtimes\mathrm{ev}$ is defined via

$$0\to(1_1\boxtimes\mathrm{ev}_{23})(G_{12}\boxtimes W_3)\to B\otimes\mathrm{Hom}(C,W)$$
$$\xrightarrow{\sum f_l\otimes\mathrm{Hom}(g_l,W)} D\otimes\mathrm{Hom}(E,W).$$

When W is injective, Hom(-, W) is exact and from the definition

$$\bigoplus_{f:Y\to Z} Y \boxtimes Z \to \bigoplus_X X \boxtimes X \to \int^{X\in\mathcal{A}} X \boxtimes X \to 0$$

we get an exact sequence

$$\bigoplus_{f:Y\to Z} Y \otimes \mathrm{Hom}(Z,W) \to \bigoplus_X X \otimes \mathrm{Hom}(X,W)$$

$$\to (1_1 \boxtimes \mathrm{ev}_{23}) \int^{X\in\mathcal{A}} X \boxtimes X \boxtimes W \to 0.$$

It follows for injective W that

$$(1_1 \boxtimes \mathrm{ev}_{23}) \int^{X\in\mathcal{A}} X \boxtimes X \boxtimes W \simeq \int^X X \otimes \mathrm{Hom}(X,W)$$

functorially in W.

Recall that the composition of morphisms

$$\mathrm{Hom}(W, X \otimes \mathrm{Hom}(X,Y)) \simeq \mathrm{Hom}(W,X) \otimes_{\Bbbk} \mathrm{Hom}(X,Y) \xrightarrow{\circ} \mathrm{Hom}(W,Y)$$

determines the canonical morphism denoted

$$\mathrm{ev}_{X,Y} : X \otimes \mathrm{Hom}_{\mathcal{A}}(X,Y) \to Y \in \mathcal{A}.$$

1.8.7. Lemma. *The system of morphisms* $(\mathrm{ev}_{X,Y})_{X\in \mathrm{Ob}\,\mathcal{A}}$ *determines an isomorphism*

$$\mathrm{ev}_Y : \int^{X\in\mathcal{A}} X \otimes \mathrm{Hom}_{\mathcal{A}}(X,Y) \to Y.$$

PROOF. Indeed, assume that a system of morphisms is given $(t_X : X \otimes \mathrm{Hom}_{\mathcal{A}}(X,Y) \to T \in \widehat{\mathcal{A}})_X$ such that the diagram

$$\begin{array}{ccc} X \otimes \mathrm{Hom}(Z,Y) & \xrightarrow{X\otimes \mathrm{Hom}(f,Y)} & X \otimes \mathrm{Hom}(X,Y) \\ {\scriptstyle f\otimes \mathrm{Hom}(Z,Y)}\downarrow & & \downarrow{\scriptstyle t_X} \\ Z \otimes \mathrm{Hom}(Z,Y) & \xrightarrow{t_Z} & T \end{array}$$

is commutative for any $f : X \to Z \in \mathcal{A}$. The morphism t_Y represents the natural map

$$\mathrm{Hom}(W,Y) \otimes_{\Bbbk} \mathrm{Hom}(Y,Y) \to \mathrm{Hom}(W,T),$$

which being restricted to $g \otimes 1_Y$, gives a natural map $\mathrm{Hom}(W,Y) \to \mathrm{Hom}(W,T)$ and therefore a morphism $s : Y \to T$.

For any $f : X \to Y = Z$ the diagram

$$\begin{array}{ccc} X \hookrightarrow X \otimes \mathrm{Hom}(Y,Y) & \xrightarrow{X\otimes \mathrm{Hom}(f,Y)} & X \otimes \mathrm{Hom}(X,Y) \\ {\scriptstyle f}\downarrow \qquad {\scriptstyle f\otimes 1}\downarrow & & \downarrow{\scriptstyle t_X} \\ Y \hookrightarrow Y \otimes \mathrm{Hom}(Y,Y) & \xrightarrow{t_Y} & T \end{array}$$

(or rather its Hom(W,-) version) shows that

$$t_X = (X \otimes \mathrm{Hom}(X,Y) \xrightarrow{\mathrm{ev}} Y \xrightarrow{s} T).$$

Uniqueness of such s follows from the particular case $X = Y$. Therefore, the morphism $\mathrm{ev}_Y : F(Y) \to Y$ is an isomorphism. $\square$

Coming back to the proof of Theorem 1.8.6 we deduce that (1.8.4) holds for injective W.

Using an injective resolution $0 \to Y \to V \to W$ of an arbitrary object $Y \in \mathcal{A}$ we get the diagram

$$\begin{array}{ccccccc} 0 & \longrightarrow & (1 \boxtimes \mathrm{ev}) \int^X X \boxtimes X \boxtimes Y & \longrightarrow & (1 \boxtimes \mathrm{ev}) \int^X X \boxtimes X \boxtimes V & \longrightarrow & (1 \boxtimes \mathrm{ev}) \int^X X \boxtimes X \boxtimes W \\ & & & & \downarrow \simeq & & \downarrow \simeq \\ 0 & \longrightarrow & Y & \longrightarrow & V & \longrightarrow & W \end{array}$$

This implies a functorial isomorphism

$$(1_1 \boxtimes \mathrm{ev}_{23}) \int^{X \in \mathcal{A}} X \boxtimes X \boxtimes Y \simeq Y.$$

The case of (1.8.2) is similar ($\mathcal{A}$ and $\mathcal{A}^{\mathrm{op}}$ change the places). Hence, $\mathcal{A}$ is a rigid object of $\mathfrak{Ab}^l_{\Bbbk}$.

Assume now that $\mathcal{A}$ is a rigid object of $\mathfrak{Ab}^l_{\Bbbk}$. Define $\mathrm{Hom} : \mathcal{A}^{\mathrm{op}} \boxtimes \mathcal{A} \to \Bbbk\text{-vect}$ as a left exact functor extending the usual Hom (see (1.8.3)) and $\mathrm{coend} : \Bbbk\text{-vect} \to \widehat{\mathcal{A} \boxtimes \mathcal{A}^{\mathrm{op}}}$, $\Bbbk \mapsto \int^X X \boxtimes X$. Then, using the presentation $\mathcal{A} = \mathrm{colim}\, \mathcal{A}_i$, $\mathcal{A}_i = \langle X_i \rangle$, we see that

$$\mathcal{A} \simeq \mathcal{I} \boxtimes \mathcal{A} \xrightarrow{\mathrm{coend} \boxtimes \mathcal{A}} \widehat{\mathcal{A} \boxtimes \mathcal{A}^{\mathrm{op}}} \boxtimes \mathcal{A}$$
$$\hookrightarrow (\mathcal{A} \boxtimes \mathcal{A}^{\mathrm{op}} \boxtimes \mathcal{A})^{\wedge} \xrightarrow{\mathcal{A} \boxtimes \mathrm{Hom}} \widehat{\mathcal{A} \boxtimes \mathcal{I}} \simeq \widehat{\mathcal{A}} \quad =$$

$$\mathcal{A}^{\mathrm{op}} \simeq \mathcal{A}^{\mathrm{op}} \boxtimes \mathcal{I} \xrightarrow{\mathcal{A}^{\mathrm{op}} \boxtimes \mathrm{coend}} \mathcal{A}^{\mathrm{op}} \boxtimes \widehat{\mathcal{A} \boxtimes \mathcal{A}^{\mathrm{op}}}$$
$$\hookrightarrow (\mathcal{A}^{\mathrm{op}} \boxtimes \mathcal{A} \boxtimes \mathcal{A}^{\mathrm{op}})^{\wedge} \xrightarrow{\mathrm{Hom} \boxtimes \mathcal{A}^{\mathrm{op}}} \widehat{\mathcal{I} \boxtimes \mathcal{A}^{\mathrm{op}}} \simeq \widehat{\mathcal{A}^{\mathrm{op}}} \quad =$$

are natural embeddings.

As $\mathcal{A}$ is a rigid object of $\mathfrak{Ab}^l_{\Bbbk}$, it has a dual $\mathcal{A}^{\vee}$ and adjunctions

$$\mathrm{ev} : \mathcal{A}^{\vee} \boxtimes \mathcal{A} \to \mathcal{I}, \qquad \mathrm{coev} : \mathcal{I} \to \mathcal{A} \boxtimes \mathcal{A}^{\vee}.$$

Then there are functors

$$F = \big(\mathcal{A}^{\mathrm{op}} \simeq \mathcal{A}^{\mathrm{op}} \boxtimes \mathcal{I} \xrightarrow{\mathcal{A}^{\mathrm{op}} \boxtimes \mathrm{coev}} \mathcal{A}^{\mathrm{op}} \boxtimes \mathcal{A} \boxtimes \mathcal{A}^{\vee} \xrightarrow{\mathrm{Hom} \boxtimes \mathcal{A}^{\vee}}$$
$$\mathcal{I} \boxtimes \mathcal{A}^{\vee} \simeq \mathcal{A}^{\vee}\big) \quad =$$

$$G = \big(\mathcal{A}^{\vee} \simeq \mathcal{A}^{\vee} \boxtimes \mathcal{I} \xrightarrow{\mathcal{A}^{\vee} \boxtimes \mathrm{coend}} \mathcal{A}^{\vee} \boxtimes \widehat{\mathcal{A} \boxtimes \mathcal{A}^{\mathrm{op}}}$$
$$\hookrightarrow (\mathcal{A}^{\vee} \boxtimes \mathcal{A} \boxtimes \mathcal{A}^{\mathrm{op}})^{\wedge} \xrightarrow{\mathrm{ev} \boxtimes \mathcal{A}^{\mathrm{op}}} \widehat{\mathcal{I} \boxtimes \mathcal{A}^{\mathrm{op}}} \simeq \widehat{\mathcal{A}^{\mathrm{op}}}\big) \quad =$$

They satisfy

$$(\mathcal{A}^{\mathrm{op}} \xrightarrow{F} \mathcal{A}^{\vee} \xrightarrow{G} \widehat{\mathcal{A}^{\mathrm{op}}}) \simeq (\mathcal{A}^{\mathrm{op}} \hookrightarrow \widehat{\mathcal{A}^{\mathrm{op}}}),$$

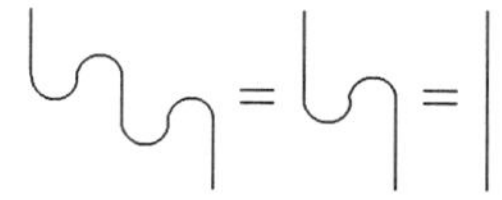

$$(\mathcal{A}^\vee \xrightarrow{G} \widehat{\mathcal{A}^{\mathrm{op}}} \xrightarrow{\hat{F}} \widehat{\mathcal{A}^\vee}) \simeq (\mathcal{A}^\vee \hookrightarrow \widehat{\mathcal{A}^\vee}),$$

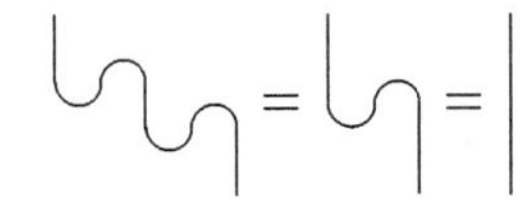

Therefore, F, G are full and faithful.

There is a commutative diagram

$$\begin{array}{ccc} \mathcal{I} & \xrightarrow{\text{coend}} & \widehat{\mathcal{A} \boxtimes \mathcal{A}^{\mathrm{op}}} \\ {\scriptstyle \text{coev}}\big\downarrow & & \big\downarrow {\scriptstyle \widehat{\mathcal{A}\boxtimes F}} \\ \mathcal{A} \boxtimes \mathcal{A}^\vee & \hookrightarrow & \widehat{\mathcal{A} \boxtimes \mathcal{A}^\vee} \end{array},$$

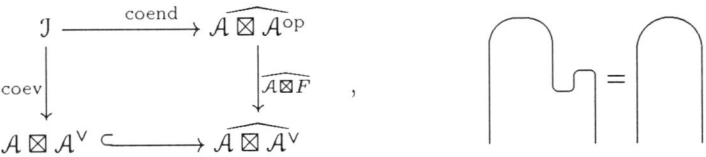

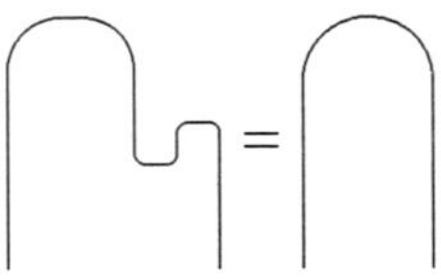

Let $(C_i)_{i \in I}$ be a filtered system of subobjects of $C = \int^{X \in \mathcal{A}} X \boxtimes X \in \widehat{\mathcal{A} \boxtimes \mathcal{A}^{\mathrm{op}}}$, such that $C_i \in \mathcal{A} \boxtimes \mathcal{A}^{\mathrm{op}}$ and $C = \operatorname{colim}_i C_i$. Applying the full faithful functor $\widehat{\mathcal{A} \boxtimes F}$ we get a filtered system of subobjects $D_i = (\mathcal{A} \boxtimes F)C_i$ of the object $D = \operatorname{coev}(\Bbbk) \in \mathcal{A} \boxtimes \mathcal{A}^\vee$. Hence, $D_l \simeq D$ for some l and for any pair (k, m) such that $l < k < m$ the morphism $D_k \to D_m$ is invertible. Therefore, $C_k \to C_m$ is an isomorphism and $C \simeq C_l \in \mathcal{A} \boxtimes \mathcal{A}^{\mathrm{op}}$. By Proposition 1.8.5 $\mathcal{A}$ is equivalent to A-mod. □

Now we shall extend our results to the case with a parameter.

1.8.8. Proposition. *Let $\mathcal{A}, \mathcal{B} \in \mathfrak{Ab}^l_{\Bbbk}$. Denote*

$$F(M, X) = \operatorname{Hom}^1(M_*, X_{*1}) \in \mathcal{B} \tag{1.8.5}$$

the value of the composite functor

$$\begin{aligned} F : \mathcal{A}^{\mathrm{op}} \times (\mathcal{A} \boxtimes \mathcal{B}) &\xrightarrow{\boxtimes} \mathcal{A}^{\mathrm{op}} \boxtimes \mathcal{A} \boxtimes \mathcal{B} \xrightarrow{\operatorname{Hom} \boxtimes \mathcal{B}} \Bbbk\text{-vect} \boxtimes \mathcal{B} \ \simeq \mathcal{B} \\ (M, Y \boxtimes Z) &\mapsto \quad M \boxtimes Y \boxtimes Z \mapsto \operatorname{Hom}_{\mathcal{A}}(M, Y) \boxtimes Z \mapsto \operatorname{Hom}_{\mathcal{A}}(M, Y) \otimes Z \end{aligned}$$

on the pair $M \in \mathcal{A}^{\mathrm{op}}$, $X \in \mathcal{A} \boxtimes \mathcal{B}$. Then (1.8.5) is the representing object for the functor

$$\mathcal{B}^{\mathrm{op}} \to \Bbbk\text{-vect}, \qquad N \mapsto \operatorname{Hom}_{\mathcal{A} \boxtimes \mathcal{B}}(M \boxtimes N, X).$$

PROOF. For objects of the form $X = Y \boxtimes Z$ this follows from the natural isomorphism

$$\begin{aligned} \operatorname{Hom}_{\mathcal{B}}(N, F(M, Y \boxtimes Z)) &= \operatorname{Hom}_{\mathcal{B}}(N, \operatorname{Hom}_{\mathcal{A}}(M, Y) \otimes Z) \\ &\simeq \operatorname{Hom}_{\mathcal{A}}(M, Y) \otimes_{\Bbbk} \operatorname{Hom}_{\mathcal{B}}(N, Z) \\ &\simeq \operatorname{Hom}_{\mathcal{A} \boxtimes \mathcal{B}}(M \boxtimes N, Y \boxtimes Z). \end{aligned}$$

The functor F is left exact in X. Using short sequences of the type (1.6.2) we deduce that the left exact in X functors $\operatorname{Hom}_{\mathcal{B}}(N, F(M, X))$ and $\operatorname{Hom}_{\mathcal{A} \boxtimes \mathcal{B}}(M \boxtimes N, X)$ are isomorphic not only for $X = Y \boxtimes Z$, but for any $X \in \mathcal{A} \boxtimes \mathcal{B}$. □

Functor (1.8.5) extends to $M \in \mathcal{A}$, $X \in \widehat{\mathcal{A} \boxtimes \mathcal{B}}$ as

$$F = (\mathcal{A}^{\mathrm{op}} \times \widehat{\mathcal{A} \boxtimes \mathcal{B}} \xrightarrow{\boxtimes} (\mathcal{A}^{\mathrm{op}} \boxtimes \mathcal{A} \boxtimes \mathcal{B})^\wedge \xrightarrow{\operatorname{Hom} \boxtimes \mathcal{B}} (\Bbbk\text{-vect} \boxtimes \mathcal{B})^\wedge \simeq \widehat{\mathcal{B}}).$$

Namely, if $X = \operatorname{colim}_I X^i$ is a filtered inductive limit, $X^i \in \mathcal{A} \boxtimes \mathcal{B}$, then

$$F(M, X) = \operatorname{Hom}^1(M_*, X_{*1}) \overset{\text{def}}{=} \varinjlim_I \operatorname{Hom}^1(M_*, X^i_{*1}) = \varinjlim_I F(M, X^i). \tag{1.8.6}$$

1.8.9. Corollary. *Formula (1.8.6) gives the representing object for the functor*

$$\mathcal{B}^{\mathrm{op}} \to \Bbbk\text{-Vect}, \qquad N \mapsto \operatorname{Hom}_{\widehat{\mathcal{A} \boxtimes \mathcal{B}}}(M \boxtimes N, X).$$

PROOF. Applying inductive limit to the adjunction of Proposition 1.8.8 we get

$$\operatorname{Hom}_{\widehat{\mathcal{B}}}(N, F(M,X)) = \operatorname{Hom}_{\widehat{\mathcal{B}}}(N, \varinjlim_I F(M,X^i)) = \varinjlim_I \operatorname{Hom}_{\mathcal{B}}(N, F(M,X^i)) \simeq$$

$$\varinjlim_I \operatorname{Hom}_{\mathcal{A}\boxtimes\mathcal{B}}(M\boxtimes N, X^i) = \operatorname{Hom}_{\widehat{\mathcal{A}\boxtimes\mathcal{B}}}(M\boxtimes N, \varinjlim_I X^i) = \operatorname{Hom}_{\widehat{\mathcal{A}\boxtimes\mathcal{B}}}(M\boxtimes N, X).$$

□

In particular, for $N = \operatorname{Hom}^1(M_*, X_{*1})$ we get a canonical morphism

$$M_1 \boxtimes \operatorname{Hom}^1(M_*, X_{*2}) \to X_{12}$$

corresponding to 1_N. Lemma 1.8.7 can be generalised to:

1.8.10. Lemma. *The induced morphism*

$$\int^{M\in\mathcal{A}} M_1 \boxtimes \operatorname{Hom}^1(M_*, X_{*2}) \to X_{12}$$

is an isomorphism.

1.8.11. Proposition. *Let $G : \mathcal{B} \to \mathcal{C}$ be a left exact functor, $M \in \mathcal{A}$, $X \in \mathcal{A}\boxtimes\mathcal{B}$. Then*

$$\operatorname{Hom}^1(M_*, (1\boxtimes G)X_{*1}) \simeq G(\operatorname{Hom}^1(M_*, X_{*1})). \tag{1.8.7}$$

PROOF. Since both functors are left exact in X it suffices to consider $X = Y\boxtimes Z$ and then use resolutions (1.6.2). By Proposition 1.8.8 the left hand side is the representing object for

$$\begin{aligned} N \mapsto\ & \operatorname{Hom}_{\mathcal{A}\boxtimes\mathcal{C}}(M\boxtimes N, (1\boxtimes G)X_{*1}) \\ =\ & \operatorname{Hom}_{\mathcal{A}\boxtimes\mathcal{C}}(M\boxtimes N, Y\boxtimes GZ) \\ \simeq\ & \operatorname{Hom}_{\mathcal{A}}(M,Y) \otimes_{\Bbbk} \operatorname{Hom}_{\mathcal{C}}(N, GZ) \\ \simeq\ & \operatorname{Hom}_{\mathcal{C}}(N, \operatorname{Hom}_{\mathcal{A}}(M,Y)\otimes GZ) \\ \simeq\ & \operatorname{Hom}_{\mathcal{C}}(N, G(\operatorname{Hom}_{\mathcal{A}}(M,Y)\otimes Z)) \\ =\ & \operatorname{Hom}_{\mathcal{C}}(N, G(\operatorname{Hom}^1(M_*, X_{*1}))). \end{aligned}$$

whence the required isomorphism. □

1.8.12. Corollary. *Let $G : \mathcal{B} \to \widehat{\mathcal{C}}$ be a filtered inductive limit of left exact functors $G_i : \mathcal{B} \to \mathcal{C}$, and let $M \in \mathcal{A}$, $X \in \mathcal{A}\boxtimes\mathcal{B}$. Then (1.8.7) holds for $1_{\mathcal{A}}\boxtimes G \simeq \operatorname{colim}_I 1_{\mathcal{A}}\boxtimes G_i$.*

PROOF. Indeed, by (1.8.6) and Proposition 1.8.11

$$\begin{aligned} \operatorname{Hom}^1(M_*, (1\boxtimes G)X_{*1}) &= \varinjlim_I \operatorname{Hom}^1(M_*, (1\boxtimes G_i)X_{*1}) \\ &\simeq \varinjlim_I G_i(\operatorname{Hom}^1(M_*, X_{*1})) = G(\operatorname{Hom}^1(M_*, X_{*1})). \end{aligned}$$ □

CHAPTER 2

Squared coalgebras

2.1. Definitions

Let $\mathcal{V}$ be a $\Bbbk$-linear abelian rigid monoidal category with length.

2.1.1. Definition. A *squared coalgebra* $C = (C, \Delta, \varepsilon)$ in $\widehat{\mathcal{V}}$ is an object $C \in \widehat{\mathcal{V} \boxtimes \mathcal{V}}$ equipped with a comultiplication $\Delta_{123} : C_{13} \boxtimes \mathbb{I}_2 \to C_{12'} \otimes C_{2''3} \in \widehat{\mathcal{V}^{\boxtimes 3}}$ and a counit $\varepsilon : C_{1'1''} \to \mathbb{I} \in \widehat{\mathcal{V}}$, such that coassociativity holds:

$$\begin{array}{ccc} C_{14} \boxtimes \mathbb{I}_2 \boxtimes \mathbb{I}_3 & \xrightarrow{\Delta_{124} \boxtimes \mathbb{I}_3} & C_{12'} \otimes C_{2''4} \boxtimes \mathbb{I}_3 \\ {\scriptstyle \Delta_{134} \boxtimes \mathbb{I}_2} \big\downarrow & & \big\downarrow {\scriptstyle C_{12'} \otimes \Delta_{2''34}} \\ C_{13'} \otimes C_{3''4} \boxtimes \mathbb{I}_2 & \xrightarrow{\Delta_{123'} \otimes C_{3''4}} & C_{12'} \otimes C_{2''3'} \otimes C_{3''4} \end{array} \tag{2.1.1}$$

and ε is the counit:

$$\big(C_{12} \xrightarrow{\sim} C_{1'2} \otimes \mathbb{I}_{1''} \xrightarrow{\Delta_{1'1''2}} C_{1'1''} \otimes C_{1'''2} \xrightarrow{\varepsilon \otimes C} \mathbb{I}_{1'} \otimes C_{1''2} \xrightarrow{\sim} C_{12}\big) = \mathrm{id}_C \tag{2.1.2a}$$

$$\big(C_{12} \xrightarrow{\sim} \mathbb{I}_{2'} \otimes C_{12''} \xrightarrow{\Delta_{12'2''}} C_{12'} \otimes C_{2''2'''} \xrightarrow{C \otimes \varepsilon} C_{12'} \otimes \mathbb{I}_{2''} \xrightarrow{\sim} C_{12}\big) = \mathrm{id}_C \tag{2.1.2b}$$

2.1.2. Example. Let $C = (C, \Delta, \varepsilon, m, \eta, \gamma)$ be an ordinary Hopf $\Bbbk$-algebra, $\mathcal{V} = C$-comod. Then (C, Δ, ε) is an example of a squared coalgebra in $\widehat{\mathcal{V}}$, where we view C as an object of $\widehat{\mathcal{V} \boxtimes \mathcal{V}} = C \otimes_{\Bbbk} C$-Comod with the coaction

$$\delta : C \to C^{\otimes 2} \otimes C, \quad c \mapsto c_{(1)} \otimes \gamma(c_{(3)}) \otimes c_{(2)}.$$

Indeed, $\Delta : C = C_{13} \otimes \Bbbk_2 \to C_{12'} \otimes C_{2''3} = C \otimes C$ is a homomorphism of $C^{\otimes 3}$-comodules, where the first comodule is equipped with the coaction

$$\delta(c) = c_{(1)} \otimes 1 \otimes \gamma(c_{(3)}) \otimes c_{(2)} \in C^{\otimes 3} \otimes C$$

and the second with

$$\delta(c \otimes d) = c_{(1)} \otimes \gamma(c_{(3)}) d_{(1)} \otimes \gamma(d_{(3)}) \otimes c_{(2)} \otimes d_{(2)} \in C^{\otimes 3} \otimes C \otimes C;$$

$\varepsilon : C = C_{1'1''} \to \Bbbk$ is a homomorphism of left C-comodules, where the first comodule carries the coaction

$$\delta(c) = c_{(1)} \gamma(c_{(3)}) \otimes c_{(2)} \in C \otimes C$$

and the second comodule is trivial. Equations (2.1.1), (2.1.2a) and (2.1.2b) are just ordinary axioms of coassociativity and counitality.

Graphical notations partially explain the choice of indices and help to memorise them. Comultiplication is denoted

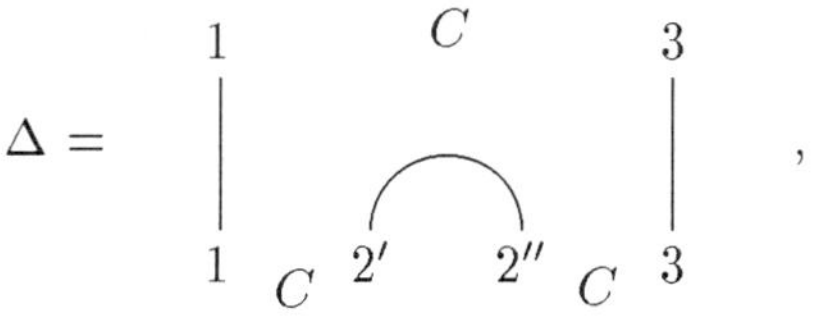

the counit is denoted

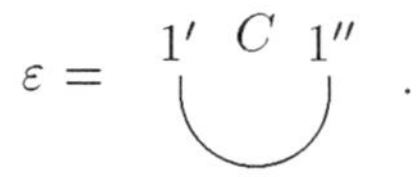

The coassociativity equation is

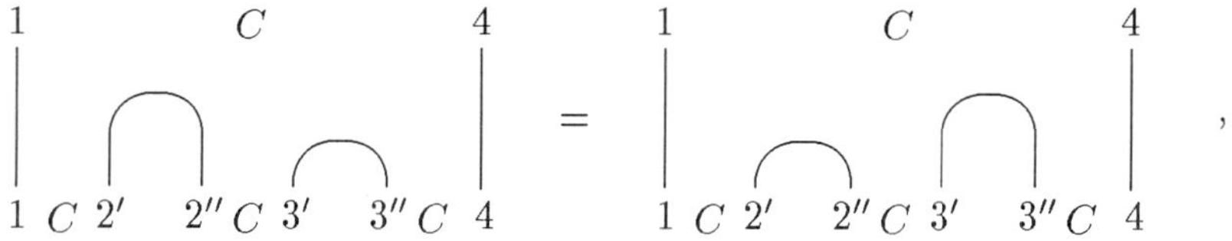

the equations for the counit are

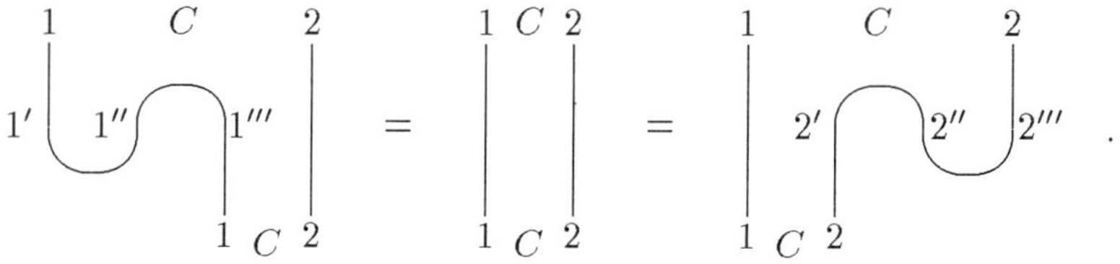

2.1.3. Definition. A *squared coalgebra homomorphism* $(C, \Delta, \varepsilon_C) \to (D, \Delta, \varepsilon_D)$ is a morphism $f : C \to D \in \operatorname{Mor} \widehat{\mathcal{V} \boxtimes \mathcal{V}}$ such that

$$\begin{array}{ccc} C_{13} \boxtimes \mathbb{1}_2 & \xrightarrow{\Delta_{123}} & C_{12'} \otimes C_{2''3} \\ {\scriptstyle f_{13} \boxtimes \mathbb{1}_2} \downarrow & & \downarrow {\scriptstyle f_{12'} \otimes f_{2''3}} \\ D_{13} \boxtimes \mathbb{1}_2 & \xrightarrow{\Delta_{123}} & D_{12'} \otimes D_{2''3} \end{array}$$

commutes and equation

$$(\circledast C \xrightarrow{\circledast f} \circledast D \xrightarrow{\varepsilon_D} \mathbb{1}) = \varepsilon_C \tag{2.1.3}$$

holds.

Squared coalgebras form a category denoted $\mathrm{Coalgsq}(\widehat{\mathcal{V}})$. Its full subcategory consisting of squared coalgebras, which are objects of $\mathcal{V} \boxtimes \mathcal{V}$, is denoted $\mathrm{Coalgsq}(\mathcal{V})$. A *subcoalgebra* in D is a squared coalgebra morphism $C \to D$, which is a monomorphism in $\widehat{\mathcal{V} \boxtimes \mathcal{V}}$. Clearly, each squared coalgebra homomorphism $f : C \to D$ has the image $\operatorname{im} f$, which is a subcoalgebra in D.

2.1.4. Definition. Let C be a squared coalgebra in $\widehat{\mathcal{V}}$. A subobject $j : J \hookrightarrow C \in \widehat{\mathcal{V} \boxtimes \mathcal{V}}$ is a *coideal*, if

$$(\circledast J \xrightarrow{\circledast j} \circledast C \xrightarrow{\varepsilon} \mathbb{1}) = 0 \tag{2.1.4}$$

and there is a morphism

$$r : W \to C_{12'} \otimes J_{2''3} \oplus J_{12'} \otimes C_{2''3} \in \widehat{\mathcal{V}^{\otimes 3}}$$

and an epimorphism

$$s : W \longrightarrow\!\!\triangleright\; J_{13} \boxtimes \mathbb{1}_2 \in \widehat{\mathcal{V}^{\boxtimes 3}}$$

such that

$$\begin{array}{ccccc} W & & \xrightarrow{\quad r \quad} & & C_{12'} \otimes J_{2''3} \oplus J_{12'} \otimes C_{2''3} \\ {\scriptstyle s}\big\downarrow & & & & \big\downarrow{\scriptstyle C_{12'}\otimes j_{2''3}\oplus j_{12'}\otimes C_{2''3}} \\ J_{13} \boxtimes \mathbb{1}_2 & \xrightarrow{j_{13}\boxtimes \mathbb{1}_2} & C_{13} \boxtimes \mathbb{1}_2 & \xrightarrow{\Delta_{123}} & C_{12'} \otimes C_{2''3} \end{array} \tag{2.1.5}$$

2.1.5. Remark. For any morphism $j : J \to C \in \widehat{\mathcal{V} \boxtimes \mathcal{V}}$, which satisfies to all hypotheses of the definition except being monic, its image $\operatorname{im} j$ is a coideal in C.

2.1.6. Proposition. *For any squared coalgebra morphism $f : C \to D$ the subobject* $\operatorname{Ker} f$ *is a coideal in C.*

PROOF. Let us denote $\mathcal{W} = \mathcal{V} \boxtimes \mathcal{V}$ and $J = \operatorname{Ker} f$. There is an exact sequence $0 \to J \xrightarrow{j} C \xrightarrow{f} D$ in $\hat{\mathcal{W}}$. First of all, the sequence

$$C \boxtimes J \oplus J \boxtimes C \xrightarrow{C\boxtimes j \oplus j \boxtimes C} C \boxtimes C \xrightarrow{f\boxtimes f} D \boxtimes D \tag{2.1.6}$$

is exact in $\widehat{\mathcal{W} \boxtimes \mathcal{W}}$. Indeed, filter C by subobjects C_i from $\mathcal{W}$ and take the induced filtration $J_i = J \cap C_i$ on J. Include $\operatorname{Coker}(J_i \to C_i)$ into a subobject $D_i \in \mathcal{W}$ of D and the question is reduced to the case $J, C, D \in \mathcal{W}$. We may restrict the consideration to the subcategory $\langle C, D\rangle \subset \mathcal{W}$. Therefore, we may assume that $J, C, D \in \text{mod-}A$ for a finite dimensional $\Bbbk$-algebra A and we have to prove that the sequence

$$C \otimes_{\Bbbk} J \oplus J \otimes_{\Bbbk} C \xrightarrow{C\otimes j \oplus j \otimes C} C \otimes_{\Bbbk} C \xrightarrow{f\otimes f} D \otimes_{\Bbbk} D$$

is exact in $\text{mod-}A \otimes_{\Bbbk} A$. This is obvious, since it is true in $\Bbbk$-vect. Therefore, (2.1.6) is exact.

Applying the exact functor $\mathrm{Id}_{\mathcal{V}} \boxtimes \circledast \boxtimes \mathrm{Id}_{\mathcal{V}} : \widehat{\mathcal{V}^{\boxtimes 4}} \to \widehat{\mathcal{V}^{\boxtimes 3}}$ to (2.1.6) we deduce that the sequence

$$C_{12'} \otimes J_{2''3} \oplus J_{12'} \otimes C_{2''3} \xrightarrow{C\otimes j \oplus j \otimes C} C_{12'} \otimes C_{2''3} \xrightarrow{f\otimes f} D_{12'} \otimes D_{2''3}$$

is exact. Therefore, there is an epimorphism $q \in \widehat{\mathcal{V}^{\boxtimes 3}}$ such that

$$C \otimes j \oplus j \otimes C$$
$$= \big(C_{12'} \otimes J_{2''3} \oplus J_{12'} \otimes C_{2''3} \xrightarrow{q} \operatorname{Ker}(f_{12'} \otimes f_{2''3}) \hookrightarrow C_{12'} \otimes C_{2''3}\big).$$

There is also a unique morphism $l \in \widehat{\mathcal{V}^{\boxtimes 3}}$ such that the diagram

$$\begin{array}{ccc}
J_{13} \boxtimes \mathbb{I}_2 & \xrightarrow{\exists l} & \operatorname{Ker}(f_{12'} \otimes f_{2''3}) \\
{\scriptstyle j_{13} \boxtimes \mathbb{I}_2}\downarrow & & \downarrow \\
C_{13} \boxtimes \mathbb{I}_2 & \xrightarrow{\Delta_{123}} & C_{12'} \otimes C_{2''3} \\
{\scriptstyle f_{13} \boxtimes \mathbb{I}_2}\downarrow & & \downarrow{\scriptstyle f_{12'} \otimes f_{2''3}} \\
D_{13} \boxtimes \mathbb{I}_2 & \xrightarrow{\Delta_{123}} & D_{12'} \otimes D_{2''3}
\end{array}$$

commutes.

Now define W as a pull back of q and l,

$$\begin{array}{ccc}
W & \xrightarrow{r} & C_{12'} \otimes J_{2''3} \oplus J_{12'} \otimes C_{2''3} \\
{\scriptstyle s}\downarrow & & \downarrow{\scriptstyle q} \\
J_{13} \boxtimes \mathbb{I}_2 & \xrightarrow{l} & \operatorname{Ker}(f_{12'} \otimes f_{2''3})
\end{array},$$

that is,

$$W = \operatorname{Ker}\bigl(C_{12'} \otimes J_{2''3} \oplus J_{12'} \otimes C_{2''3} \oplus J_{13} \boxtimes \mathbb{I}_2 \xrightarrow{q \oplus -l} \operatorname{Ker}(f_{12'} \otimes f_{2''3})\bigr).$$

Clearly, s is an epimorphism and the diagram (2.1.5) commutes. □

2.1.7. Proposition. *If $j : J \hookrightarrow C$ is a coideal, then C/J has a unique squared coalgebra structure such that the canonical projection $C \to C/J$ is a homomorphism of squared coalgebras.*

PROOF. We use notations and results from the proof of the previous proposition. Denote $D = C/J$. The columns of the following diagram are exact.

$$\begin{array}{ccccc}
J_{13} \boxtimes \mathbb{I}_2 & \xleftarrow{s} & W & \xrightarrow{r} & C_{12'} \otimes J_{2''3} \oplus J_{12'} \otimes C_{2''3} \\
{\scriptstyle j_{13} \boxtimes \mathbb{I}_2}\downarrow & & & & \downarrow{\scriptstyle C \otimes j \oplus j \otimes C} \\
C_{13} \boxtimes \mathbb{I}_2 & & \xrightarrow{\Delta^C_{123}} & & C_{12'} \otimes C_{2''3} \\
{\scriptstyle f_{13} \boxtimes \mathbb{I}_2}\downarrow & & & & \downarrow{\scriptstyle f_{12'} \otimes f_{2''3}} \\
D_{13} \boxtimes \mathbb{I}_2 & & \xrightarrow{\exists \Delta^D_{123}} & & D_{12'} \otimes D_{2''3} \\
\downarrow & & & & \downarrow \\
0 & & & & 0
\end{array}$$

Indeed, exactness of the right column in the middle term was shown in the proof of Proposition 2.1.6. Since $f \boxtimes f$ is epi, $f_{12'} \otimes f_{2''3}$ is also epi.

This implies existence and uniqueness of the morphism Δ^D_{123} making the diagram commutative. Coassociativity of Δ^D follows from that of Δ^C.

The condition (2.1.4) implies existence of $\varepsilon_D : \circledast D \to \mathbb{I}$ such that (2.1.3) holds. Since ε_C is the counit for C, ε_D is the counit for D. □

2.1.8. Corollary. *Let $f : C \to D$ be a squared coalgebra homomorphism and $f = (C \xrightarrow{h} E \xrightarrow{g} D)$ in $\widehat{\mathcal{V} \boxtimes \mathcal{V}}$, where h is epi and g is mono. Then there is a unique squared coalgebra structure on E such that h, g are squared coalgebra homomorphisms.*

2.2. Comodules

2.2.1. Definition. A *left comodule* $X \in \widehat{\mathcal{V}}$ over a squared coalgebra C is an object X of $\widehat{\mathcal{V}}$ equipped with the coaction

$$\delta = \delta_X : X_1 \boxtimes \mathbb{1}_2 \to C_{12'} \otimes X_{2''} \in \widehat{\mathcal{V} \boxtimes \mathcal{V}},$$

which is coassociative:

$$\begin{array}{ccc} X_1 \boxtimes \mathbb{1}_2 \boxtimes \mathbb{1}_3 & \xrightarrow{\delta_{12} \boxtimes \mathbb{1}_3} & C_{12'} \otimes X_{2''} \boxtimes \mathbb{1}_3 \\ {\scriptstyle \delta_{13} \boxtimes \mathbb{1}_2} \downarrow & & \downarrow {\scriptstyle C_{12'} \otimes \delta_{2''3}} \\ C_{13'} \boxtimes \mathbb{1}_2 \otimes X_{3''} & \xrightarrow{\Delta_{123'} \otimes X_{3''}} & C_{12'} \otimes C_{2''3'} \otimes X_{3''} \end{array} \tag{2.2.1}$$

and counital:

$$\left(X \xrightarrow{\sim} X \otimes \mathbb{1} \xrightarrow{\circledast \delta} (\circledast C) \otimes X \xrightarrow{\varepsilon \otimes X} \mathbb{1} \otimes X \xrightarrow{\sim} X \right) = \mathrm{id}_X \, . \tag{2.2.2}$$

Graphical notation for the coaction is

$\delta =$ 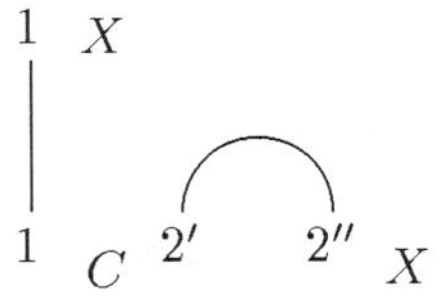

and it will be explained later. Coassociativity takes the form

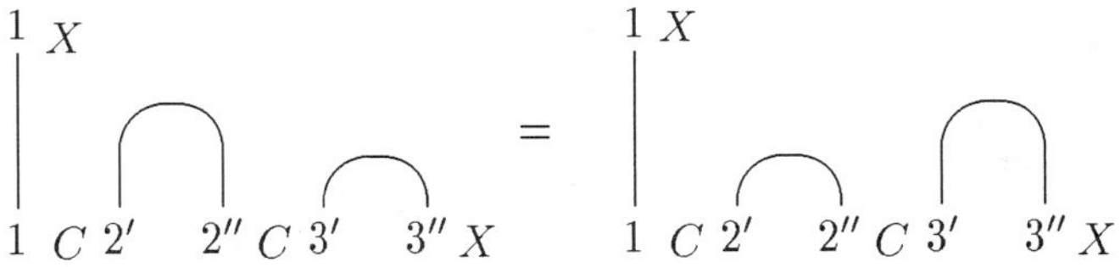

and counitality is

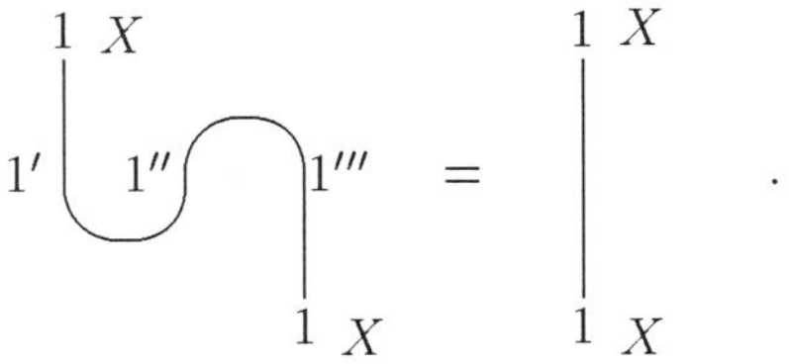

The definition above should be generalised. Let $\mathcal{A}$ be a $\Bbbk$-linear abelian category with length.

2.2.2. Definition. A *left comodule* $X \in \widehat{\mathcal{V} \boxtimes \mathcal{A}}$ over a squared coalgebra $C \in \mathrm{Coalgsq}(\widehat{\mathcal{V}})$ is an object $X = X_{10} \in \widehat{\mathcal{V} \boxtimes \mathcal{A}}$ equipped with the coaction

$$\delta = \delta_X : X_{10} \boxtimes \mathbb{1}_2 \to C_{12'} \otimes X_{2''0} \in (\mathcal{V} \boxtimes \mathcal{V} \boxtimes \mathcal{A})\hat{}$$

such that coassociativity (2.2.1) in $(\mathcal{V}^{\boxtimes 3} \boxtimes \mathcal{A})\hat{}$ and counitality (2.2.2) in $\widehat{\mathcal{V} \boxtimes \mathcal{A}}$ hold.

When $\mathcal{A} = \Bbbk$-vect this reduces to the previous definition.

2.2.3. Definition. A morphism of left C-comodules $(X,\delta_X) \to (Y,\delta_Y)$ in $\widehat{\mathcal{V}}$ (resp. in $\widehat{\mathcal{V}\boxtimes\mathcal{A}}$) is $f : X \to Y \in \widehat{\mathcal{V}}$ (resp. $f : X_{10} \to Y_{10} \in \widehat{\mathcal{V}\boxtimes\mathcal{A}}$) such that

$$\begin{array}{ccc} X_1 \boxtimes \mathbb{1}_2 & \xrightarrow{\delta_X} & C_{12'} \otimes X_{2''} \\ {\scriptstyle f_1\boxtimes\mathbb{1}_2}\downarrow & & \downarrow{\scriptstyle C_{12'}\otimes f_{2''}} \\ Y_1 \boxtimes \mathbb{1}_2 & \xrightarrow{\delta_Y} & C_{12'} \otimes Y_{2''} \end{array} \tag{2.2.3}$$

(resp. the same diagram in $(\mathcal{V}\boxtimes\mathcal{V}\boxtimes\mathcal{A})^\frown$) commutes.

Left C-comodules form a category, which is denoted ${}^C\widehat{\mathcal{V}}$ (resp. ${}^C\widehat{\mathcal{V}\boxtimes\mathcal{A}}$). It has a full subcategory ${}^C\mathcal{V}$ (resp. ${}^C(\mathcal{V}\boxtimes\mathcal{A})$) formed by objects from $\mathcal{V}$ (resp. $\mathcal{V}\boxtimes\mathcal{A}$).

It is easy to show that if $f : X \to Y$ is a morphism of comodules in ${}^C\widehat{\mathcal{V}}$ (resp. ${}^C\widehat{\mathcal{V}\boxtimes\mathcal{A}}$, ${}^C\mathcal{V}$, ${}^C(\mathcal{V}\boxtimes\mathcal{A})$), and

$$\operatorname{Ker} f \xrightarrow{\ker f} X \xrightarrow{\operatorname{coim} f} \operatorname{Coim} f \simeq \operatorname{Im} f \xrightarrow{\operatorname{im} f} Y \xrightarrow{\operatorname{coker} f} \operatorname{Coker} f$$

is its canonical decomposition in $\widehat{\mathcal{V}}$ (resp. $\widehat{\mathcal{V}\boxtimes\mathcal{A}}$, $\mathcal{V}$, $\mathcal{V}\boxtimes\mathcal{A}$), then the objects $\operatorname{Ker} f$, $\operatorname{Coim} f$, $\operatorname{Im} f$, $\operatorname{Coker} f$ have unique C-comodule structure such that the morphisms above are morphisms of comodules. It follows:

2.2.4. Proposition. *The category ${}^C\widehat{\mathcal{V}}$ (resp. ${}^C\widehat{\mathcal{V}\boxtimes\mathcal{A}}$, ${}^C\mathcal{V}$, ${}^C(\mathcal{V}\boxtimes\mathcal{A})$) is $\Bbbk$-linear and abelian and the underlying functor $\mathcal{U} : {}^C\widehat{\mathcal{V}} \to \widehat{\mathcal{V}}$ (resp. $\mathcal{U} : {}^C\widehat{\mathcal{V}\boxtimes\mathcal{A}} \to \widehat{\mathcal{V}\boxtimes\mathcal{A}}$, $\mathcal{U} : {}^C\mathcal{V} \to \mathcal{V}$, $\mathcal{U} : {}^C(\mathcal{V}\boxtimes\mathcal{A}) \to \mathcal{V}\boxtimes\mathcal{A}$) is exact and faithful.*

Right C-comodules are defined as objects $X \in \widehat{\mathcal{V}}$ (resp. $X_{01} \in \widehat{\mathcal{A}\boxtimes\mathcal{V}}$) equipped with a coaction $\delta : \mathbb{1}_1\boxtimes X_2 \to X_{1'}\otimes C_{1''2} \in \widehat{\mathcal{V}\boxtimes\mathcal{V}}$ (resp. $\delta : \mathbb{1}_1\boxtimes X_{02} \to X_{01'}\otimes C_{1''2} \in (\mathcal{A}\boxtimes\mathcal{V}\boxtimes\mathcal{V})^\frown$) such that the properties similar to (2.2.1)–(2.2.2) hold. An analogue of Proposition 2.2.4 holds for $\widehat{\mathcal{V}}^C$, $\widehat{\mathcal{A}\boxtimes\mathcal{V}}^C$, $\mathcal{V}^C$, $(\mathcal{A}\boxtimes\mathcal{V})^C$.

Any $\Bbbk$-linear exact functor $F : \mathcal{A} \to \mathcal{B}$ induces a $\Bbbk$-linear exact functor $\mathcal{V}\boxtimes F : {}^C\widehat{\mathcal{V}\boxtimes\mathcal{A}} \to {}^C\widehat{\mathcal{V}\boxtimes\mathcal{B}}$.

2.2.5. Example. (C,Δ) is a C-comodule from ${}^C\widehat{\mathcal{V}\boxtimes\mathcal{V}}$ (resp. $\widehat{\mathcal{V}\boxtimes\mathcal{V}}^C$). It is called the left (resp. right) regular comodule.

2.3. The fundamental theorem on coalgebras

2.3.1. Theorem. *Any comodule from ${}^C\widehat{\mathcal{V}}$ (resp. ${}^C\widehat{\mathcal{V}\boxtimes\mathcal{A}}$) is a union, i.e. filtered inductive limit, of its subcomodules from ${}^C\mathcal{V}$ (resp. ${}^C(\mathcal{V}\boxtimes\mathcal{A})$).*

PROOF. Let us prove for ${}^C\widehat{\mathcal{V}}$. The case of ${}^C\widehat{\mathcal{V}\boxtimes\mathcal{A}}$ is considered similarly.

Let $X \in {}^C\widehat{\mathcal{V}}$ be a comodule, and let $Y \subset X$ be a subobject, $Y \in \mathcal{V}$. There exist subobjects $D \subset C$, $D \in \mathcal{V}\boxtimes\mathcal{V}$ and $Z \subset X$, $Z \in \mathcal{V}$ such that $\delta(Y_1\boxtimes\mathbb{1}_2) \subset D_{12'}\otimes Z_{2''}$. Since $\mathcal{V}\boxtimes\mathcal{V}$ is rigid by Theorem 1.6.1, there is an object $D^\vee \in \mathcal{V}\boxtimes\mathcal{V}$. Let us decompose the morphism

$$D^\vee_{1''2'} \otimes Y_{1'} \otimes \mathbb{1}_{2''} \xrightarrow{D^\vee\otimes\delta_{1'2''}} D^\vee_{1''2'} \otimes D_{1'2''} \otimes Z_{2'''} \xrightarrow{\operatorname{ev}\otimes Z} \mathbb{1}_1 \boxtimes Z_2 \tag{2.3.1}$$

into an epimorphism and a monomorphism. By Propositions 1.1.10 and 1.1.11 the subcategory $\langle\mathbb{1}\rangle\boxtimes\mathcal{V} \simeq \Bbbk\text{-vect}\boxtimes\mathcal{V} \simeq \mathcal{V}$ of $\mathcal{V}\boxtimes\mathcal{V}$ is stable with respect to subquotients. Therefore, any subobject of $\mathbb{1}_1\boxtimes Z_2$ has the form $\mathbb{1}_1\boxtimes N_2$ for some $N \in \mathcal{V}$. Hence (2.3.1) equals to

$$D^\vee_{1''2'} \otimes Y_{1'} \otimes \mathbb{1}_{2''} \xrightarrow{p} \mathbb{1}_1 \boxtimes N_2 \hookrightarrow \mathbb{1}_1 \boxtimes Z_2$$

and p is epi.

By the properties of rigid categories we deduce that $\delta(Y_1 \boxtimes \mathbb{1}_2) \subset D_{12'} \otimes N_{2''}$. The composite

$$Y \simeq Y \otimes \mathbb{1} \xrightarrow{\circledast\delta} D_{1'1''} \otimes N_{1'''} \xrightarrow{\varepsilon\otimes N} \mathbb{1}_{1'} \otimes N_{1''} \simeq N$$

is coherent with the embeddings $Y \subset X$, $N \subset X$. Therefore, it is an embedding $Y \subset N$. It remains to prove that N is a subcomodule of X.

2.3.2. Lemma. *Let $\Delta(D_{13} \boxtimes \mathbb{1}_2) \subset L_{12'} \otimes M_{2''3}$ for some $L, M \subset C$, $L, M \in \mathcal{V} \boxtimes \mathcal{V}$. Then $\delta(N_1 \boxtimes \mathbb{1}_2) \subset M_{12'} \otimes N_{2''}$.*

PROOF. Let us construct an embedding $j : D \hookrightarrow L$,

$$j = \big(D_{12} \simeq D_{12''} \otimes \mathbb{1}_{2'} \xrightarrow{\Delta_{12'2''}} L_{12'} \otimes M_{2''2'''} \xrightarrow{L\otimes\varepsilon} L_{12'} \otimes \mathbb{1}_{2''} \simeq L_{12}\big).$$

Then $j^t : L^\vee \to D^\vee$ is an epimorphism.

The following equations hold:

$$\begin{aligned}
&\big(L^\vee_{1''2'} \otimes Y_{1'} \otimes \mathbb{1}_{2''} \boxtimes \mathbb{1}_3 \xrightarrow{j^t\otimes Y\otimes\mathbb{1}\boxtimes\mathbb{1}} D^\vee_{1''2'} \otimes Y_{1'} \otimes \mathbb{1}_{2''} \boxtimes \mathbb{1}_3 \\
&\qquad \xrightarrow{p\boxtimes\mathbb{1}_3} \mathbb{1}_1 \boxtimes N_2 \boxtimes \mathbb{1}_3 \xrightarrow{\mathbb{1}_1\boxtimes\delta_{23}} \mathbb{1}_1 \boxtimes C_{23'} \otimes X_{3''}\big) \\
&= \big(L^\vee_{1''2'} \otimes Y_{1'} \otimes \mathbb{1}_{2''} \boxtimes \mathbb{1}_3 \xrightarrow{j^t_{1''2'}\otimes\delta_{1'2''}\boxtimes\mathbb{1}_3} D^\vee_{1''2'} \otimes D_{1'2''} \otimes N_{2'''} \boxtimes \mathbb{1}_3 \\
&\qquad \xrightarrow{\mathrm{ev}\otimes N\boxtimes\mathbb{1}} \mathbb{1}_1 \boxtimes N_2 \boxtimes \mathbb{1}_3 \xrightarrow{\mathbb{1}_1\boxtimes\delta_{23}} \mathbb{1}_1 \boxtimes C_{23'} \otimes X_{3''}\big) \\
&= \big(L^\vee_{1''2'} \otimes Y_{1'} \otimes \mathbb{1}_{2''} \boxtimes \mathbb{1}_3 \xrightarrow{L^\vee\otimes\delta_{1'2''}\boxtimes\mathbb{1}_3} L^\vee_{1''2'} \otimes D_{1'2''} \otimes N_{2'''} \boxtimes \mathbb{1}_3 \\
&\qquad \xrightarrow{L^\vee\otimes j\otimes\delta_{2'''3}} L^\vee_{1''2'} \otimes L_{1'2''} \otimes C_{2'''3'} \otimes X_{3''} \xrightarrow{\mathrm{ev}\otimes C\otimes X} \mathbb{1}_1 \boxtimes C_{23'} \otimes X_{3''}\big) \\
&= \big(L^\vee_{1''2'} \otimes Y_{1'} \otimes \mathbb{1}_{2''} \boxtimes \mathbb{1}_3 \xrightarrow{L^\vee\otimes\delta_{1'3}\otimes\mathbb{1}_{2''}} L^\vee_{1''2'} \otimes D_{1'3'} \otimes \mathbb{1}_{2''} \otimes N_{3''} \\
&\qquad \xrightarrow{L^\vee\otimes\Delta_{1'2''3'}\otimes N_{3''}} L^\vee_{1''2'} \otimes L_{1'2''} \otimes M_{2'''3'} \otimes N_{3''} \xrightarrow{\mathrm{ev}\otimes M\otimes N} \mathbb{1}_1 \boxtimes M_{23'} \otimes N_{3''}\big)
\end{aligned}$$

Since the first two arrows in the first row are epimorphisms, we deduce that $\mathbb{1}_1 \boxtimes \delta_{23}(N_2 \boxtimes \mathbb{1}_3) \subset \mathbb{1}_1 \boxtimes M_{23'} \otimes N_{3''}$, and the lemma follows. □

Therefore, any subobject $Y \in \mathcal{V}$ is included into a subcomodule $N \in {}^C\mathcal{V}$ and the theorem is proved. □

2.3.3. Corollary. *$\widehat{{}^C\mathcal{V}} \simeq {}^C\widehat{\mathcal{V}}$ and $\big({}^C(\mathcal{V} \boxtimes \mathcal{A})\big)^\wedge \simeq {}^C\widehat{\mathcal{V}\boxtimes\mathcal{A}}$.*

Similar results hold for right comodules.

2.3.4. Theorem (fundamental theorem on coalgebras). *A squared coalgebra $C \in \mathrm{Coalgsq}(\widehat{\mathcal{V}})$ is a filtered inductive limit of its subcoalgebras from $\mathrm{Coalgsq}(\mathcal{V})$.*

PROOF. The right regular comodule (C, Δ) is a filtered inductive limit of its subcomodules (right coideals) from $(\mathcal{V} \boxtimes \mathcal{V})^C$ by Theorem 2.3.1. Therefore, any subobject $X \subset C$, $X \in \mathcal{V} \boxtimes \mathcal{V}$ is contained in a right subcomodule $Y \subset C$, $Y \in \mathcal{V} \boxtimes \mathcal{V}$, $\Delta(Y_{13} \boxtimes \mathbb{1}_2) \subset Y_{12'} \otimes C_{2''3}$.

Let us apply the procedure of Theorem 2.3.1 to get a left coideal N (a subcomodule of the left regular comodule) containing Y. For some $N \in \mathcal{V} \boxtimes \mathcal{V}$, $Y \subset N \subset C$ we have $\Delta(Y_{13} \boxtimes \mathbb{1}_2) \subset Y_{12'} \otimes N_{2''3}$ and the morphism (2.3.1) with $D = Y$, $Z = N$,

$$Y^\vee_{1''2'} \otimes Y_{1'3} \boxtimes \mathbb{1}_{2''} \xrightarrow{Y^\vee\otimes\Delta_{1'2''3}} Y^\vee_{1''2'} \otimes Y_{1'2''} \otimes N_{2'''3} \xrightarrow{\mathrm{ev}\otimes N} \mathbb{1}_1 \boxtimes N_{23},$$

is epi.

Since $D = Y$ we can choose $L = Y$, $M = N$ in hypotheses of Lemma 2.3.2. And the lemma gives $\Delta(N_{13} \boxtimes \mathbb{1}_2) \subset N_{12'} \otimes N_{2''3}$. Therefore, N is a subcoalgebra and the theorem is proved. □

2.4. Relationship between categories of comodules

Let $\mathcal{A} \in \mathfrak{Ab}_{\Bbbk}$ and let C be a squared coalgebra in $\mathcal{V}$.

2.4.1. Theorem. *The functor* $\Psi : ({}^C\mathcal{V}) \boxtimes \mathcal{A} \to {}^C(\mathcal{V} \boxtimes \mathcal{A})$ *induced by* $({}^C\mathcal{V}) \times \mathcal{A} \to {}^C(\mathcal{V} \boxtimes \mathcal{A})$, $(X, M) \mapsto X \boxtimes M$, *is an equivalence.*

PROOF. Let us construct a quasiinverse functor $\Phi : {}^C(\mathcal{V} \boxtimes \mathcal{A}) \to ({}^C\mathcal{V}) \boxtimes \mathcal{A}$. Let $X \in {}^C(\mathcal{V} \boxtimes \mathcal{A})$. According to Lemma 1.8.10

$$\int^{M \in \mathcal{A}} \operatorname{Hom}^2(M_*, X_{1*}) \boxtimes M_2 \to X_{12}$$

is an isomorphism in $\mathcal{V} \boxtimes \mathcal{A}$. The object $\operatorname{Hom}^2(M_*, X_{1*}) \in \mathcal{V}$ has a C-comodule structure since it comes from (M, X) via the composite functor

$$\mathcal{A}^{\mathrm{op}} \boxtimes {}^C(\mathcal{V} \boxtimes \mathcal{A}) \xrightarrow{\boxtimes} \mathcal{A}^{\mathrm{op}} \boxtimes {}^C(\mathcal{V} \boxtimes \mathcal{A}) \to {}^C(\mathcal{V} \boxtimes \mathcal{A}^{\mathrm{op}} \boxtimes \mathcal{A})$$
$$\xrightarrow{{}^C(\mathcal{V} \boxtimes \mathrm{Hom})} {}^C(\mathcal{V} \boxtimes \Bbbk\text{-vect}) \xrightarrow{\sim} {}^C\mathcal{V}.$$

Therefore,

$$X \mapsto \Phi(X) = \int^{M \in \mathcal{A}} \operatorname{Hom}^2(M_*, X_{1*}) \boxtimes M_2 \in \widehat{({}^C\mathcal{V}) \boxtimes \mathcal{A}} \cap (\mathcal{V} \boxtimes \mathcal{A}) = ({}^C\mathcal{V}) \boxtimes \mathcal{A}$$

is the required functor.

We have $\Phi(\Psi(X)) \simeq X$ for $X \in ({}^C\mathcal{V}) \boxtimes \mathcal{A}$. Indeed, since Φ and Ψ are exact, it suffices to check this for $X = Y \boxtimes Z \in ({}^C\mathcal{V}) \boxtimes \mathcal{A}$ and then use resolutions (1.6.2). For such X we find $\operatorname{Hom}^2(M_*, X_{1*}) = Y \otimes \operatorname{Hom}_{\mathcal{A}}(M, Z)$ and Lemma 1.8.7 gives the isomorphism in question.

Now we show that $\Psi(\Phi(X)) \simeq X$ for $X \in {}^C(\mathcal{V} \boxtimes \mathcal{A})$. By Lemma 1.8.10 all we have to do is to prove that

$$\operatorname{Hom}^2(M_*, X_{1*}) \boxtimes M_2 \to X_{12} \tag{2.4.1}$$

is a homomorphism of C-comodules. It is easier to prove more: for any $N \in {}^C\mathcal{V}$ the image of the inclusion

$$\operatorname{Hom}_{{}^C\mathcal{V}}(N, \operatorname{Hom}^2(M_*, X_{1*})) \hookrightarrow \operatorname{Hom}_{\mathcal{V}}(N, \operatorname{Hom}^2(M_*, X_{1*}))$$
$$\simeq \operatorname{Hom}_{\mathcal{V} \boxtimes \mathcal{A}}(N_1 \boxtimes M_2, X_{12})$$

is contained in $\operatorname{Hom}_{{}^C(\mathcal{V} \boxtimes \mathcal{A})}(N_1 \boxtimes M_2, X_{12})$. Then, set $N = \operatorname{Hom}^2(M_*, X_{1*})$ and the image (2.4.1) of 1_N is in ${}^C(\mathcal{V} \boxtimes \mathcal{A})$.

The comodule structure of $\operatorname{Hom}^2(M_*, X_{1*})$ is

$$\delta : \operatorname{Hom}^2(M_*, X_{1*}) \boxtimes \mathbb{1}_2 = \operatorname{Hom}^3(M_*, X_{1*} \boxtimes \mathbb{1}_2) \xrightarrow{\operatorname{Hom}(M, \delta)}$$
$$\operatorname{Hom}^3(M_*, C_{12'} \otimes X_{2''*}) \simeq C_{12'} \otimes \operatorname{Hom}(M_*, X_{2''*}),$$

where the last isomorphism is that one of Corollary 1.8.12 for the functor $G : \mathcal{V} \to \widehat{\mathcal{V} \boxtimes \mathcal{V}}$, $W \mapsto C_{12'} \otimes W_{2''}$, $G = \operatorname{colim} G_i$, $G_i : \mathcal{V} \to \mathcal{V} \boxtimes \mathcal{V}$, $W \mapsto C^i_{12'} \otimes W_{2''}$, where $C = \operatorname{colim} C^i$, $C^i \in \mathcal{V} \boxtimes \mathcal{V}$.

So, assume that $f : N \to \operatorname{Hom}(M_*, X_{1*})$ is a morphism of C-comodules. By Proposition 1.8.8 it is given by a morphism $g : N_1 \boxtimes M_2 \to X_{12} \in \mathcal{V} \boxtimes \mathcal{A}$. Let us prove that $g \in {}^C(\mathcal{V} \boxtimes \mathcal{A})$. Commutativity of diagram (2.2.3) implies the equation

$$\begin{aligned}
&\big(\operatorname{Hom}(Q_1 \boxtimes P_2, N_1 \boxtimes \mathbb{1}_2) \xrightarrow{\operatorname{Hom}(Q\boxtimes P,\delta_N)} \operatorname{Hom}(Q_1 \boxtimes P_2, C_{12'} \otimes N_{2''}) \\
&\quad \xrightarrow{\boxtimes 1_M} \operatorname{Hom}(Q_1 \boxtimes P_2 \boxtimes M_3, C_{12'} \otimes N_{2''} \boxtimes M_3) \\
&\quad \xrightarrow{\operatorname{Hom}(Q\boxtimes P\boxtimes M, C_{12'}\otimes g_{2''3})} \operatorname{Hom}(Q_1 \boxtimes P_2 \boxtimes M_3, C_{12'} \otimes X_{2''3}\big) \\
= &\big(\operatorname{Hom}(Q_1 \boxtimes P_2, N_1 \boxtimes \mathbb{1}_2) \xrightarrow{\boxtimes 1_M} \operatorname{Hom}(Q_1 \boxtimes P_2 \boxtimes M_3, N_1 \boxtimes \mathbb{1}_2 \boxtimes M_3) \\
&\quad \xrightarrow{\operatorname{Hom}(Q\boxtimes P\boxtimes M, g_{13}\boxtimes \mathbb{1}_2)} \operatorname{Hom}(Q_1 \boxtimes P_2 \boxtimes M_3, X_{13} \boxtimes \mathbb{1}_2) \\
&\quad \xrightarrow{\operatorname{Hom}(Q\boxtimes P\boxtimes M, \delta_X)} \operatorname{Hom}(Q_1 \boxtimes P_2 \boxtimes M_3, C_{12'} \otimes X_{2''3}\big)
\end{aligned}$$

for any $Q, P \in \mathcal{V}$. In particular, it holds for $Q = N$, $P = \mathbb{1}$. Following $1_{N\boxtimes\mathbb{1}}$ at the right hand side we get

$$1_{N\boxtimes\mathbb{1}} \mapsto 1_{N\boxtimes\mathbb{1}\boxtimes M} \mapsto g_{13} \boxtimes \mathbb{1}_2 \mapsto \delta_X \circ (g_{13} \boxtimes \mathbb{1}_2).$$

Following it at the left hand side we get

$$1_{N\boxtimes\mathbb{1}} \mapsto \delta_N \mapsto \delta_N \boxtimes M \mapsto (C_{12'} \otimes g_{2''3}) \circ (\delta_N \boxtimes M).$$

Coincidence of the results means that g is a homomorphism of comodules. □

2.5. The category of fibre functors

Let $\mathcal{V}$ be a $\Bbbk$-linear abelian category. Extending the definition of Saavedra [**26**] let us call a *fibre functor* to $\mathcal{V}$ a $\Bbbk$-linear exact faithful functor $a : \mathcal{A} \to \mathcal{V}$, where $\mathcal{A}$ is a $\Bbbk$-linear abelian essentially small category. Following Schauenburg [**27**] we define the category of fibre functors.

2.5.1. Definition. Let the category $\mathfrak{A} = \mathfrak{A}(\mathcal{V})$ have fibre functors to $\mathcal{V}$ as objects and let morphisms from $a : \mathcal{A} \to \mathcal{V}$ to $b : \mathcal{B} \to \mathcal{V}$ be equivalence classes of pairs (F, ϕ), where $F : \mathcal{A} \to \mathcal{B}$ is a functor and $\phi : a \xrightarrow{\sim} bF$ is a functorial isomorphism. Two such pairs (F, ϕ) and (G, γ) are equivalent if there is a functorial isomorphism $\zeta : F \to G$ such that

$$\gamma = \big(a \xrightarrow{\phi} bF \xrightarrow{b\zeta} bG\big). \tag{2.5.1}$$

The composite of two morphisms represented by (F, ϕ) and (G, γ) is represented by $(GF, \gamma F \circ \phi)$. Clearly, $\operatorname{Hom}_{\mathfrak{A}}(a, b)$ is a set.

2.5.2. Lemma. *Let $a : \mathcal{A} \to \mathcal{V}$ and $b : \mathcal{B} \to \mathcal{V}$ be fibre functors, let $F : \mathcal{A} \to \mathcal{B}$ be a functor, and let $\phi : a \xrightarrow{\sim} bF$. Then F is $\Bbbk$-linear and exact.*

PROOF. The functor bF isomorphic to the $\Bbbk$-linear functor a is $\Bbbk$-linear as well. Since b is $\Bbbk$-linear and faithful, F is $\Bbbk$-linear.

Denote by ξ an exact sequence $0 \to A'' \to A \to A' \to 0$ in $\mathcal{A}$. The homology $H(bF\xi) \simeq bH(F\xi)$, since b is exact. On the other hand, $a\xi \simeq bF\xi$ is exact, so $bH(F\xi) = 0$. Since b is faithful, $H(F\xi) = 0$ and F is exact. □

Now let $\mathcal{V}$ be a $\Bbbk$-linear abelian rigid monoidal category with length.

There is a functor $\Phi : \operatorname{Coalgsq}(\widehat{\mathcal{V}}) \to \mathfrak{A}(\mathcal{V})$, $C \mapsto (\mathcal{U} : {}^C\mathcal{V} \to \mathcal{V})$, where $\mathcal{U}$ is the underlying functor. To a squared coalgebra morphism $f : C \to D$ corresponds the equivalence class of the pair (F, ϕ), where

$$F(X, \delta_X) = (X, X_1 \boxtimes \mathbb{1}_2 \xrightarrow{\delta_X} C_{12'} \otimes X_{2''} \xrightarrow{f \otimes X} D_{12'} \otimes X_{2''})$$

and ϕ is the identity automorphism $\mathcal{U} \to \mathcal{U}$.

Our primary goal is to show that the functor Φ is an equivalence. And the second aim is to construct abelian categories equipped with fibre functors to $\mathcal{V}$ via generators and relations. To achieve this, we use the notion of a precategory, which is a technical synonym for a system of generators and relations. It already appeared in Joyal and Street study of monoidal categories [**16**] under the name of tensor scheme or valuation.

2.5.3. Definition. A *precategory* $\mathcal{P}$ is a class of objects $\mathcal{O} = \operatorname{Ob} \mathcal{P}$ with a set of morphisms $\operatorname{Hom}_{\mathcal{P}}(X, Y)$ given for each pair $X, Y \in \mathcal{O}$, and a distinguished morphism $\mathrm{id}_X \in \operatorname{Hom}_{\mathcal{P}}(X, X)$ for each $X \in \mathcal{O}$, called the identity morphism. A *prefunctor* from a precategory $\mathcal{P}$ to a precategory (or a category) $\mathcal{S}$ is a pair of maps $\operatorname{Ob} \mathcal{P} \to \operatorname{Ob} \mathcal{S}$, $\operatorname{Mor} \mathcal{P} \to \operatorname{Mor} \mathcal{S}$ coherent with the functions id, source and target. *(Re)construction data* is a prefunctor from a small precategory $\mathcal{P}$ ($\operatorname{Ob} \mathcal{P}$ is a set) to a $\Bbbk$-linear abelian category.

As the name suggests, (re)construction data are used to construct abelian categories equipped with an exact faithful functor to the given category $\mathcal{V}$. We shall do this later. At the moment we concentrate on monoidal $\mathcal{V}$ and squared coalgebras.

2.5.4. The coend. Let $\mathcal{C}$ be a $\Bbbk$-linear abelian category with length. We define the coend C of a bifunctor $B : \mathcal{P} \times \mathcal{P}^{\mathrm{op}} \to \mathcal{C}$, where $\mathcal{P}$ is a precategory, as an object of $\widehat{\mathcal{C}}$ which is the inductive limit of the diagram

$$B(X, X) \xleftarrow{B(X,f)} B(X, Y) \xrightarrow{B(f,Y)} B(Y, Y),$$

where $f : X \to Y$ runs over $\operatorname{Mor} \mathcal{P}$. That is, C is equipped with a morphism $i_X : B(X, X) \to C \in \widehat{\mathcal{C}}$ for each $X \in \operatorname{Ob} \mathcal{P}$, the diagram

$$\begin{array}{ccc} B(X,Y) & \xrightarrow{B(f,Y)} & B(Y,Y) \\ {\scriptstyle B(X,f)}\big\downarrow & & \big\downarrow{\scriptstyle i_Y} \\ B(X,X) & \xrightarrow{i_X} & C \end{array}$$

is commutative for any $f : X \to Y \in \operatorname{Mor} \mathcal{P}$, and C is universal between such objects. If $\mathcal{P}$ is small, we can say that the sequence

$$\bigoplus_{f : X \to Y \in \operatorname{Mor} \mathcal{P}} B(X, Y) \xrightarrow{B(X,f) - B(f,Y)} \bigoplus_{X \in \operatorname{Ob} \mathcal{P}} B(X, X) \xrightarrow{\oplus i_X} C \to 0 \qquad (2.5.2)$$

is exact. So in this case the coend exists. It is denoted $C = \int^{X \in \mathcal{P}} B(X, X)$.

Let us consider the particular case. Let $p : \mathcal{P} \to \mathcal{V}$ be a prefunctor from a small precategory $\mathcal{P}$, let $\mathcal{C} = \mathcal{V} \boxtimes \mathcal{V}$ and let $B : \mathcal{P} \times \mathcal{P}^{\mathrm{op}} \to \mathcal{C}$, $B(X, Y) = pX \boxtimes (pY)^{\vee}$. The coend is denoted

$$C = \int^{X \in \mathcal{P}} pX \boxtimes (pX)^{\vee}. \qquad (2.5.3)$$

The object $M \boxtimes M^\vee \in \mathcal{V} \boxtimes \mathcal{V}$ has a canonical squared coalgebra structure for any $M \in \mathcal{V}$. Namely,

$$\Delta_{123} = M \boxtimes \operatorname{coev} \boxtimes M^\vee : M \boxtimes \mathbb{1} \boxtimes M^\vee \to M \boxtimes M^\vee \otimes M \boxtimes M^\vee,$$
$$\varepsilon = \operatorname{ev} : M \otimes M^\vee \to \mathbb{1}.$$

Compare these formulae with the graphical notations on page 44. The object M has a canonical structure of a left $M \boxtimes M^\vee$-comodule, namely,

$$\delta = M \boxtimes \operatorname{coev} : M_1 \boxtimes \mathbb{1}_2 \to M_1 \boxtimes M^\vee_{2'} \otimes M_{2''}.$$

Compare with the graphical notations on page 47. In particular, this holds for $M = pX$.

2.5.5. Proposition. *If the coend* (2.5.3) *exists, it has a unique squared coalgebra structure such that the structure morphisms $i_X : pX \boxtimes (pX)^\vee \to C$ are coalgebra morphisms.*

PROOF. We have the squared coalgebra $C^{\mathcal{O}} = \oplus_{X \in \operatorname{Ob} \mathcal{P}} pX \boxtimes (pX)^\vee$. Since a sum of coideals is a coideal, we only have to show that the image of

$$\operatorname{rel} f = \bigl(pX \boxtimes (pX)^\vee \xrightarrow{1\boxtimes(pf)^t \oplus -pf\boxtimes 1} pX \boxtimes (pX)^\vee \oplus pY \boxtimes (pY)^\vee \xrightarrow{i_X \oplus i_Y} C^{\mathcal{O}}\bigr) \tag{2.5.4}$$

is a coideal in $C^{\mathcal{O}}$ for any $f : X \to Y \in \mathcal{P}$. Set $J = pX \boxtimes (pY)^\vee$, $j = \operatorname{rel} f$, $W = J_{13} \boxtimes \mathbb{1}_2 = pX \boxtimes \mathbb{1} \boxtimes (pY)^\vee$, $s = \mathrm{id}_W$ and

$$r = \bigl(pX_1 \boxtimes \mathbb{1}_2 \boxtimes pY_3^\vee \xrightarrow{1\boxtimes \operatorname{coev}_{pX} \boxtimes 1 \oplus 1 \boxtimes \operatorname{coev}_{pY} \boxtimes 1}$$
$$pX_1 \boxtimes pX^\vee_{2'} \otimes pX_{2''} \boxtimes pY_3^\vee \oplus pX_1 \boxtimes pY_{2'} \otimes pY_{2''} \boxtimes pY_3^\vee$$
$$\xrightarrow{i_X \otimes 1 \boxtimes 1 \oplus 1 \boxtimes 1 \otimes i_Y} C^{\mathcal{O}}_{12'} \otimes pX_{2''} \boxtimes pY_3^\vee \oplus pX_1 \boxtimes pY^\vee_{2'} \otimes C^{\mathcal{O}}_{2''3}\bigr).$$

Then the diagram (2.1.5) commutes. Indeed, it reduces to the following equations, where pX, pY, pf are abbreviated to X, Y, f:

$$[1 \boxtimes 1 \boxtimes f^t, 1 \boxtimes \operatorname{coev}_X \boxtimes 1] = 0,$$
$$[f \boxtimes 1 \boxtimes 1, 1 \boxtimes \operatorname{coev}_Y \boxtimes 1] = 0,$$

$$\bigl(X_1 \boxtimes \mathbb{1}_2 \boxtimes Y_3^\vee \xrightarrow{X \boxtimes \operatorname{coev}_X \boxtimes Y^\vee \oplus X \boxtimes \operatorname{coev}_Y \boxtimes Y^\vee}$$
$$X_1 \boxtimes X_2^\vee \otimes X_{2''} \boxtimes Y_3^\vee \oplus X_1 \boxtimes Y^\vee_{2'} \otimes Y_{2''} \boxtimes Y_3^\vee$$
$$\xrightarrow{-X \boxtimes X^\vee \otimes f \boxtimes Y^\vee \oplus X \boxtimes f^t \otimes Y \boxtimes Y^\vee} X_1 \boxtimes X^\vee_{2'} \otimes Y_{2''} \boxtimes Y_3^\vee\bigr) = 0,$$

which are obvious. By Remark 2.1.5 we get that $\operatorname{Im} \operatorname{rel} f$ is a coideal. □

Since pX is a $pX \boxtimes (pX)^\vee$-comodule, it is a C-comodule as well for any $X \in \operatorname{Ob} \mathcal{P}$.

2.5.6. Proposition. *If the coend C exists, the map $\operatorname{Ob} \mathcal{P} \to \operatorname{Ob} {}^C\mathcal{V}$, $X \mapsto (pX, \delta)$ extends to the prefunctor $F : \mathcal{P} \to {}^C\mathcal{V}$, $f \mapsto pf$ such that $p = \bigl(\mathcal{P} \xrightarrow{F} {}^C\mathcal{V} \xrightarrow{\mathcal{U}} \mathcal{V}\bigr)$. If p is a functor from the category $\mathcal{P}$, so is F.*

PROOF. We have to prove that for any $g : X \to Y \in \operatorname{Mor} \mathcal{P}$ the morphism pg is a comodule homomorphism. Denote $f = pg$, $M = pX$, $N = pY$. The required conclusion follows from the diagram

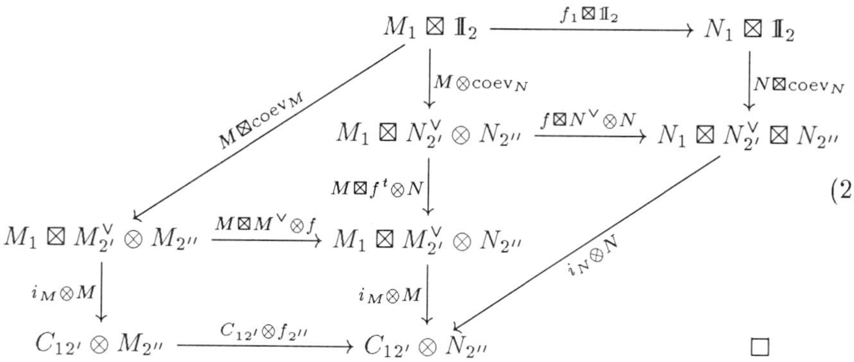

(2.5.5) □

2.5.7. An exact sequence of comodules. Let C be a squared coalgebra and $M \in {}^C\widehat{\mathcal{V}}$. Applying the functor $\mathrm{Id}_{\mathcal{V}} \boxtimes \circledast$ to the diagram (2.2.1) we get a sequence

$$0 \to M_1 \boxtimes \mathbb{I}_2 \xrightarrow{\delta} C_{12'} \otimes M_{2''} \xrightarrow{f-g} C_{12'} \otimes C_{2''2'''} \otimes M_{2^4}, \tag{2.5.6}$$

where

$$f = \left(C_{12'} \otimes M_{2''} \simeq C_{12'} \otimes M_{2''} \otimes \mathbb{I}_{2'''} \xrightarrow{C_{12'} \otimes \delta_{2''2'''}} C_{12'} \otimes C_{2''2'''} \otimes M_{2^4}\right),$$

$$g = \left(C_{12'} \otimes M_{2''} \simeq C_{12''} \otimes \mathbb{I}_{2'} \otimes M_{2'''} \xrightarrow{\Delta_{12'2''} \otimes M_{2'''}} C_{12'} \otimes C_{2''2'''} \otimes M_{2^4}\right).$$

2.5.8. Proposition. *Sequence* (2.5.6) *is an exact sequence of C-comodules from* ${}^C\widehat{\mathcal{V} \boxtimes \mathcal{V}}$.

PROOF. The functor $\circledast$ is exact and faithful by Proposition 1.3.3. Applying it to (2.5.6) we get a sequence

$$0 \to M \xrightarrow{\bar\delta} \bar C \otimes M \xrightarrow{\bar C \otimes \bar\delta - \bar\Delta \otimes M} \bar C \otimes \bar C \otimes M, \tag{2.5.7}$$

where $(\bar C = \circledast C, \bar\Delta = \circledast(\mathcal{V} \otimes \circledast)\Delta)$ is an ordinary coalgebra in $\widehat{\mathcal{V}}$ and $\left(M, \bar\delta = (M \simeq M \otimes \mathbb{I} \xrightarrow{\circledast\delta} \bar C \otimes M)\right)$ is the canonical $\bar C$-comodule structure on M. The functor $\circledast$ takes cohomology of (2.5.6) to the cohomology of (2.5.7), therefore, exactness of (2.5.7) implies exactness of (2.5.6). However, it is well known that (2.5.7) is exact.

2.5.9. Lemma (cf. Saavedra [26] Section 2.0.4). *Sequence* (2.5.7) *is exact for any comodule* $(M, \bar\delta)$ *over any ordinary coalgebra* $\bar C \in \widehat{\mathcal{V}}$.

PROOF. Indeed, $\bar\delta$ is monic since $\left(M \xrightarrow{\bar\delta} \bar C \otimes M \xrightarrow{\varepsilon \otimes X} \mathbb{I} \otimes M \simeq M\right) = \mathrm{id}_M$. Composing the last arrow in (2.5.7) with $\varepsilon \otimes \bar C \otimes M$ we get a sequence

$$0 \to M \xrightarrow{\bar\delta} \bar C \otimes M \xrightarrow{\varepsilon \otimes \bar\delta - 1} \bar C \otimes M, \tag{2.5.8}$$

whose cohomology contains the cohomology of (2.5.7). It remains to notice that sequence (2.5.8) is exact. □

The C-comodule structure of $C_{12'} \otimes M_{2''}$ and $C_{12'} \otimes C_{2''2'''} \otimes M_{2^4}$ is obtained from the left regular comodule C via the functor $\mathrm{Id}_{\mathcal{V}} \boxtimes \circledast : {}^C\widehat{\mathcal{V}^{\boxtimes 3}} \to {}^C\widehat{\mathcal{V}^{\boxtimes 2}}$, $C_{12} \boxtimes Z_3 \mapsto C_{12'} \otimes Z_{2''}$. The coassociativity of the coaction (2.2.1) implies that $\delta \in$

${}^{C}\widehat{\mathcal{V}\boxtimes\mathcal{V}}$. Clearly, f is a homomorphism of C-comodules as well. The coassociativity of the comultiplication (2.1.1) implies that $\Delta_{123} : C_{13}\boxtimes \mathbb{I}_2 \to C_{12'}\otimes C_{2''3} \in \widehat{\mathcal{V}^{\boxtimes 3}}$ is a homomorphism of C-comodules. Therefore,

$$\Delta_{12'2''} : C_{12''}\boxtimes \mathbb{I}_{2'} \to C_{12'}\otimes C_{2''2'''} \in \widehat{\mathcal{V}\boxtimes\mathcal{V}}$$

is a homomorphism of C-comodules. This implies that $g \in {}^{C}\widehat{\mathcal{V}\boxtimes\mathcal{V}}$. □

2.6. Reconstruction theorems

Now we come back to the case of a fibre functor $a : \mathcal{A}\to\mathcal{V}$. Assume that $\mathcal{P}\subset\mathcal{A}$ is a full subcategory equivalent to $\mathcal{A}$ and $\mathcal{O} = \operatorname{Ob}\mathcal{P}$ is a set. Denote $p = a|_{\mathcal{P}} : \mathcal{P}\to\mathcal{V}$. Notice, that coend (2.5.3) serves as the coend of the bifunctor $\mathcal{A}^{\mathrm{op}}\times\mathcal{A}\to\mathcal{V}\boxtimes\mathcal{V}$, $(X,Y)\mapsto aX\boxtimes(aY)^{\vee}$ as well. Similarly to the coalgebra C given by (2.5.3) we consider the coend

$$B = \int^{X\in\mathcal{P}} X\boxtimes(pX)^{\vee} \in \widehat{\mathcal{A}\boxtimes\mathcal{V}}.$$

Clearly, $(a\boxtimes 1_{\mathcal{V}})B = C \in \widehat{\mathcal{V}\boxtimes\mathcal{V}}$.

Let $M\in {}^{C}\mathcal{V}$. We shall lift the exact sequence (2.5.6) to $\widehat{\mathcal{A}\boxtimes\mathcal{V}}$. Consider the morphisms in $\widehat{\mathcal{A}\boxtimes\mathcal{V}}$

$$\tilde f = \big(B_{12'}\otimes M_{2''} \simeq B_{12'}\otimes M_{2''}\otimes\mathbb{I}_{2'''} \xrightarrow{B_{12'}\otimes\delta_{2''2'''}} B_{12'}\otimes C_{2''2'''}\otimes M_{2^4}\big),$$

$$\tilde g = \big(B_{12'}\otimes M_{2''} \simeq B_{12''}\otimes\mathbb{I}_{2'}\otimes M_{2'''} \xrightarrow{\tilde\Delta_{12'2''}\otimes M_{2'''}} B_{12'}\otimes C_{2''2'''}\otimes M_{2^4}\big),$$

where $\tilde\Delta_{12'2''}$ comes from

$$\tilde\Delta_{123} : B_{13}\boxtimes\mathbb{I}_2 \to B_{12'}\otimes C_{2''3} \in (\mathcal{A}\boxtimes\mathcal{V}\boxtimes\mathcal{V})\widehat{\ },$$

which is determined by the system of morphisms

$$X_1\boxtimes\mathbb{I}_2\boxtimes pX_3^{\vee} \xrightarrow{1\boxtimes\mathrm{coev}\boxtimes 1} X_1\boxtimes pX_{2'}^{\vee}\otimes pX_{2''}\boxtimes pX_3^{\vee} \to B_{12'}\otimes C_{2''3}.$$

The necessary check is left to the reader. Clearly, $(a\boxtimes\mathcal{V})\tilde f = f$, $(a\boxtimes\mathcal{V})\tilde g = g \in \widehat{\mathcal{V}\boxtimes\mathcal{V}}$.

Moreover, applying $(F\boxtimes\mathcal{V})$ to $B_{12'}\otimes M_{2''}$ and $B_{12'}\otimes C_{2''2'''}\otimes M_{2^4}$ we get C-comodules $C_{12'}\otimes M_{2''}$ and $C_{12'}\otimes C_{2''2'''}\otimes M_{2^4}$, where $F:\mathcal{A}\to {}^{C}\mathcal{V}$ was introduced in Proposition 2.5.6. This follows from the lemma.

2.6.1. Lemma. *$(F\boxtimes\mathcal{V})B = C$ in ${}^{C}\widehat{\mathcal{V}\boxtimes\mathcal{V}}$.*

PROOF. The comodule structure δ_{123} of $(F\boxtimes\mathcal{V})B \in (({}^{C}\mathcal{V})\boxtimes\mathcal{V})\widehat{\ }\subset ({}^{C}(\mathcal{V}\boxtimes\mathcal{V}))\widehat{\ } = {}^{C}\widehat{\mathcal{V}\boxtimes\mathcal{V}}$ is given by the same formulae

$$\begin{array}{ccc}
pX_1\boxtimes\mathbb{I}_2\boxtimes pX_3^{\vee} & \xrightarrow{i_{X\,13}\boxtimes\mathbb{I}_2} & (F\boxtimes\mathcal{V})B_{13}\boxtimes\mathbb{I}_2\\
{\scriptstyle 1\boxtimes\mathrm{coev}\boxtimes 1}\downarrow & & \downarrow{\scriptstyle\delta_{123}}\\
pX_1\boxtimes pX_{2'}^{\vee}\otimes pX_{2''}\boxtimes pX_3^{\vee} & \xrightarrow{i_X\otimes i_X} & C_{12'}\otimes(F\boxtimes\mathcal{V})B_{2''3}
\end{array}$$

as the comultiplication in C. □

Let $W = \operatorname{Ker}(\tilde f-\tilde g) \in \widehat{\mathcal{A}\boxtimes\mathcal{V}}$. Then $(F\boxtimes\mathcal{V})W \simeq \operatorname{Ker}(f-g) = M\boxtimes\mathbb{I}$ by Proposition 2.5.8. Now we may apply the following lemma (set $\mathcal{B} = {}^{C}\mathcal{V}$).

2.6.2. Lemma. *Let $\mathcal{A}$, $\mathcal{B}$, $\mathcal{V}$ be $\Bbbk$-linear abelian categories with length, assume that $\mathcal{V}$ is rigid monoidal with $\operatorname{End}_{\mathcal{V}} \mathbb{I} = \Bbbk$, and let $F : \mathcal{A} \to \mathcal{B}$ be a $\Bbbk$-linear exact faithful functor. If $(F \boxtimes \mathcal{V})W = M \boxtimes \mathbb{I}$ for some $W \in \widehat{\mathcal{A} \boxtimes \mathcal{V}}$ and $M \in \mathcal{B}$, then there exists such $Y \in \mathcal{A}$ that $W \simeq Y \boxtimes \mathbb{I}$ and $F(Y) \simeq M$.*

PROOF. Standard reasoning allows to consider the special case $\mathcal{A} = \text{mod-}A$, $\mathcal{B} = \text{mod-}B$, $\mathcal{V} = \text{mod-}V$, where A, B, V are finite dimensional $\Bbbk$-algebras. Then $W \in \mathcal{A} \boxtimes \mathcal{V} = \text{mod-}A \otimes_{\Bbbk} V$, $M \in \mathcal{B}$, $\mathbb{I}$ is a simple V-module such that $\operatorname{End}_V \mathbb{I} = \Bbbk$, $M \boxtimes \mathbb{I} = M \otimes_{\Bbbk} \mathbb{I} \in \mathcal{B} \boxtimes \mathcal{V} = \text{mod-}B \otimes_{\Bbbk} V$. Any exact faithful functor $F : \mathcal{A} \to \mathcal{B}$ is given by a bimodule $P = {}_A P_B \in A\text{-}B\text{-bimod}$, $F(E) = E \otimes_A P$, such that ${}_A P$ is a direct summand of ${}_A A^n$ and ${}_A A$ is a direct summand of ${}_A P^r$ for some $n, r \in \mathbb{N}$. The functor $F \boxtimes \mathcal{V} : \mathcal{A} \boxtimes \mathcal{V} \to \mathcal{B} \boxtimes \mathcal{V}$ is also given by $Z_{AV} \mapsto Z_{AV} \otimes_A {}_A P_B$.

Since ${}_A A \oplus {}_A L \simeq {}_A P^r$ for some $L \in A\text{-mod}$, the V-module W is a direct summand of

$$W_{AV} \otimes_A A \oplus W_{AV} \otimes_A L \simeq W_{AV} \otimes_A {}_A P^r = (F \boxtimes \mathcal{V})(W^r)_V \simeq M^r \otimes_{\Bbbk} \mathbb{I}_V .$$

Since $\mathbb{I}_V$ is simple, the V-module W is an isotypic semisimple module of the type $\mathbb{I}_V$, that is, $W_V \simeq Y \otimes_{\Bbbk} \mathbb{I}_V$ for some $\Bbbk$-vector space Y.

The homomorphism of action $\phi : V \to \operatorname{End}_{\Bbbk} \mathbb{I}$ is surjective (its image is the centraliser of $\Bbbk$). Any $a \in A$ is represented by a linear map $\pi(a) \in \operatorname{End}_{\Bbbk}(Y \otimes_{\Bbbk} \mathbb{I}) \simeq \operatorname{End}_{\Bbbk} W$ which commutes with any $1_Y \otimes \phi(v)$, $v \in V$. Thus $\pi(a)$ has the form $\psi(a) \otimes 1_{\mathbb{I}}$ for some $\psi(a) \in \operatorname{End}_{\Bbbk} Y$. Therefore, the algebra A acts in Y via the homomorphism $\psi : A \to \operatorname{End}_{\Bbbk} Y$ and the $A \otimes_{\Bbbk} V$-module W is isomorphic to $Y_A \otimes_{\Bbbk} \mathbb{I}_V$. Finally, the isomorphism $F(Y) \boxtimes \mathbb{I} \simeq M \boxtimes \mathbb{I}$ implies $F(Y) \simeq M$ by Proposition 1.1.11. □

We deduce from the above lemma that $W = \operatorname{Ker}(\tilde{f} - \tilde{g}) \simeq Y \boxtimes \mathbb{I}$ for some $Y \in \mathcal{A}$ and $F(Y) \simeq M$. That is, any C-comodule M is isomorphic to a comodule which lies in the image of $F : \mathcal{A} \to {}^C\mathcal{V}$. All is ready for the following theorem.

2.6.3. Theorem. *The functor $F : \mathcal{A} \to {}^C\mathcal{V}$ is an equivalence of categories.*

PROOF. We already know that F is faithful and essentially surjective on objects. We shall prove that F is full. Let $M = F(X)$, $N = F(Y)$ and let $h : M \to N$ be any C-comodule homomorphism. We have a morphism of exact sequences (2.5.6) in $\widehat{\mathcal{V} \boxtimes \mathcal{V}}$

$$\begin{array}{ccccccc}
0 \longrightarrow & M_1 \boxtimes \mathbb{I}_2 & \xrightarrow{\delta} & C_{12'} \otimes M_{2''} & \xrightarrow{f_M - g_M} & C_{12'} \otimes C_{2''2'''} \otimes M_{2^4} \\
& \Big\downarrow{\scriptstyle f_1 \boxtimes \mathbb{I}_2} & & \Big\downarrow{\scriptstyle C_{12'} \otimes f_{2''}} & & \Big\downarrow{\scriptstyle C \otimes C \otimes f} \\
0 \longrightarrow & N_1 \boxtimes \mathbb{I}_2 & \xrightarrow{\delta} & C_{12'} \otimes N_{2''} & \xrightarrow{f_N - g_N} & C_{12'} \otimes C_{2''2'''} \otimes N_{2^4}
\end{array} \tag{2.6.1}$$

and a morphism of exact sequences in $\widehat{\mathcal{A} \boxtimes \mathcal{V}}$

$$\begin{array}{ccccccc}
0 \longrightarrow & X_1 \boxtimes \mathbb{I}_2 & \longrightarrow & B_{12'} \otimes M_{2''} & \xrightarrow{\tilde{f}_M - \tilde{g}_M} & B_{12'} \otimes C_{2''2'''} \otimes M_{2^4} \\
& \exists \Big\downarrow{\scriptstyle j} & & \Big\downarrow{\scriptstyle B_{12'} \otimes f_{2''}} & & \Big\downarrow{\scriptstyle B \otimes C \otimes f} \\
0 \longrightarrow & Y_1 \boxtimes \mathbb{I}_2 & \xrightarrow{\delta} & B_{12'} \otimes N_{2''} & \xrightarrow{\tilde{f}_N - \tilde{g}_N} & B_{12'} \otimes C_{2''2'''} \otimes N_{2^4}
\end{array} \tag{2.6.2}$$

The last diagram implies existence of a morphism $j : X_1 \boxtimes \mathbb{I}_2 \to Y_1 \boxtimes \mathbb{I}_2 \in \mathcal{A} \boxtimes \mathcal{V}$. Since $\mathcal{A} \boxtimes \langle \mathbb{I} \rangle \simeq \mathcal{A} \boxtimes \Bbbk\text{-vect} \simeq \mathcal{A}$, any such morphism has the form $j = t \boxtimes \mathbb{I}$ for some $t : X \to Y \in \mathcal{A}$. Applying $F \boxtimes \mathcal{V}$ to (2.6.2) we should get (2.6.1), which implies, in particular, that $f = F(t)$. □

In particular case $\mathcal{V} = \Bbbk$-vect this theorem was proved by Saavedra [**26**] (see also Schauenburg [**27**]). Notice that they also used exact sequence (2.5.6) which in this case coincides with (2.5.7).

2.6.4. Proposition. *The map $(a : \mathcal{A} \to \mathcal{V}) \longmapsto C_a$ constructed in Section 2.5.4 extends to a functor*

$$\Psi : \mathfrak{A}(\mathcal{V}) \to \mathrm{Coalgsq}(\widehat{\mathcal{V}}).$$

PROOF. For a given $a \in \mathfrak{A}(\mathcal{V})$ we constructed a squared coalgebra C_a together with canonical morphisms $i_X : aX \boxtimes (aX)^\vee \to C_a$ for each $X \in \mathrm{Ob}\,\mathcal{A}$. Suppose that (G, γ) represents a morphism $a \to b \in \mathfrak{A}(\mathcal{V})$, that is, $G : \mathcal{A} \to \mathcal{B}$ is a functor and $\gamma : a \xrightarrow{\sim} bG$ is an isomorphism of functors. There is a unique morphism $g : C_a \to C_b \in \widehat{\mathcal{V} \boxtimes \mathcal{V}}$ which makes the diagram

$$\begin{array}{ccc} aX \boxtimes (aX)^\vee & \xrightarrow{i_X} & C_a \\ {\scriptstyle \gamma_X \boxtimes \gamma_X^{-1\,t}} \downarrow & & \exists \downarrow {\scriptstyle g} \\ bGX \boxtimes (bGX)^\vee & \xrightarrow{i_{GX}} & C_b \end{array} \tag{2.6.3}$$

commutative for any $X \in \mathcal{A}$. This follows by universality from the diagram

$$\begin{array}{ccccc} aY \boxtimes (aY)^\vee & \xleftarrow{af \boxtimes 1} & aX \boxtimes (aY)^\vee & \xrightarrow{1 \boxtimes (af)^t} & aX \boxtimes (aX)^\vee \\ {\scriptstyle \gamma_Y \boxtimes \gamma_Y^{-1\,t}} \downarrow & & {\scriptstyle \gamma_X \boxtimes \gamma_Y^{-1\,t}} \downarrow & & \downarrow {\scriptstyle \gamma_X \boxtimes \gamma_X^{-1\,t}} \\ bGY \boxtimes (bGY)^\vee & \xleftarrow{bGf \boxtimes 1} & bGX \boxtimes (bGY)^\vee & \xrightarrow{1 \boxtimes (bGf)^t} & bGX \boxtimes (bGX)^\vee \\ & {\scriptstyle i_{GY}} \searrow & & \swarrow {\scriptstyle i_{GX}} & \\ & & C_b & & \end{array}$$

which commutes for any $f : X \to Y \in \mathcal{A}$. One can check that $\gamma_X \boxtimes \gamma_X^{-1\,t}$ in diagram (2.6.3) is an isomorphism of coalgebras and i_X, i_{GX} are coalgebra homomorphisms by Proposition 2.5.5. Using presentation of C_a of the type (2.5.2) we deduce that g is a coalgebra homomorphism. The remaining check is left to the reader. □

2.6.5. Example. Let $C = (C, \Delta, \varepsilon, m, \eta, \gamma)$ be an ordinary Hopf $\Bbbk$-algebra, $\mathcal{A} = \mathcal{V} = C$-comod, $a = \mathrm{Id}_{\mathcal{V}}$. Then the squared coalgebra C_a reconstructed from $\mathrm{Id}_{\mathcal{V}} : \mathcal{V} \to \mathcal{V}$ is that of Example 2.1.2. That is, $C_a = (C, \Delta, \varepsilon)$. Proof is given in Example 3.3.20.

2.6.6. Proposition. *The functor $F_a : \mathcal{A} \to \Phi(\Psi(\mathcal{A})) = {}^{C_a}\mathcal{V}$ constructed in Theorem 2.6.3 together with $\mathrm{id} : a \to \mathcal{U} \circ F_a$ gives an isomorphism of functors*

$$F : \mathrm{Id}_{\mathfrak{A}(\mathcal{V})} \xrightarrow{\sim} \Phi\Psi.$$

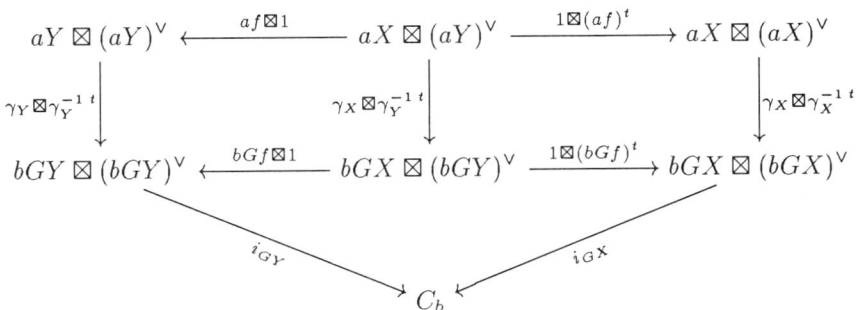

PROOF. We already noticed in Proposition 2.5.6 that $a = \mathcal{U} \circ F_a$. Let (G,γ) represent a morphism $a \to b \in \mathfrak{A}(\mathcal{V})$. Consider the following diagram

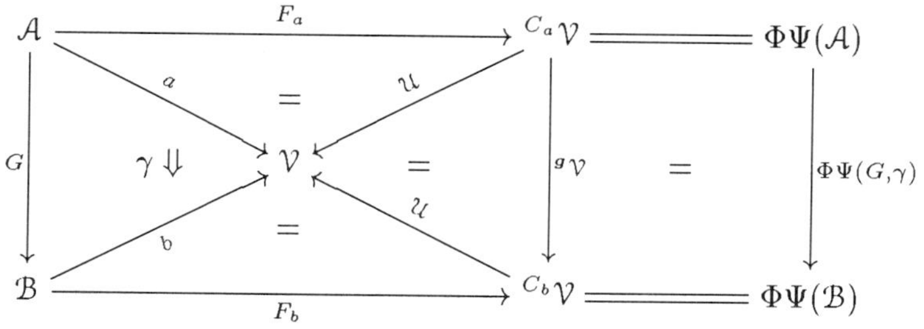

For any $X \in \mathcal{A}$ we have

$$
\begin{aligned}
{}^{g}\mathcal{V}(F_a X) &= \big(aX, aX \boxtimes \mathbb{1} \xrightarrow{1\boxtimes\text{coev}} aX \boxtimes (aX)^\vee \otimes aX \\
&\qquad \xrightarrow{i_X \otimes 1} C_{a\,12'} \otimes aX_{2''} \xrightarrow{g\otimes 1} C_{b\,12'} \otimes aX_{2''}\big), \\
F_b(GX) &= \big(bGX, bGX \boxtimes \mathbb{1} \xrightarrow{1\boxtimes\text{coev}} bGX \boxtimes (bGX)^\vee \otimes bGX \\
&\qquad \xrightarrow{i_{GX}\otimes 1} C_{b\,12'} \otimes bGX_{2''}\big).
\end{aligned}
$$

By definition of $\boldsymbol{g}$ (see diagram (2.6.3)) it follows that $\zeta = \gamma : aX \to bGX$ is an isomorphism of C_b-comodules. Thus, $\zeta : {}^{g}\mathcal{V} \circ F_a \to F_b \circ G$ is a morphism of functors. Therefore, the pairs $({}^{g}\mathcal{V} \circ F_a, \mathrm{id})$ and $(F_b \circ G, \gamma)$ are equivalent by ζ and define the same morphism in $\mathfrak{A}(\mathcal{V})$. Hence, $F : \mathrm{Id}_{\mathfrak{A}(\mathcal{V})} \to \Phi\Psi$ is a morphism of functors. By Theorem 2.6.3 F_a is an equivalence, therefore (F_a, id) is an isomorphism in $\mathfrak{A}(\mathcal{V})$ (see Schauenburg [**27**]), that is, F is an isomorphism. □

2.6.7. Theorem. *The functors* $\Phi : \mathrm{Coalgsq}(\widehat{\mathcal{V}}) \to \mathfrak{A}(\mathcal{V})$, $C \mapsto {}^{C}\mathcal{V}$, *and* $\Psi : \mathfrak{A}(\mathcal{V}) \to \mathrm{Coalgsq}(\widehat{\mathcal{V}})$, $a \to C_a$, *are equivalences, quasiinverse to each other.*

PROOF. By Theorem 2.6.3 we know that Φ is essentially surjective on objects and Proposition 2.6.6 implies that Φ is full. It remains to show that Φ is faithful.

Let $f, g : C \to D$ be squared coalgebra homomorphisms which give equivalent pairs $({}^{f}\mathcal{V}, \mathrm{id})$ and $({}^{g}\mathcal{V}, \mathrm{id})$, so that $\Phi(f) = \Phi(g)$. It follows from the definition (2.5.1) that, in fact, ${}^{f}\mathcal{V} = {}^{g}\mathcal{V} : {}^{C}\mathcal{V} \to {}^{D}\mathcal{V}$. Thus for any $X \in {}^{C}\mathcal{V}$ the equation

$$
\begin{aligned}
&\big(X_1 \boxtimes \mathbb{1}_2 \xrightarrow{\delta} C_{12'} \otimes X_{2''} \xrightarrow{f\otimes X} D_{12'} \otimes X_{2''}\big) \\
&\qquad\qquad = \big(X_1 \boxtimes \mathbb{1}_2 \xrightarrow{\delta} C_{12'} \otimes X_{2''} \xrightarrow{g\otimes X} D_{12'} \otimes X_{2''}\big)
\end{aligned}
$$

holds. It holds also for any comodule $X \in {}^{C}(\mathcal{V} \boxtimes \mathcal{A})$ since this category is equivalent to $({}^{C}\mathcal{V}) \boxtimes \mathcal{A}$ by Theorem 2.4.1. The same equation holds for $X \in {}^{C}\widehat{\mathcal{V} \boxtimes \mathcal{A}}$ by Theorem 2.3.1. In particular, it holds for the left regular comodule:

$$
\begin{aligned}
&\big(C_{13} \boxtimes \mathbb{1}_2 \xrightarrow{\Delta_{123}} C_{12'} \otimes C_{2''3} \xrightarrow{f_{12'}\otimes C_{2''3}} D_{12'} \otimes C_{2''3}\big) \\
&\qquad\qquad = \big(C_{13} \boxtimes \mathbb{1}_2 \xrightarrow{\Delta_{123}} C_{12'} \otimes C_{2''3} \xrightarrow{g_{12'}\otimes C_{2''3}} D_{12'} \otimes C_{2''3}\big).
\end{aligned}
$$

Applying the functor $\mathrm{Id}_{\mathcal{V}} \boxtimes \circledast$ and composing with $D_{12'} \otimes C_{2''2'''} \xrightarrow{D\otimes\varepsilon} D_{12'} \otimes \mathbb{1}_{2''}$ we get

$$
f_{12'} \boxtimes \mathbb{1}_{2''} = g_{12'} \boxtimes \mathbb{1}_{2''} : C_{12''} \otimes \mathbb{1}_{2'} \to D_{12'} \otimes \mathbb{1}_{2''}.
$$

Therefore, $f = g$ and Φ is an equivalence. Hence, the functor Ψ adjoint to Φ by Proposition 2.6.6 is also an equivalence, and Φ and Ψ are quasiinverse to each other. □

In the case of $\mathcal{V} = \Bbbk$-vect this theorem was proved by Schauenburg [**27**].

Constructing another adjunction $\Psi\Phi \xrightarrow{\sim} \mathrm{Id}$, is also of practical interest.

2.6.8. Proposition. *Let C be a squared coalgebra in $\widehat{\mathcal{V}}$ and let $X \in \mathcal{V}$. The structures of C-comodules in X and squared coalgebra homomorphisms $X \boxtimes X^\vee \to C$ are in bijective correspondence.*

PROOF. If $\delta : X_1 \boxtimes \mathbb{1}_2 \to C_{12'} \otimes X_{2''}$ is a comodule structure, then

$$\ddot{\imath} = \ddot{\imath}_X = \big(X_1 \boxtimes X_2^\vee \xrightarrow{\sim} X_1 \boxtimes \mathbb{1}_{2'} \otimes X_{2''}^\vee \xrightarrow{\delta_{12'} \otimes X_{2''}^\vee} C_{12'} \otimes X_{2''} \otimes X_{2'''}^\vee \xrightarrow{C_{12'} \otimes \mathrm{ev}_{2''}} C_{12'} \otimes \mathbb{1}_{2''} \xrightarrow{\simeq} C_{12}\big) \qquad (2.6.4)$$

is a homomorphism of squared coalgebras. The verification is straightforward and it is left to the reader. On the other hand, if $\ddot{\imath}_X : X_1 \boxtimes X_2^\vee \to C_{12}$ is a squared coalgebra homomorphism, then

$$\delta = \big(X_1 \boxtimes \mathbb{1}_2 \xrightarrow{X_1 \boxtimes \mathrm{coev}_2} X_1 \boxtimes X_{2'}^\vee \otimes X_{2''} \xrightarrow{\ddot{\imath}_{12'} \otimes X_{2''}} C_{12'} \otimes X_{2''}\big)$$

is a C-comodule structure on X by Section 2.5.4. The properties of the evaluation and coevaluation imply that the constructed maps $\delta \mapsto \ddot{\imath}$ and $\ddot{\imath} \mapsto \delta$ are inverse to each other. □

2.6.9. Proposition. *There is a unique squared coalgebra homomorphism h_C : $C' \overset{\mathrm{def}}{=} \Psi(\Phi C) \to C$ such that for any $X \in {}^C\mathcal{V}$ the composite coalgebra morphism $\ddot{\imath}_X : X \boxtimes X^\vee \xrightarrow{i_X} C' \xrightarrow{h_C} C$ is the canonical* (2.6.4).

PROOF. First we prove that the system of morphisms (2.6.4) determine a unique morphism $h_C : C' \to C \in \widehat{\mathcal{V} \boxtimes \mathcal{V}}$. This follows from the commutative diagram

$$\begin{array}{ccccc}
Y_1 \boxtimes \mathbb{1}_{2'} \otimes Y_{2''}^\vee & \xleftarrow{f \boxtimes \mathbb{1} \otimes Y^\vee} & X_1 \boxtimes \mathbb{1}_{2'} \otimes Y_{2''}^\vee & \xrightarrow{X \boxtimes \mathbb{1} \otimes f^t} & X_1 \boxtimes \mathbb{1}_{2'} \otimes X_{2''}^\vee \\
\downarrow{\scriptstyle \delta_Y \otimes Y^\vee} & & \downarrow{\scriptstyle \delta_X \otimes Y^\vee} & & \downarrow{\scriptstyle \delta_X \otimes X^\vee} \\
C_{12'} \otimes Y_{2''} \otimes Y_{2'''}^\vee & \xleftarrow{C \otimes f \otimes Y^\vee} & C_{12'} \otimes X_{2''} \otimes Y_{2'''}^\vee & \xrightarrow{C \otimes X \otimes f^t} & C_{12'} \otimes X_{2''} \otimes X_{2'''}^\vee \\
& \searrow{\scriptstyle C \otimes \mathrm{ev}_Y} & & \swarrow{\scriptstyle C \otimes \mathrm{ev}_X} & \\
& & C_{12'} \otimes \mathbb{1}_{2''} & &
\end{array} \qquad (2.6.5)$$

Propositions 2.5.5 and 2.6.8 imply that h_C is a squared coalgebra morphism. □

2.6.10. Proposition. *The family $h_C : \Psi(\Phi C) \to C$ gives an isomorphism of functors $h : \Psi\Phi \to \mathrm{Id}_{\mathrm{Coalgsq}(\widehat{\mathcal{V}})}$.*

PROOF. To prove that h is a functorial morphism, consider the following diagram, where $f : C \to D \in \mathrm{Coalgsq}(\widehat{\mathcal{V}})$ and $f' = \Psi\Phi f : C' = \Psi\Phi C \to D' = \Psi\Phi D$,

$$\begin{array}{ccccc} X \boxtimes X^\vee & \xrightarrow{i_X} & C' & \xrightarrow{h_C} & C \\ \| & & \downarrow{\scriptstyle f'} & & \downarrow{\scriptstyle f} \\ X \boxtimes X^\vee & \xrightarrow{i_X} & D' & \xrightarrow{h_D} & D \end{array}$$

The left square commutes by (2.6.3). The exterior commutes by Proposition 2.6.9 and (2.6.4). Hence, the right square commutes.

Proposition 2.6.9 implies that the composite

$$^C\mathcal{V} \xrightarrow{F_\mathcal{U}} {}^{C'}\mathcal{V} \xrightarrow{^{h_C}\mathcal{V}} {}^C\mathcal{V}$$

is the identity functor. Since $F_\mathcal{U}$ is an equivalence, $^{h_C}\mathcal{V}$ is also an equivalence quasiinverse to $F_\mathcal{U}$. As $\Phi(h_C)$ is an isomorphism, so is h_C by Theorem 2.6.7. □

2.6.11. Corollary. *Any squared coalgebra C is the union of the images of the canonical morphisms* (2.6.4) *for all comodules X.*

This is a detailed form of the fundamental theorem on coalgebras 2.3.4.

2.7. Comodules over ordinary coalgebras

Let C be a squared coalgebra in $\mathcal{V}$. Then $\bar{C} = \circledast C \in \mathcal{V}$ is an ordinary coalgebra in $\mathcal{V}$. How big is the category of $\bar{C}$-comodules in comparison with C-comodules?

First of all, any $\bar{C}$-comodule is a union of its rigid subcomodules, $^{\bar{C}}\widehat{\mathcal{V}} = \widehat{^{\bar{C}}\mathcal{V}}$. This is a general fact. It follows from the embedding $\delta : X \hookrightarrow \bar{C} \otimes X$ of $\bar{C}$-comodules and the fundamental theorem on coalgebras. In our case, for any rigid subobject Y of a $\bar{C}$-comodule $(X, \delta) \in {}^{\bar{C}}\widehat{\mathcal{V}}$ there exist rigid subobjects $D \subset \bar{C}$ and $Z \subset X$ such that $\delta Y \subset D \otimes Z$. Due to Theorem 2.3.4 the subobject D can be chosen as a subcoalgebra of $\bar{C}$. Hence, δY is contained in a subcomodule $D \otimes Z$ of the left regular comodule $\bar{C}$ tensored with X. Therefore, the inverse image $\overset{-1}{\delta}(D \otimes Z) \supset Y$ is a subcomodule of X.

2.7.1. Theorem. *The functor*

$$^C\mathcal{V} \times \mathcal{V} \xrightarrow{\otimes} {}^{\bar{C}}\mathcal{V}, \qquad (X, \delta_X) \times Y \mapsto (X \otimes Y, \bar{\delta}_X \otimes Y)$$

$$\bar{\delta}_X = \left(X \simeq X \otimes \mathbb{1} \xrightarrow{\circledast\delta_X} C_{1'1''} \otimes X_{1'''} = \bar{C} \otimes X\right)$$

makes $^{\bar{C}}\mathcal{V}$ into $^C\mathcal{V} \boxtimes \mathcal{V}$.

PROOF. By definition this functor decomposes as $^C\mathcal{V} \times \mathcal{V} \xrightarrow{\boxtimes} {}^C\mathcal{V} \boxtimes \mathcal{V} \xrightarrow{\circledast} {}^{\bar{C}}\mathcal{V}$. We want to construct a quasiinverse $K : {}^{\bar{C}}\mathcal{V} \to {}^C\mathcal{V} \boxtimes \mathcal{V}$ to $\circledast$.

Let $(M, \delta) \in {}^{\bar{C}}\mathcal{V}$. Lemma 2.5.9 shows that the sequence of $\bar{C}$-comodules

$$0 \to M \xrightarrow{\delta} \bar{C} \otimes M \xrightarrow{\bar{C}\otimes\delta - \bar{\Delta}\otimes M} \bar{C} \otimes \bar{C} \otimes M \tag{2.7.1}$$

is exact. The second arrow lifts to a morphism $C_{12'} \otimes \delta_{2''} - \Delta_{12'2''} \otimes M$ in $\widehat{^C\mathcal{V} \boxtimes \mathcal{V}}$ and we set

$$K(M) = \mathrm{Ker}(C_{12'} \otimes \delta_{2''} - \Delta_{12'2''} \otimes M : C_{12'} \otimes M_{2''} \to C_{12'} \otimes C_{2''2'''} \otimes M_{2^4}).$$

This is an object of $\widehat{{}^C\mathcal{V}\boxtimes\mathcal{V}}$, whose image $\circledast K(M)$ in $\widehat{\mathcal{V}}$ is isomorphic to M by exactness of sequence (2.7.1). Since $\circledast$ is faithful and exact by Proposition 1.3.3, we have, in fact, $K(M) \in {}^C\mathcal{V}\boxtimes\mathcal{V}$.

If $f : M \to N \in {}^{\bar{C}}\mathcal{V}$, there is a commutative diagram

$$\begin{array}{ccc} C_{12'}\otimes M_{2''} & \xrightarrow{C_{12'}\otimes\delta_{2''}-\Delta_{12'2''}\otimes M} & C_{12'}\otimes C_{2''2'''}\otimes M_{2^4} \\ {\scriptstyle 1\otimes f}\downarrow & & \downarrow{\scriptstyle 1\otimes 1\otimes f} \\ C_{12'}\otimes N_{2''} & \xrightarrow{C_{12'}\otimes\delta_{2''}-\Delta_{12'2''}\otimes N} & C_{12'}\otimes C_{2''2'''}\otimes N_{2^4} \end{array} \tag{2.7.2}$$

which induces a morphism of kernels of rows $K(f) : K(M) \to K(N)$. Clearly, it gives a functor $K : {}^{\bar{C}}\mathcal{V} \to {}^C\mathcal{V}\boxtimes\mathcal{V}$.

Comparing $\circledast$-image of diagram (2.7.2) with

$$\begin{array}{ccccccc} 0 & \longrightarrow & M & \xrightarrow{\delta} & \bar{C}\otimes M & \xrightarrow{\bar{C}\otimes\delta-\bar{\Delta}\otimes M} & \bar{C}\otimes\bar{C}\otimes M \\ & & {\scriptstyle f}\downarrow & & \downarrow{\scriptstyle 1\otimes f} & & \downarrow{\scriptstyle 1\otimes 1\otimes f} \\ 0 & \longrightarrow & N & \xrightarrow{\delta} & \bar{C}\otimes N & \xrightarrow{\bar{C}\otimes\delta-\bar{\Delta}\otimes N} & \bar{C}\otimes\bar{C}\otimes N \end{array}$$

we see that ${}^{\bar{C}}\mathcal{V} \xrightarrow{K} {}^C\mathcal{V}\boxtimes\mathcal{V} \xrightarrow{\circledast} {}^{\bar{C}}\mathcal{V}$ is isomorphic to the identity functor.

Let us compute the composite

$${}^C\mathcal{V}\times\mathcal{V} \xrightarrow{\boxtimes} {}^C\mathcal{V}\boxtimes\mathcal{V} \xrightarrow{\circledast} {}^{\bar{C}}\mathcal{V} \xrightarrow{K} {}^C\mathcal{V}\boxtimes\mathcal{V}. \tag{2.7.3}$$

Let $X \in {}^C\mathcal{V}$, $Y \in \mathcal{V}$, $\circledast(X\boxtimes Y) = (X\otimes Y, \bar{\delta}_X\otimes Y)$. Tensoring exact sequence (2.5.6) for $M = X$ with Y, we get an exact sequence in ${}^C\mathcal{V}\boxtimes\mathcal{V}$

$$0 \to X_1\boxtimes Y_2 \xrightarrow{\delta_X\otimes Y} C_{12'}\otimes X_{2''}\otimes Y_{2'''}$$
$$\xrightarrow{C\otimes\bar{\delta}_X\otimes Y-\Delta\otimes X\otimes Y} C_{12'}\otimes C_{2''2'''}\otimes X_{2^4}\otimes Y_{2^5}.$$

Thus, $K(\circledast(X\boxtimes Y)) \simeq X\boxtimes Y$, and functor (2.7.3) is isomorphic to $\boxtimes$. Hence, by definition of tensor product of categories, ${}^C\mathcal{V}\boxtimes\mathcal{V} \xrightarrow{\circledast} {}^{\bar{C}}\mathcal{V} \xrightarrow{K} {}^C\mathcal{V}\boxtimes\mathcal{V}$ is isomorphic to the identity functor. Therefore, $\circledast$ and K are quasiinverse to each other. □

2.7.2. Corollary. *Let $\omega : \mathcal{A} \to \mathcal{V} \in \mathfrak{Ab}_{\Bbbk}$ be a $\Bbbk$-linear exact faithful functor, and $\bar{C} = \int^{X\in\mathcal{A}} \omega X\otimes(\omega X)^\vee$. Then the category of $\bar{C}$-comodules ${}^{\bar{C}}\mathcal{V}$ is equivalent to $\mathcal{A}\boxtimes\mathcal{V}$.*

This is closely related with results of Pareigis [**23**].

2.8. Construction data

The main purpose of introducing precategories is the possibility of constructing fibre functors. Let $\mathcal{C}$ be a $\Bbbk$-linear abelian category. Similarly to the category $\mathfrak{A}(\mathcal{C})$ one can consider morphisms from a prefunctor $p : \mathcal{P} \to \mathcal{C}$ to a fibre functor $b : \mathcal{B} \to \mathcal{C}$, defined as equivalence classes of pairs (F, ϕ), where $F : \mathcal{P} \to \mathcal{B}$ is a prefunctor and $\phi : p \to bF$ is an isomorphism of prefunctors (this makes sense since the target $\mathcal{C}$ is a category). The equivalence relation is defined with the help of isomorphisms of prefunctors $\zeta : F \to G$ as in Definition 2.5.1. The following explains how an abelian category and a functor are constructed from a prefunctor.

2.8.1. Theorem. *Any prefunctor $p : \mathcal{P} \to \mathcal{C}$ from a small precategory $\mathcal{P}$ has a canonical morphism (F, id) to a fibre functor $a : \mathcal{A} \to \mathcal{C}$ such that the induced map*

$$\mathrm{Hom}_{\mathfrak{A}(\mathcal{C})}(a, b) \to \mathrm{Hom}(p, b)$$

is a bijection for any fibre functor $b : \mathcal{B} \to \mathcal{C}$.

PROOF. Let us add new objects and morphisms to $\mathcal{P}$ turning it into an abelian category. We shall do it step by step, transforming a prefunctor $p : \mathcal{P} \to \mathcal{C}$ into a prefunctor $p' : \mathcal{P}' \to \mathcal{C}$, such that $\mathcal{P}'$ contains $\mathcal{P}$. The result p' denoted again by p is used for the next step and so on. At each step we produce faithful $p' : \mathcal{P}' \to \mathcal{C}$. In particular, at Steps 2–4 we define a subset $\mathrm{Hom}_{\mathcal{P}'}(X, Y) \subset \mathrm{Hom}_{\mathcal{C}}(pX, pY)$, which contains $\mathrm{Hom}_{\mathcal{P}}(X, Y)$. At each step we prove that the restriction map $\mathrm{Hom}(p', b) \to \mathrm{Hom}(p, b)$ is bijective.

Step 1. For all finite families of objects $X : S \to \mathrm{Ob}\,\mathcal{P}$, $i \mapsto X_i$, add to $\mathcal{P}$ the object $\oplus_{i \in S} X_i$ and morphisms $p_j : \oplus_{i \in S} X_i \to X_j$, $i_j : X_j \to \oplus_{i \in S} X_i$, thus forming $\mathcal{P}'$. The prefunctor p' sends them to the direct sum $\oplus_{i \in S} pX_i$, the canonical projections and injections respectively.

Let us prove that $[(F, \phi)] : p \to b$ extends uniquely to $[(F', \phi')] : p' \to b$. There is a freedom in choosing $F'(\oplus X_i)$, $F' i_j$, $F' p_j$. However, the isomorphism $\phi' : p' \to bF'$ implies that $bF' i_j : bFX_j \to bF'(\oplus X_i)$, $bF' p_j : bF'(\oplus X_i) \to bFX_j$ make $bF'(\oplus X_i)$ into a direct sum $\oplus bFX_i$, and $\phi'_{\oplus X_i} : \oplus pX_i \to bF'(\oplus X_i)$ is determined uniquely. Since b is linear and faithful, $F' i_j : FX_j \to F'(\oplus X_i)$, $F' p_j : F'(\oplus X_i) \to FX_j$ make $F'(\oplus X_i)$ into a direct sum $\oplus FX_i$. Any other choice $\bar{F}'(\oplus X_i)$, $\bar{F}' i_j$, $\bar{F}' p_j$ of the direct sum of FX_i is isomorphic to it by the unique $\zeta_{\oplus X_i} : F'(\oplus X_i) \to \bar{F}'(\oplus X_i)$. Set $\zeta_X = \mathrm{id} : FX \to FX$ for any $X \in \mathrm{Ob}\,\mathcal{P}$. This gives an isomorphism $\zeta : F' \to \bar{F}'$ which satisfies (2.5.1), hence, the extensions (F', ϕ') and $(\bar{F}', \bar{\phi}')$ are equivalent.

Step 2. For all pairs of objects $X, Y \in \mathcal{P}$ set $\mathrm{Hom}_{\mathcal{P}'}(X, Y)$ to be the $\Bbbk$-linear span of $p(\mathrm{Hom}_{\mathcal{P}}(X, Y)) \subset \mathrm{Hom}_{\mathcal{C}}(pX, pY)$. Let us prove that $F : \mathcal{P} \to \mathcal{B}$ extends uniquely to sums of morphisms. If $f, g : X \to Y \in \mathcal{P}$, $\alpha, \beta \in \Bbbk$, then $p'(\alpha f + \beta g) = \alpha pf + \beta pg$ by definition. Since bF' is isomorphic to p' through ϕ, the equation $bF'(\alpha f + \beta g) = \alpha bFf + \beta bFg$ holds as well. The functor b is $\Bbbk$-linear and faithful, hence, $F'(\alpha f + \beta g) = \alpha Ff + \beta Fg$. Thus (F', ϕ) is the unique extension of (F, ϕ).

Step 3. If pf is invertible in $\mathcal{C}$ for $f : X \to Y \in \mathrm{Mor}\,\mathcal{P}$, add $f^{-1} : Y \to X$ to $\mathrm{Mor}\,\mathcal{P}'$. The functor p' sends it to $p'(f^{-1}) = (pf)^{-1}$. The map $\mathrm{Hom}(p', b) \to \mathrm{Hom}(p, b)$ is bijective similarly to Step 2.

Step 4. $\mathrm{Mor}\,\mathcal{P}'$ is made of products of morphisms from $\mathcal{P}$. The functor p' sends such product to the composite morphism. The map $\mathrm{Hom}(p', b) \to \mathrm{Hom}(p, b)$ is bijective similarly to Step 2.

Step 5. For all $u : M \to N \in \mathcal{P}$ choose a morphism $i : K \to pM$ which is the kernel of pu in $\mathcal{C}$. Add K, i to $\mathcal{P}'$ with $p'K = K$, $p'i = i$. For any $X \in \mathcal{P}$ set

$$\mathrm{Hom}_{\mathcal{P}'}(X, K) = \{g : pX \to K \in \mathcal{C} \mid \exists f \in \mathcal{P} \quad pf = \bigl(pX \xrightarrow{g} K \xrightarrow{i} pM\bigr)\}.$$

Let us prove that $[(F, \phi)] : p \to b$ extends uniquely to $[(F', \phi')] : p' \to b$. There is a freedom in choosing $F'K$, $F'i$. However, the sequence $0 \to bF'K \xrightarrow{bF'i} bFM \xrightarrow{bFu} bFN$ isomorphic to $0 \to K \xrightarrow{i} pM \xrightarrow{pu} pN$ is exact. By exactness and faithfulness of b we get that $0 \to F'K \xrightarrow{F'i} FM \xrightarrow{Fu} FN$ is exact. Any other choice $\bar{F}'i$:

$\bar{F}'K \to FM$ of $\ker Fu$ is isomorphic to $F'i : F'K \to FM$ via a unique isomorphism $\zeta_K : F'K \to \bar{F}'K$. Set $\zeta_X = \mathrm{id} : FX \to FX$ for $X \in \mathrm{Ob}\,\mathcal{P}$. Notice that $F'g$ are determined uniquely for $g : X \to K \in \mathcal{P}'$ such that $pf = i \circ g$, namely, $F'g : FX \to F'K$ is the unique morphism such that $Ff = F'i \circ F'g$. Notice, that Ff is uniquely determined by g. Therefore, $\zeta : F' \to \bar{F}'$ is an isomorphism of functors which makes (F', ϕ') and $(\bar{F}', \bar{\phi}')$ equivalent.

Step 6. For all $u : A \to B \in \mathcal{P}$ choose a morphism $j : pB \to L$ which is the cokernel of pu in $\mathcal{C}$. Add L, j to $\mathcal{P}'$ with $p'L = L$, $p'j = j$. For any $X \in \mathcal{P}$ set

$$\mathrm{Hom}_{\mathcal{P}'}(L, X) = \{g : L \to pX \in \mathcal{C} \mid \exists f \in \mathcal{P} \quad pf = \big(pB \xrightarrow{j} L \xrightarrow{g} pX\big)\}.$$

The proof of bijectivity of $\mathrm{Hom}(p', b) \to \mathrm{Hom}(p, b)$ is dual to that of Step 5.

Repeat cyclically Steps 1–6. This gives a sequence of precategories and prefunctors $\mathcal{P} \to \mathcal{P}_1 \to \mathcal{P}_2 \to \mathcal{P}_3 \to \dots$ such that $\mathrm{Hom}(p_{i+1}, b) \to \mathrm{Hom}(p_i, b) \to \mathrm{Hom}(p, b)$ are bijective. Denote $\mathcal{P}_\infty$ the inductive limit. Then $p = \big(\mathcal{P} \to \mathcal{P}_\infty \xrightarrow{p_\infty} \mathcal{C}\big)$. The category $\mathcal{P}_\infty$ is abelian. Indeed, Steps 4, 1, 2 make sure that $\mathcal{P}_\infty$ is a category, additive and $\Bbbk$-linear. Steps 5, 6 guarantee that any morphism in $\mathcal{P}_\infty$ has the kernel and the cokernel, and p_∞ preserves kernels and cokernels. Step 3 implies that the canonical morphism $\mathrm{Coker}(\ker u) \to \mathrm{Ker}(\mathrm{coker}\, u)$, which is an isomorphism in $\mathcal{C}$, is an isomorphism in $\mathcal{P}_\infty$ as well. In the limit we have $\mathrm{Hom}(p_\infty, b) \xrightarrow{\sim} \mathrm{Hom}(p, b)$. Hence, the category $\mathcal{A} = \mathcal{P}_\infty$, $a = p_\infty$ has the required universal properties. □

2.8.2. Remark. Assume that a $\Bbbk$-linear exact faithful functor $F : \mathcal{C} \to \mathcal{D}$ and a prefunctor $p : \mathcal{P} \to \mathcal{C}$ are given where $\mathcal{P}$ is small. Following Steps 1–6 it is easy to see that the abelian category and the fibre functor constructed from the composite $F \circ p$ are $\mathcal{A}$ and $\mathcal{A} \xrightarrow{a} \mathcal{C} \xrightarrow{F} \mathcal{D}$, where $\mathcal{A}$, a are constructed from p.

2.8.3. Theorem. *Let $\mathcal{C} = \mathcal{V}$ be monoidal and let it satisfy the assumptions of Section 1.3. Let $\mathcal{P}$ be a small precategory, let $p : \mathcal{P} \to \mathcal{V}$ be a prefunctor, and let C_p be the squared coalgebra (2.5.3) associated with p. The category $\mathcal{A} = {}^{C_p}\mathcal{V}$, the functor $a = \mathcal{U} : {}^{C_p}\mathcal{V} \to \mathcal{V}$, and the functor $F_p : \mathcal{P} \to {}^{C_p}\mathcal{V}$ are the canonical ones associated with p in Theorem 2.8.1.*

PROOF. Let us prove that each step in the proof of Theorem 2.8.1 does not change the associated category of comodules and the induced morphism of coalgebras $C_p \to C_{p'}$ is an isomorphism. Then, passing to the limit, we shall get that the coalgebra $C_\infty = C_{p_\infty}$ corresponding to the limit category $\mathcal{P}_\infty$ is isomorphic to the initial C_p. On the other hand, $\mathcal{P}_\infty$ is abelian and the functor $F_\infty : \mathcal{P}_\infty \to {}^{C_\infty}\mathcal{V}$ is an equivalence by Theorem 2.6.3.

Step 1. Let $pX = \oplus_{i \in S} pX_i$. The morphisms p_j, i_j induce relations which tell precisely that the morphism i_X factorises as

$$pX \boxtimes pX^\vee = \oplus_{i,j} pX_i \boxtimes pX_j^\vee \xrightarrow{\mathrm{diag(id)}} \oplus_j pX_j \boxtimes pX_j^\vee \xrightarrow{\sum i_{X_j}} C_p.$$

Therefore, $C_{p'} \simeq C_p$.

Step 2. Since $\mathrm{rel}(\lambda f + \mu g) = \lambda \,\mathrm{rel}\, f + \mu \,\mathrm{rel}\, g$ (see (2.5.4)), $\mathrm{Im}\,\mathrm{rel}(\lambda f + \mu g) \subset \mathrm{Im}\,\mathrm{rel}\, f + \mathrm{Im}\,\mathrm{rel}\, g$.

Step 3. We have

$$\mathrm{rel}\, f^{-1} = \mathrm{rel}\, f \circ (pf^{-1} \boxtimes pf^{-1\,t}),$$

so $\operatorname{Im}\operatorname{rel} f^{-1}$ coincides with $\operatorname{Im}\operatorname{rel} f$.

Step 4. Let $f : X \to Y$, $g : Y \to Z \in \mathcal{P}$. Then $\operatorname{rel}(g \circ f)$ coincides with

$$pX \boxtimes pZ^\vee \xrightarrow{1\boxtimes pg^t \oplus pf\boxtimes 1} pX \boxtimes pY^\vee \oplus pY \boxtimes pZ^\vee \xrightarrow{\operatorname{rel} f \oplus \operatorname{rel} g} C^{\mathcal{O}}. \tag{2.8.1}$$

Therefore, the coideal in $C^{\mathcal{O}}$ does not increase with adding $g \circ f$ to $\operatorname{Mor}\mathcal{P}$.

Step 5. We have $C^{\mathcal{O}'} = C^{\mathcal{O}} \oplus K \boxtimes K^\vee$. However, in the quotient $C_{p'}$ the image of $K \boxtimes K^\vee$ is identified with a subquotient of $pM \boxtimes pM^\vee$. Indeed, the relation $\operatorname{rel} i$ gives

$$\big(K \boxtimes pM^\vee \xrightarrow{1\boxtimes i^t \oplus -i\boxtimes 1} K \boxtimes K^\vee \oplus pM \boxtimes pM^\vee \xrightarrow{i_K \oplus i_M} C_{p'}\big) = 0$$

and i^t is epi. To check that no extra factorisation happens in $pM \boxtimes pM^\vee$, we have to deduce that

$$\big(\operatorname{Ker}(K \boxtimes i^t) \xrightarrow{i\boxtimes 1} pM \boxtimes pM^\vee \xrightarrow{i_M} C_p\big) = 0$$

from the relations already existing in C_p. So, covering $\operatorname{Ker}(K \boxtimes i^t)$ with $K \boxtimes pN^\vee \xrightarrow{1\boxtimes pu^t} K \boxtimes pM^\vee$ we get

$$\begin{aligned}\big(K \boxtimes pN^\vee &\xrightarrow{i\boxtimes 1} pM \boxtimes pN^\vee \xrightarrow{1\boxtimes pu^t} pM \boxtimes pM^\vee \xrightarrow{i_M} C_p\big) \\ &= \big(K \boxtimes pN^\vee \xrightarrow{i\boxtimes 1} pM \boxtimes pN^\vee \xrightarrow{pu\boxtimes 1} pN \boxtimes pN^\vee \xrightarrow{i_N} C_p\big) = 0.\end{aligned}$$

If $pf = \big(pX \xrightarrow{g} K \xrightarrow{i} pM\big)$ we deduce from (2.8.1) that

$$\begin{aligned}&\big(pX \boxtimes pM^\vee \xrightarrow{1\boxtimes i^t} pX \boxtimes K^\vee \xrightarrow{\operatorname{rel} g} C^{\mathcal{O}'}\big) \\ &\qquad = \operatorname{rel} f - \big(pX \boxtimes pM^\vee \xrightarrow{pg\boxtimes 1} K \boxtimes pM^\vee \xrightarrow{\operatorname{rel} i} C^{\mathcal{O}'}\big).\end{aligned}$$

Since $1 \boxtimes i^t$ is epi the image of $\operatorname{rel} g$ is contained in the sum of $\operatorname{Im}\operatorname{rel} f$ and $\operatorname{Im}\operatorname{rel} i$. Therefore, $C_{p'} \simeq C_p$.

Step 6. The summand $L \boxtimes L^\vee$ is identified with a subquotient of $pB \boxtimes pB^\vee$ in $C_{p'}$. Indeed,

$$\big(pB \boxtimes L^\vee \xrightarrow{1\boxtimes j^t \oplus -j\boxtimes 1} pB \boxtimes pB^\vee \oplus L \boxtimes L^\vee \xrightarrow{i_B \oplus i_L} C_{p'}\big) = 0.$$

This does not impose extra factorisation in $\operatorname{Im} i_B$, since

$$\big(\operatorname{Ker}(j \boxtimes L^\vee) \xrightarrow{1\boxtimes j^t} pB \boxtimes pB^\vee \xrightarrow{i_B} C_p\big) = 0$$

follows from

$$\big(A \boxtimes L^\vee \xrightarrow{u\boxtimes j^t} pB \boxtimes pB^\vee \xrightarrow{i_B} C_p\big) = 0$$

The possibility to add $g : L \to pX$ to $\operatorname{Mor}\mathcal{P}'$ without changing $C_{p'}$ follows from (2.8.1) quite similarly to Step 5.

The theorem is proved. □

CHAPTER 3

Squared bicoalgebras

Before we discuss bicoalgebras, let us consider the operation of tensor product of squared coalgebras. Notice that a braiding in $\mathcal{V}$ is not required.

3.1. Tensor product of squared coalgebras

Recall that $\mathcal{V} \boxtimes \mathcal{V}$ has a rigid monoidal structure $(A \bar{\otimes} B)_{12} = A_{1'2''} \otimes B_{1''2'}$ (Theorem 1.6.1).

3.1.1. Proposition. *Let $A, B \in \operatorname{Coalgsq}(\widehat{\mathcal{V}})$. The tensor product $A \bar{\otimes} B$ has a coalgebra structure*

$$\Delta^{A\bar{\otimes}B} : A_{1'3''} \otimes B_{1''3'} \boxtimes \mathbb{1}_2 \xrightarrow{A_{1'3''} \otimes \Delta_{1''23'}} A_{1'3''} \otimes B_{1''2'} \otimes B_{2''3'} \simeq$$
$$A_{1'3''} \otimes B_{1''2'} \otimes \mathbb{1}_{2''} \otimes B_{2'''3'} \xrightarrow{\Delta_{1'2''3''} \otimes B_{1''2'} \otimes B_{2'''3'}} A_{1'2''} \otimes B_{1''2'} \otimes A_{2'''3''} \otimes B_{2^4 3'}, \tag{3.1.1}$$

$$\varepsilon^{A\bar{\otimes}B} : A^{14} \boxtimes B^{23} \xrightarrow{A \boxtimes \varepsilon} A^{13} \boxtimes \mathbb{1}^2 \simeq A^{12} \xrightarrow{\varepsilon} \mathbb{1}^1. \tag{3.1.2}$$

The tensor product $A \bar{\otimes} B$ is associative and turn $\operatorname{Coalgsq}(\widehat{\mathcal{V}})$ into a monoidal category.

PROOF. Coassociativity of $\Delta^{A\bar{\otimes}B}$ is proved in Diagram A on the following page. Counitality of $A \bar{\otimes} B$ follows from the equation

$$\big(A_{1'2''} \otimes B_{1''2'} \simeq A_{1'2'''} \otimes B_{1''2''} \otimes \mathbb{1}_{2'} \xrightarrow{A \otimes \Delta_{1''2'2''}} A_{1'2^4} \otimes B_{1''2'} \otimes B_{2''2'''}$$
$$\simeq A_{1'2^5} \otimes \mathbb{1}_{2''} \otimes B_{1''2'} \otimes B_{2'''2^4} \xrightarrow{\Delta_{1'2''2^5} \otimes B \otimes B} A_{1'2''} \otimes A_{2'''2^6} \otimes B_{1''2'} \otimes B_{2^4 2^5}$$
$$\xrightarrow{A \otimes A \otimes B \otimes \varepsilon} A_{1'2''} \otimes A_{2'''2^5} \otimes B_{1''2'} \otimes \mathbb{1}_{2^4} \simeq A_{1'2''} \otimes A_{2'''2^4} \otimes B_{1''2'}$$
$$\xrightarrow{A \otimes \varepsilon \otimes B} A_{1'2''} \otimes \mathbb{1}_{2'''} \otimes B_{1''2'} \simeq A_{1'2''} \otimes B_{1''2'}\big)$$
$$= \big(A_{1'2''} \otimes B_{1''2'} \simeq A_{1'2'''} \otimes \mathbb{1}_{2''} \otimes B_{1''2'} \xrightarrow{\Delta_{1'2''2'''} \otimes B} A_{1'2''} \otimes A_{2'''2^4} \otimes B_{1''2'}$$
$$\xrightarrow{A \otimes \varepsilon \otimes B} A_{1'2''} \otimes \mathbb{1}_{2'''} \otimes B_{1''2'} \simeq A_{1'2''} \otimes B_{1''2'}\big) = 1_{A\bar{\otimes}B},$$

and another similar one.

The associativity isomorphism is that one of $(\mathcal{V} \boxtimes \mathcal{V}, \bar{\otimes})$. Clearly, both ways to get a comultiplication on the product of three coalgebras $A_{1'2'''} \otimes B_{1''2''} \otimes C_{1'''2'}$ coincide. The coalgebra $\mathbb{1} \boxtimes \mathbb{1}$ with the standard comultiplication $\Delta = \mathbb{1} \boxtimes r_{\mathbb{1}}^{-1} \boxtimes \mathbb{1} : \mathbb{1} \boxtimes \mathbb{1} \boxtimes \mathbb{1} \to \mathbb{1} \boxtimes \mathbb{1} \otimes \mathbb{1} \boxtimes \mathbb{1}$ and the standard counit $\varepsilon = r_{\mathbb{1}} : \mathbb{1} \otimes \mathbb{1} \to \mathbb{1}$ is the unit object of the monoidal category $(\operatorname{Coalgsq}(\widehat{\mathcal{V}}), \bar{\otimes})$. □

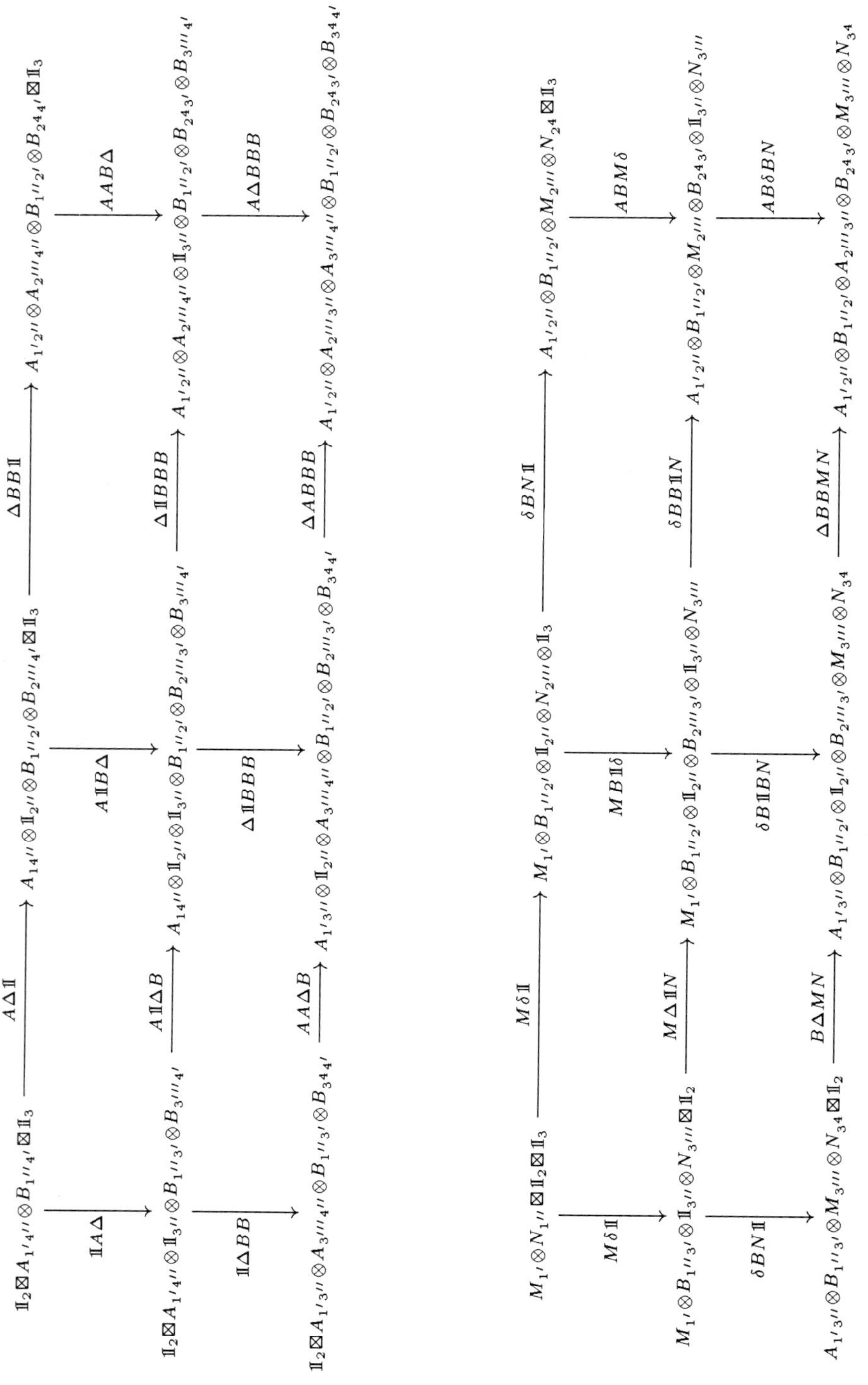

FIGURE 3.1. Diagrams A and B

Graphical notation and explanation of the comultiplication (3.1.1) and the counit (3.1.2) is the following

$\Delta =$

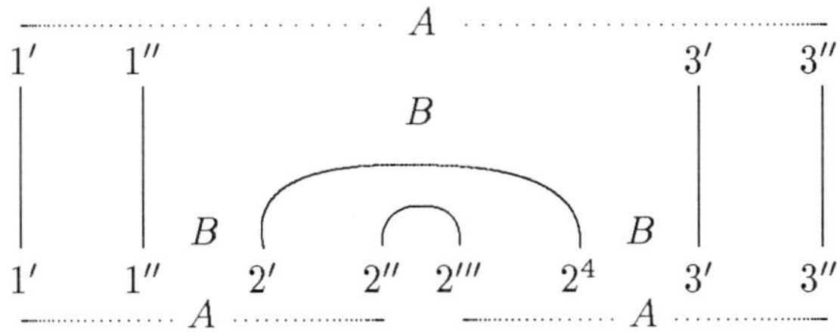

,

$\varepsilon =$

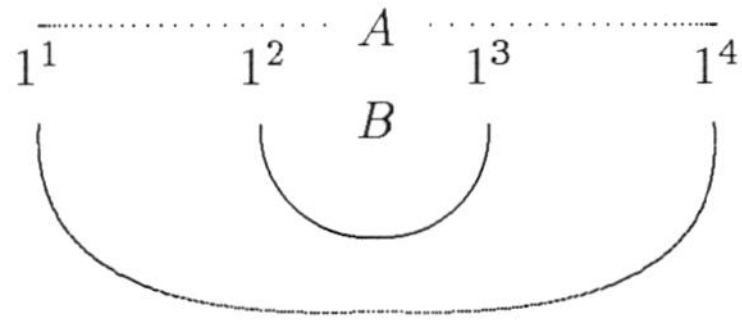

.

It reflects the canonical isomorphism $j_+ : Y^\vee \otimes X^\vee \to (X \otimes Y)^\vee$ which is described in Section 1.2.6 on page 14.

3.1.2. Remark. Let $A = X \boxtimes X^\vee$, $B = Y \boxtimes Y^\vee$. Then $A \bar{\otimes} B = (X \otimes Y) \boxtimes (Y^\vee \otimes X^\vee) \xrightarrow{1 \boxtimes j_+} (X \otimes Y) \boxtimes (X \otimes Y)^\vee$ is a squared coalgebra isomorphism.

3.1.3. Proposition. *Let $M \in {}^A\widehat{\mathcal{V}}$, $N \in {}^B\widehat{\mathcal{V}}$. Then $M \otimes N$ has the structure of an $A \bar{\otimes} B$-comodule*

$$\delta^{M \otimes N} = \bigl(M_{1'} \otimes N_{1''} \boxtimes \mathbb{1}_2 \xrightarrow{M \otimes \delta^N} M_{1'} \otimes B_{1''2'} \otimes N_{2''}$$
$$\simeq M_{1'} \otimes B_{1''2'} \otimes \mathbb{1}_{2''} \otimes N_{2'''} \xrightarrow{\delta^M \otimes B \otimes N} A_{1'2''} \otimes B_{1''2'} \otimes M_{2'''} \otimes N_{2^4}\bigr). \quad (3.1.3)$$

For $L \in {}^C\widehat{\mathcal{V}}$, both ways to construct an $A \bar{\otimes} B \bar{\otimes} C$-comodule structure on $M \otimes N \otimes L$ coincide.

Proof. Coassociativity of the coaction is proved in Diagram B on the preceding page. Counitality follows from the computation:

$$\bigl(M^1 \otimes N^2 \simeq M^1 \otimes N^2 \otimes \mathbb{1}^3 \xrightarrow{M \otimes \delta} M^1 \otimes B^{23} \otimes N^4 \simeq M^1 \otimes B^{23} \otimes \mathbb{1}^4 \otimes N^5$$
$$\xrightarrow{\delta \otimes B \otimes N} A^{14} \otimes B^{23} \otimes M^5 \otimes N^6 \xrightarrow{A \otimes \varepsilon \otimes M \otimes N} A^{13} \otimes \mathbb{1}^2 \otimes M^4 \otimes N^5$$
$$\simeq A^{12} \otimes M^3 \otimes N^4 \xrightarrow{\varepsilon \otimes M \otimes N} \mathbb{1}^1 \otimes M^2 \otimes N^3 \simeq M^1 \otimes N^2\bigr)$$
$$= \bigl(M^1 \otimes N^2 \simeq M^1 \otimes \mathbb{1}^2 \otimes N^3 \xrightarrow{\delta \otimes N} A^{12} \otimes M^3 \otimes N^4$$
$$\xrightarrow{\varepsilon \otimes M \otimes N} \mathbb{1}^1 \otimes M^2 \otimes N^3 \simeq M^1 \otimes N^2\bigr) = 1_{M \otimes N}.$$

The last assertion is clear. □

Graphical notation for the coaction $\delta^{M \otimes N}$ is

$\delta^{M \otimes N} =$

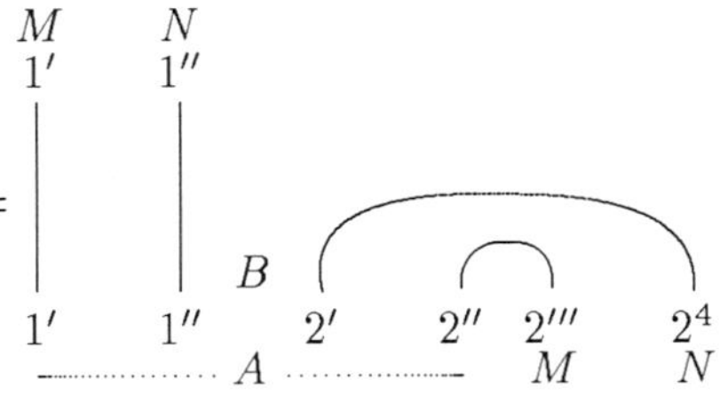

.

3.1.4. Remark. If $A = M \boxtimes M^\vee$, $B = N \boxtimes N^\vee$, the coaction $\delta^{M \otimes N}$ of $A \bar{\otimes} B$ on $M \otimes N$ is mapped by $1 \boxtimes j_+$ to the canonical coaction of $(M \otimes N) \boxtimes (M \otimes N)^\vee$ (see Remark 3.1.2).

3.1.5. Proposition. *Let $j : J \to B$ be a coideal in a squared coalgebra B, and let A be a squared coalgebra. Then $A \bar{\otimes} j : A \bar{\otimes} J \to A \bar{\otimes} B$ and $j \bar{\otimes} A : J \bar{\otimes} A \to B \bar{\otimes} A$ are coideals.*

PROOF. Proposition 2.1.7 implies that there exists an exact sequence $0 \to J \xrightarrow{j} B \xrightarrow{f} C \to 0$ in $\widehat{\mathcal{V} \boxtimes \mathcal{V}}$, where f is a homomorphism of squared coalgebras. Then $0 \to A \bar{\otimes} J \xrightarrow{A \bar{\otimes} j} A \bar{\otimes} B \xrightarrow{A \bar{\otimes} f} A \bar{\otimes} C \to 0$ and $0 \to J \bar{\otimes} A \xrightarrow{j \bar{\otimes} A} B \bar{\otimes} A \xrightarrow{f \bar{\otimes} A} C \bar{\otimes} A \to 0$ are also exact sequences in $\widehat{\mathcal{V} \boxtimes \mathcal{V}}$ since $(\mathcal{V} \boxtimes \mathcal{V}, \bar{\otimes})$ is rigid by Theorem 1.6.1. The morphisms $A \bar{\otimes} f$ and $f \bar{\otimes} A$ are squared coalgebra morphisms, hence, their kernels $A \bar{\otimes} J$ and $J \bar{\otimes} A$ are coideals by Proposition 2.1.6. □

3.2. Bicoalgebras

3.2.1. Definition. A *squared bicoalgebra* $B = (B, \Delta, \varepsilon, m, \eta)$ in $\widehat{\mathcal{V}}$ is a squared coalgebra $(B, \Delta, \varepsilon) \in \mathrm{Coalgsq}(\widehat{\mathcal{V}})$ equipped with an algebra structure (B, m, η) in $(\widehat{\mathcal{V} \boxtimes \mathcal{V}}, \bar{\otimes})$ (such that the multiplication $m : B \bar{\otimes} B \to B \in \widehat{\mathcal{V} \boxtimes \mathcal{V}}$ is associative and $\eta : \mathbb{1} \boxtimes \mathbb{1} \to B \in \widehat{\mathcal{V} \boxtimes \mathcal{V}}$ is the unit) and such that m, η are squared coalgebra homomorphisms, that is,

$$\begin{array}{ccc}
B_{1'3''} \otimes B_{1''3'} \boxtimes \mathbb{1}_2 & \xrightarrow{B_{1'3''} \otimes \Delta_{1''23'}} & B_{1'3''} \otimes B_{1''2'} \otimes B_{2''3'} \\
\downarrow{\scriptstyle m_{13} \boxtimes \mathbb{1}_2} & & \downarrow{\scriptstyle \simeq} \\
B_{13} \boxtimes \mathbb{1}_2 & & B_{1'3''} \otimes B_{1''2'} \otimes \mathbb{1}_{2''} \otimes B_{2'''3'} \\
\downarrow{\scriptstyle \Delta_{123}} & & \downarrow{\scriptstyle \Delta_{1'2''3''} \otimes B_{1''2'} \otimes B_{2'''3'}} \\
B_{12'} \otimes B_{2''3} & \xleftarrow{m \otimes m} & B_{1'2''} \otimes B_{1''2'} \otimes B_{2'''3''} \otimes B_{2^4 3'}
\end{array} \tag{3.2.1}$$

$$\begin{array}{ccccc}
B^{14} \otimes B^{23} & \xrightarrow{B \otimes \varepsilon} & B^{13} \otimes \mathbb{1}^2 & \xrightarrow{\sim} & B^{12} \\
\downarrow{\scriptstyle \circledast m} & & & & \downarrow{\scriptstyle \varepsilon} \\
B^{12} & & \xrightarrow{\varepsilon} & & \mathbb{1}
\end{array} \tag{3.2.2}$$

$$\begin{array}{ccc}
\mathbb{1}_1 \boxtimes \mathbb{1}_2 \boxtimes \mathbb{1}_3 & \xrightarrow{\mathbb{1}_1 \boxtimes r^{-1}_{\mathbb{1}\,2} \boxtimes \mathbb{1}_3} & \mathbb{1}_1 \boxtimes \mathbb{1}_{2'} \otimes \mathbb{1}_{2''} \boxtimes \mathbb{1}_3 \\
\downarrow{\scriptstyle \eta_{13} \boxtimes \mathbb{1}_2} & & \downarrow{\scriptstyle \eta_{12'} \otimes \eta_{2''3}} \\
B_{13} \boxtimes \mathbb{1}_2 & \xrightarrow{\Delta_{123}} & B_{12'} \otimes B_{2''3}
\end{array}$$

$$\left(\mathbb{1} \otimes \mathbb{1} \xrightarrow{\circledast \eta} \circledast B \xrightarrow{\varepsilon} \mathbb{1}\right) = r_{\mathbb{1}}$$

hold.

Morphisms of bicoalgebras are those preserving algebra and squared coalgebra structures. The category of bicoalgebras in $\widehat{\mathcal{V}}$ is denoted $\mathrm{Bicoalg}(\widehat{\mathcal{V}})$.

3.2.2. Remark. Definition of a squared bicoalgebra is not self-dual. One can define dually squared bialgebras which are ordinary coalgebras and squared algebras.

For $M, N \in {}^B\widehat{\mathcal{V}}$ let us define a B-comodule structure on $M \otimes N \in \widehat{\mathcal{V}}$ via (3.1.3):

$$M_{1'} \otimes N_{1''} \boxtimes \mathbb{I}_2 \xrightarrow{\delta^{M\otimes N}} B_{1'2''} \otimes B_{1''2'} \otimes M_{2''} \otimes N_{24} \xrightarrow{m\otimes M\otimes N} B_{12'} \otimes M_{2''} \otimes N_{2'''}. \quad (3.2.3)$$

Proposition 3.1.3 shows that the tensor product of B-comodules is associative. The associativity isomorphism coincides with that one of $(\widehat{\mathcal{V}}, \otimes)$. Therefore, the category $({}^B\widehat{\mathcal{V}}, \otimes)$ is monoidal with the unit object $(\mathbb{I}, \delta_{\mathbb{I}})$,

$$\delta_{\mathbb{I}} = \big(\mathbb{I}_1 \boxtimes \mathbb{I}_2 \xrightarrow{\eta} B_{12} \simeq B_{12'} \otimes \mathbb{I}_{2''}\big).$$

3.2.3. Example. Let $C = (C, \Delta, \varepsilon, m, \eta, \gamma)$ be an ordinary Hopf $\Bbbk$-algebra, $\mathcal{V} = C$-comod. Then $(C, \Delta, \varepsilon, m, \eta, \gamma)$ is an example of a squared bicoalgebra in $\widehat{\mathcal{V}}$. The underlying squared coalgebra (C, Δ, ε) is described in Example 2.1.2. In particular, C is a left $C \otimes_{\Bbbk} C$-comodule with the coaction

$$\delta : C \to C^{\otimes 2} \otimes C, \quad c \mapsto c_{(1)} \otimes \gamma(c_{(3)}) \otimes c_{(2)}.$$

The remaining structure maps are, indeed, morphisms of $C \otimes_{\Bbbk} C$-comodules: the unit $\eta : \Bbbk \to C = H$, $1 \mapsto 1$ and the multiplication $m : C \otimes C = C_{1'2''} \otimes C_{1''2'} \to C$, where the first comodule is equipped with the coaction

$$\delta(c \otimes d) = c_{(1)}d_{(1)} \otimes \gamma(d_{(3)})\gamma(c_{(3)}) \otimes c_{(2)} \otimes d_{(2)} \in C^{\otimes 2} \otimes C \otimes C.$$

3.2.4. Example. Let C be a squared coalgebra in $\widehat{\mathcal{V}}$. Then $T(C) = \oplus_{k\geqslant 0} C^{\bar{\otimes} k}$ is a bicoalgebra in $\widehat{\mathcal{V}}$. Indeed, $C^{\bar{\otimes} k}$ has a unique coalgebra structure by Proposition 3.1.1. For any $n, l \geqslant 0$ the squared coalgebra structure of $C^{\bar{\otimes}(n+l)}$ is that of $(C^{\bar{\otimes} n})\bar{\otimes}(C^{\bar{\otimes} l})$. Therefore, the multiplication morphism $m = \mathrm{id} : C^{\bar{\otimes} n}\bar{\otimes}C^{\bar{\otimes} l} \to C^{\bar{\otimes} n+l}$ is a coalgebra morphism.

Let B be a bicoalgebra and let $f : C \to B$ be a squared coalgebra morphism. Then it extends to a unique bicoalgebra morphism $h : T(C) \to B$. Indeed, the unique algebra morphism $h : T(C) \to B$ extending f is given by

$$C^{\bar{\otimes} k} \xrightarrow{f^{\bar{\otimes} k}} B^{\bar{\otimes} k} \xrightarrow{m} B$$

and both morphisms here are coalgebra morphisms.

3.2.5. Proposition. *Let B be a bicoalgebra, and let $J \subset B$ be a coideal. Then the two-sided (resp. left, resp. right) ideal generated by J is a coideal as well. Thus (J) is a biideal and $B/(J)$ is a bicoalgebra.*

PROOF. By Proposition 3.1.5 $B\bar{\otimes}J$ and $J\bar{\otimes}B$ are coideals in $B\bar{\otimes}B$. Since $m : B\bar{\otimes}B \to B$ is a squared coalgebra homomorphism, the image BJ of $B\bar{\otimes}J \xrightarrow{B\bar{\otimes}j} B\bar{\otimes}B \xrightarrow{m} B$ and the image JB of $J\bar{\otimes}B \xrightarrow{j\bar{\otimes}B} B\bar{\otimes}B \xrightarrow{m} B$ are coideals. Whence the assertion follows. □

3.3. Monoidal construction data

For a class $\mathcal{O}$ let Words($\mathcal{O}$) be the class of all nonassociative words built with the alphabet $\mathcal{O}$. Brackets are used in such words, but usually not shown.

3.3.1. Definition. A *premonoidal precategory* $\mathcal{P}$ is a class of objects $\mathcal{O} = \operatorname{Ob}\mathcal{P}$, a set of morphisms $\operatorname{Hom}_{\mathcal{P}}(X_1 \dots X_k, Y_1 \dots Y_n)$ for each pair of nonassociative words $X_1 \dots X_k, Y_1 \dots Y_n \in$ Words($\mathcal{O}$), and a distinguished morphism $\mathrm{id}_X \in \operatorname{Hom}_{\mathcal{P}}(X, X)$ for each object $X \in \mathcal{O}$, called the identity morphism. We denote the class of morphisms by $\mathcal{M} = \operatorname{Mor} \mathcal{P}$ and consider three maps $\mathrm{ID} : \mathcal{O} \to \mathcal{M}$, $X \mapsto \mathrm{id}_X$, and source, target : $M \to \mathrm{Words}(\mathcal{O})$. A *premonoidal prefunctor* $\mathcal{P} \to \mathcal{S}$ is a pair of maps $\operatorname{Ob} \mathcal{P} \to \operatorname{Ob} \mathcal{S}$, $Mor\mathcal{P} \to \operatorname{Mor} \mathcal{S}$ coherent with ID, source and target.

3.3.2. Example. A monoidal category $(\mathcal{C}, \otimes)$ can be made into a premonoidal precategory $\overline{\mathcal{C}}$ with $\operatorname{Ob} \overline{\mathcal{C}} = \operatorname{Ob} \mathcal{C}$, whose set of morphisms from $X_1 \dots X_k$ to $Y_1 \dots Y_n$ with some brackets is the set of morphisms $X_1 \otimes \dots \otimes X_k \to Y_1 \otimes \dots \otimes Y_n \in \operatorname{Mor} \mathcal{C}$ with the same bracketing.

With a monoidal functor $(F, f) : (\mathcal{C}, \otimes) \to (\mathcal{D}, \otimes)$ we associate a premonoidal prefunctor $\bar{F} : \overline{\mathcal{C}} \to \overline{\mathcal{D}}$,

$$\begin{aligned}\bar{F}\big(X_1 \otimes \dots \otimes X_k \xrightarrow{h} Y_1 \otimes \dots \otimes Y_n\big) \\ = \big(F(X_1) \otimes \dots \otimes F(X_k) \xrightarrow{\sim} F(X_1 \otimes \dots \otimes X_k) \xrightarrow{F(h)} F(Y_1 \otimes \dots \otimes Y_n) \\ \xrightarrow{\sim} F(Y_1) \otimes \dots \otimes F(Y_n)\big).\end{aligned}$$

3.3.3. Definition. Let $\mathcal{P}$ be a premonoidal precategory, let $\mathcal{C}$ be a monoidal category and let $F, G : \mathcal{P} \to \overline{\mathcal{C}}$ be premonoidal prefunctors. A *morphism of premonoidal prefunctors* $\lambda : F \to G$ is a family $(\lambda_X : FX \to GX \in \mathcal{C})_X$ such that for any $h : X_1 \dots X_k \to Y_1 \dots Y_n \in \operatorname{Mor} \mathcal{P}$

$$\begin{array}{ccc} FX_1 \otimes \dots \otimes FX_k & \xrightarrow{Fh} & FY_1 \otimes \dots \otimes FY_n \\ {\scriptstyle \lambda_{X_1} \otimes \dots \otimes \lambda_{X_k}} \downarrow & & \downarrow {\scriptstyle \lambda_{Y_1} \otimes \dots \otimes \lambda_{Y_n}} \\ GX_1 \otimes \dots \otimes GX_k & \xrightarrow{Gh} & GY_1 \otimes \dots \otimes GY_n \end{array}$$

holds.

3.3.4. Example. Let $(F, f, \boldsymbol{f})$, $(G, g, \boldsymbol{g}) : (\mathcal{C}, \otimes, \mathbb{1}) \to (\mathcal{D}, \otimes, \mathbb{1})$ be monoidal functors, and let $\lambda : (F, f, \boldsymbol{f}) \to (G, g, \boldsymbol{g})$ be a morphism of monoidal functors. Then $\lambda : \bar{F} \to \bar{G}$ is a morphism of associated premonoidal prefunctors as well and vice versa.

Composition of premonoidal prefunctors is defined obviously.

3.3.5. Proposition. *Let*

$$(H, h, \boldsymbol{h}) = \big((\mathcal{C}, \otimes, \mathbb{1}) \xrightarrow{(F,f,\boldsymbol{f})} (\mathcal{D}, \otimes, \mathbb{1}) \xrightarrow{(G,g,\boldsymbol{g})} (\mathcal{E}, \otimes, \mathbb{1})\big)$$

be a composite monoidal functor. Then

$$\bar{H} = \big(\overline{\mathcal{C}} \xrightarrow{\bar{F}} \overline{\mathcal{D}} \xrightarrow{\bar{G}} \overline{\mathcal{E}}\big).$$

The proof is left to the reader.

3.3.6. Definition. Let $(\mathcal{V}, \otimes)$ be a monoidal abelian category as in Section 1.3, and let $\mathcal{P}$ be a small premonoidal precategory. A premonoidal prefunctor $p : \mathcal{P} \to \overline{\mathcal{V}}$ is called *monoidal construction data.*

Let us construct a bicoalgebra from monoidal construction data $p : \mathcal{P} \to \overline{\mathcal{V}}$. For any $X \in \mathcal{O} = \operatorname{Ob} \mathcal{P}$ denote

$$C^X = pX \boxtimes (pX)^\vee, \qquad C^\mathcal{O} = \oplus_{X \in \mathcal{O}} C_X.$$

For any $f : X_1 \dots X_k \to Y_1 \dots Y_n \in \operatorname{Mor} \mathcal{P}$ define $\operatorname{rel} f$ as the difference of

$$\begin{aligned}
&(pX_1 \otimes \cdots \otimes pX_k) \boxtimes (pY_n^\vee \otimes \cdots \otimes pY_1^\vee) \\
&\qquad \xrightarrow{1 \boxtimes pf^t} (pX_1 \otimes \cdots \otimes pX_k) \boxtimes (pX_k^\vee \otimes \cdots \otimes pX_1^\vee) \\
&\qquad\qquad = C^{X_1} \bar{\otimes} \dots \bar{\otimes} C^{X_k} \hookrightarrow T^k(C^{\mathcal{O}}) \hookrightarrow T(C^{\mathcal{O}})
\end{aligned}$$

and

$$\begin{aligned}
&(pX_1 \otimes \cdots \otimes pX_k) \boxtimes (pY_n^\vee \otimes \cdots \otimes pY_1^\vee) \\
&\qquad \xrightarrow{pf \boxtimes 1} (pY_1 \otimes \cdots \otimes pY_n) \boxtimes (pY_n^\vee \otimes \cdots \otimes pY_1^\vee) \\
&\qquad\qquad = C^{Y_1} \bar{\otimes} \dots \bar{\otimes} C^{Y_n} \hookrightarrow T^n(C^{\mathcal{O}}) \hookrightarrow T(C^{\mathcal{O}}).
\end{aligned}$$

3.3.7. Proposition. *Let $J_p = \sum_{f \in \operatorname{Mor} \mathcal{P}} \operatorname{Im} \operatorname{rel} f \subset T(C^{\mathcal{O}})$. Then the quotient $B_p = T(C^{\mathcal{O}})/(J_p)$ is a bicoalgebra.*

PROOF. By Proposition 3.2.5 it suffices to show that $\operatorname{Im} \operatorname{rel} f$ is a coideal in $T(C^{\mathcal{O}})$. Consider the left coaction of $T(C^{\mathcal{O}})$ in $pX_1 \otimes \cdots \otimes pX_k$

$$\begin{aligned}
&\delta^l_{X_1 \dots X_k} = \big((pX_1 \otimes \cdots \otimes pX_k) \boxtimes \mathbb{I} \xrightarrow{\delta^{pX_1 \otimes \cdots \otimes pX_k}} \\
&(C^{X_1} \bar{\otimes} \dots \bar{\otimes} C^{X_k})_{12'} \otimes (pX_1 \otimes \cdots \otimes pX_k)_{2''} \hookrightarrow T(C^{\mathcal{O}})_{12'} \otimes (pX_1 \otimes \cdots \otimes pX_k)_{2''}\big)
\end{aligned}$$

and the right coaction of $T(C^{\mathcal{O}})$ in $pY_n^\vee \otimes \cdots \otimes pY_1^\vee$

$$\begin{aligned}
&\delta^r_{Y_n^\vee \dots Y_1^\vee} = \big(\mathbb{I} \boxtimes (pY_n^\vee \otimes \cdots \otimes pY_1^\vee) \xrightarrow{\operatorname{coev} \boxtimes 1} \\
&\qquad (pY_n^\vee \otimes \cdots \otimes pY_1^\vee \otimes pY_1 \otimes \cdots \otimes pY_n) \boxtimes (pY_n^\vee \otimes \cdots \otimes pY_1^\vee) = \\
&(pY_n^\vee \otimes \cdots \otimes pY_1^\vee)_{1'} \otimes (C^{Y_1} \bar{\otimes} \dots \bar{\otimes} C^{Y_n})_{1''2} \hookrightarrow (pY_n^\vee \otimes \cdots \otimes pY_1^\vee)_{1'} \otimes T(C^{\mathcal{O}})_{1''2}\big).
\end{aligned}$$

3.3.8. Lemma. *The diagram (2.1.5) with $C = T(C^{\mathcal{O}})$, $J = (pX_1 \otimes \cdots \otimes pX_k) \boxtimes (pY_n^\vee \otimes \cdots \otimes pY_1^\vee)$, $j = \operatorname{rel} f$, $W = J_{13} \boxtimes \mathbb{I}_2$, $s = \operatorname{id}_W$ and*

$$\begin{aligned}
r = \delta^l_{X_1 \dots X_k} \boxtimes 1 \oplus 1 \boxtimes \delta^r_{Y_n^\vee \dots Y_1^\vee} &: (pX_1 \otimes \cdots \otimes pX_k)_1 \boxtimes \mathbb{I}_2 \boxtimes (pY_n^\vee \otimes \cdots \otimes pY_1^\vee)_3 \\
&\to T(C^{\mathcal{O}})_{12'} \otimes (pX_1 \otimes \cdots \otimes pX_k)_{2''} \boxtimes (pY_n^\vee \otimes \cdots \otimes pY_1^\vee)_3 \\
&\quad \oplus (pX_1 \otimes \cdots \otimes pX_k)_1 \boxtimes (pY_n^\vee \otimes \cdots \otimes pY_1^\vee)_{2'} \otimes T(C^{\mathcal{O}})_{2''3}
\end{aligned}$$

is commutative.

PROOF. Remarks 3.1.2 and 3.1.4 permit to view $M = pX_1 \otimes \cdots \otimes pX_k$ as a left comodule over $M \boxtimes M^\vee \simeq C^{X_1} \bar{\otimes} \dots \bar{\otimes} C^{X_k}$ and to view $N^\vee = pY_n^\vee \otimes \cdots \otimes pY_1^\vee$ as a right comodule over $N \boxtimes N^\vee \simeq C^{Y_1} \bar{\otimes} \dots \bar{\otimes} C^{Y_n}$. Thus the lemma is reduced to the case $k = n = 1$, applied to M, N. This is precisely Proposition 2.5.5. □

Proposition 3.3.7 follows from the lemma by Remark 2.1.5. □

The meaning of the morphisms $\operatorname{rel} f$ is explained in the following statement.

3.3.9. Proposition. *Let B be a bicoalgebra and let $X_1, \dots, X_k, Y_1, \dots, Y_n \in {}^B\mathcal{V}$. Then $f : X_1 \otimes \dots \otimes X_k \to Y_1 \otimes \dots \otimes Y_n \in \mathcal{V}$ is a morphism of B-comodules iff*

$$\begin{aligned}
&\big((X_1 \otimes \dots \otimes X_k) \boxtimes (Y_n^\vee \otimes \dots \otimes Y_1^\vee) \xrightarrow{1 \boxtimes f^t} (X_1 \otimes \dots \otimes X_k) \boxtimes (X_k^\vee \otimes \dots \otimes X_1^\vee) \\
&\qquad \simeq C^{X_1} \bar{\otimes} \dots \bar{\otimes} C^{X_k} \xrightarrow{\ddot{\imath}_{X_1} \bar{\otimes} \dots \bar{\otimes} \ddot{\imath}_{X_k}} B^{\bar{\otimes} k} \xrightarrow{m} B\big) \\
&= \big((X_1 \otimes \dots \otimes X_k) \boxtimes (Y_n^\vee \otimes \dots \otimes Y_1^\vee) \xrightarrow{f \boxtimes 1} (Y_1 \otimes \dots \otimes Y_n) \boxtimes (Y_n^\vee \otimes \dots \otimes Y_1^\vee) \\
&\qquad \simeq C^{Y_1} \bar{\otimes} \dots \bar{\otimes} C^{Y_n} \xrightarrow{\ddot{\imath}_{Y_1} \bar{\otimes} \dots \bar{\otimes} \ddot{\imath}_{Y_n}} B^{\bar{\otimes} n} \xrightarrow{m} B\big).
\end{aligned} \tag{3.3.1}$$

PROOF. The definition of the B-comodule structure on $X_1 \otimes \dots \otimes X_k$ (see (3.1.3), (3.2.3)) is built so that the morphism

$$(X_1 \otimes \dots \otimes X_k) \boxtimes (X_k^\vee \otimes \dots \otimes X_1^\vee) = C^{X_1} \bar{\otimes} \dots \bar{\otimes} C^{X_k} \xrightarrow{\ddot{\imath}_{X_1} \bar{\otimes} \dots \bar{\otimes} \ddot{\imath}_{X_k}} B^{\bar{\otimes} k} \xrightarrow{m} B$$

coincides with the canonical morphism (2.6.4)

$$\ddot{\imath}_{X_1 \otimes \dots \otimes X_k} : (X_1 \otimes \dots \otimes X_k) \boxtimes (X_1 \otimes \dots \otimes X_k)^\vee \to B.$$

Therefore, the result follows from the particular case $k = n = 1$. Diagram (2.5.5) with i_M replaced by $\ddot{\imath}_M$ shows that f satisfying (3.3.1) is a comodule morphism. Diagram (2.6.5) shows that a comodule morphism f satisfies (3.3.1). □

3.3.10. Corollary. *Monoidal construction data $p : \mathcal{P} \to \overline{\mathcal{V}}$ factorise as*

$$p = \big(\mathcal{P} \xrightarrow{F_p} \overline{{}^{B_p}\mathcal{V}} \xrightarrow{\overline{\mathcal{U}}} \overline{\mathcal{V}}\big).$$

3.3.11. Definition. Let $(\mathcal{A}, \otimes)$ be a $\Bbbk$-linear abelian monoidal essentially small category. A *monoidal fibre functor* is a $\Bbbk$-linear exact faithful monoidal functor $(\omega_\mathcal{A}, \omega^\mathcal{A}) : \mathcal{A} \to \mathcal{V}$. Let the *category of monoidal fibre functors* $\mathfrak{M}(\mathcal{V})$ have monoidal fibre functors as objects and let morphisms from $(\omega_\mathcal{A}, \omega^\mathcal{A}) : \mathcal{A} \to \mathcal{V}$ to $(\omega_\mathcal{B}, \omega^\mathcal{B}) : \mathcal{B} \to \mathcal{V}$ be equivalence classes of triples (F, f, ϕ), where $(F, f) : \mathcal{A} \to \mathcal{B}$ is a monoidal functor and $\phi : (\omega_\mathcal{A}, \omega^\mathcal{A}) \to (\omega_\mathcal{B}, \omega^\mathcal{B}) \circ (F, f)$ is an isomorphism of monoidal functors. Two such triples (F, f, ϕ) and (G, g, γ) are equivalent if there is a functorial isomorphism $\zeta : F \to G$ such that (2.5.1) holds. The composite of two morphisms represented by (F, f, ϕ) and (G, g, γ) is represented by $(G \circ F, Gf \circ g_{F,F}, \gamma_F \circ \phi)$.

By Lemma 2.5.2 F is exact and faithful. Forgetting monoidal structure we get a functor $\mathfrak{M}(\mathcal{V}) \to \mathfrak{A}(\mathcal{V})$. It is faithful since $\omega_\mathcal{B} f_{X,Y}$ is determined uniquely by given F, ϕ from the condition $\phi : (\omega_\mathcal{A}, \omega^\mathcal{A}) \xrightarrow{\sim} (\omega_\mathcal{B}, \omega^\mathcal{B}) \circ (F, f)$, and $\omega_\mathcal{B}$ is faithful.

3.3.12. Lemma. *Let $(\omega_\mathcal{A}, \omega^\mathcal{A}) : \mathcal{A} \to \mathcal{V}$ and $(\omega_\mathcal{B}, \omega^\mathcal{B}) : \mathcal{B} \to \mathcal{V}$ be fibre functors, let (F, f), $(G, g) : \mathcal{A} \to \mathcal{B}$ be $\Bbbk$-linear monoidal functors, and let $\phi : (\omega_\mathcal{A}, \omega^\mathcal{A}) \to (\omega_\mathcal{B}, \omega^\mathcal{B}) \circ (F, f)$ and $\gamma : (\omega_\mathcal{A}, \omega^\mathcal{A}) \to (\omega_\mathcal{B}, \omega^\mathcal{B}) \circ (G, g)$ be isomorphisms of monoidal functors. If a functorial isomorphism $\zeta : F \xrightarrow{\sim} G$ satisfies (2.5.1), then it is an isomorphism of monoidal functors.*

PROOF. Let $X, Y \in \operatorname{Ob}\mathcal{A}$. Consider diagram of isomorphisms

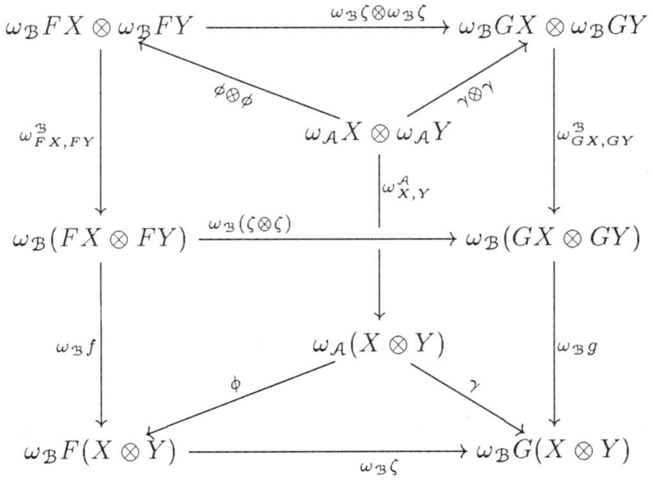

Each pentagon and triangle commutes. The upper square commutes, hence, the lower square commutes. Since $\omega_{\mathcal{B}}$ is faithful it follows that $\zeta : (F, f) \to (G, g)$ is an isomorphism of monoidal functors. $\square$

There is a functor $\Phi : \operatorname{Bicoalg}(\widehat{\mathcal{V}}) \to \mathfrak{M}(\mathcal{V})$, $B \mapsto ((\mathcal{U}, \mathrm{id}) : {}^{B}\mathcal{V} \to \mathcal{V})$. We want to prove that this is an equivalence.

Let $(\omega_{\mathcal{A}}, \omega^{\mathcal{A}}) : (\mathcal{A}, \otimes) \to (\mathcal{V}, \otimes)$ be a monoidal fibre functor. Choose a set $\mathcal{O} \subset \operatorname{Ob}\mathcal{A}$ so that any object of $\mathcal{A}$ is isomorphic to an object of $\mathcal{O}$, and $\mathcal{O}$ is closed under the operation $\otimes : \operatorname{Ob}\mathcal{A} \times \operatorname{Ob}\mathcal{A} \to \operatorname{Ob}\mathcal{A}$.

Introduce a premonoidal precategory $\mathcal{P}$ with objects $\operatorname{Ob}\mathcal{P} = \mathcal{O}$ and morphisms of two kinds

$$\operatorname{Mor}_{\mathcal{P}}(X, Y) = \operatorname{Hom}_{\mathcal{A}}(X, Y)$$
$$\operatorname{Mor}_{\mathcal{P}}(XY, X \otimes Y) = \{h_{X,Y}\}$$

for any pair $X, Y \in \mathcal{O}$. Thus, $\operatorname{source}(h_{X,Y}) = XY$ is a 2-letter word and $\operatorname{target}(h_{X,Y}) = X \otimes Y$ is a 1-letter word. In fact, $\mathcal{P}$ is a subcategory of $\overline{\mathcal{A}}$ and the inclusion $\mathcal{P} \subset \overline{\mathcal{A}}$ sends $h_{X,Y}$ to $\mathrm{id}_{X \otimes Y}$. Consider a premonoidal prefunctor $p : \mathcal{P} \to \overline{\mathcal{V}}$, $pX = \omega_{\mathcal{A}}X$, $p(f : X \to Y) = (\omega_{\mathcal{A}}f : \omega_{\mathcal{A}}X \to \omega_{\mathcal{A}}Y)$, $p(h_{X,Y}) = (\omega^{\mathcal{A}}_{X,Y} : \omega_{\mathcal{A}}X \otimes \omega_{\mathcal{A}}Y \to \omega_{\mathcal{A}}(X \otimes Y))$. Clearly, p is the restriction of $\overline{(\omega_{\mathcal{A}}, \omega^{\mathcal{A}})}$ to $\mathcal{P}$.

According to Proposition 3.3.7 $B_p = T(C^{\mathcal{O}})/(J)$ is a bicoalgebra, where $J = J_1 + J_2$, $J_1 = \sum_{f \in \operatorname{Mor}\mathcal{A}} \operatorname{Im}\operatorname{rel} f$, $J_2 = \sum_{X,Y \in \mathcal{O}} \operatorname{Im}\operatorname{rel} h_{X,Y}$. Recall that $C_{\mathcal{A}} = C^{\mathcal{O}}/J_1$ is the classifying coalgebra of the category $\mathcal{A}$, that is, $\mathcal{A} \simeq {}^{C_{\mathcal{A}}}\mathcal{V}$. The relation $\operatorname{rel} h_{X,Y}$ is the difference

$$\begin{aligned}
&((\omega_{\mathcal{A}}X \otimes \omega_{\mathcal{A}}Y) \boxtimes \omega_{\mathcal{A}}(X \otimes Y)^{\vee} \xrightarrow{1 \boxtimes \omega^{\mathcal{A}\,t}_{X,Y}} \\
&\qquad (\omega_{\mathcal{A}}X \otimes \omega_{\mathcal{A}}Y) \boxtimes (\omega_{\mathcal{A}}Y^{\vee} \otimes \omega_{\mathcal{A}}X^{\vee}) = C^{X} \bar{\otimes} C^{Y} \subset T^{2}(C^{\mathcal{O}}) \subset T(C^{\mathcal{O}})) \\
&-((\omega_{\mathcal{A}}X \otimes \omega_{\mathcal{A}}Y) \boxtimes \omega_{\mathcal{A}}(X \otimes Y)^{\vee} \xrightarrow{\omega^{\mathcal{A}}_{X,Y} \boxtimes 1} \\
&\qquad \omega_{\mathcal{A}}(X \otimes Y) \boxtimes \omega_{\mathcal{A}}(X \otimes Y)^{\vee} = C^{X \otimes Y} \subset T^{1}(C^{\mathcal{O}}) \subset T(C^{\mathcal{O}})).
\end{aligned}$$

The image of rel $h_{X,Y}$ is the same as the image of the difference

$$(C^X\bar{\otimes}C^Y \hookrightarrow T(C^{\mathcal{O}})) - (C^X\bar{\otimes}C^Y \xrightarrow{\omega^{\mathcal{A}}_{X,Y}\boxtimes\omega^{\mathcal{A}\,-1\,t}_{X,Y}} C^{X\otimes Y} \hookrightarrow T(C^{\mathcal{O}})).$$

The second summand induces a (non-associative, in general) multiplication

$$m^{\mathcal{O}} = \omega^{\mathcal{A}} \boxtimes \omega^{\mathcal{A}\,-1\,t} : C^{\mathcal{O}}\bar{\otimes}C^{\mathcal{O}} \to C^{\mathcal{O}}.$$

3.3.13. Lemma. *The coideal J_1 is an ideal of the (eventually non-associative) algebra $(C^{\mathcal{O}}, m^{\mathcal{O}})$.*

PROOF. Let us show that

$$\big(J_1\bar{\otimes}C^{\mathcal{O}} \hookrightarrow C^{\mathcal{O}}\bar{\otimes}C^{\mathcal{O}} \xrightarrow{m^{\mathcal{O}}} C^{\mathcal{O}} \to C_{\mathcal{A}}\big) = 0.$$

This follows from the commutative diagram, valid for any $f : X \to Z$, $X, Y, Z \in \mathcal{O}$, $f \in \mathcal{A}$, written with a shorthand $\omega = \omega_{\mathcal{A}}$:

$$\begin{array}{ccccc}
\omega Z\otimes\omega Y\boxtimes\omega Y^\vee\otimes\omega Z^\vee & \xleftarrow{\omega f\otimes 1} & \omega X\otimes\omega Y\boxtimes\omega Y^\vee\otimes\omega Z^\vee & \xrightarrow{1\boxtimes\omega f^t} & \omega X\otimes\omega Y\boxtimes\omega Y^\vee\otimes\omega X^\vee \\
\downarrow{\scriptstyle \omega^{\mathcal{A}}_{Z,Y}\boxtimes\omega^{\mathcal{A}\,-1\,t}_{Z,Y}} & & \downarrow{\scriptstyle \omega^{\mathcal{A}}_{X,Y}\boxtimes\omega^{\mathcal{A}\,-1\,t}_{Z,Y}} & & \downarrow{\scriptstyle \omega^{\mathcal{A}}_{X,Y}\boxtimes\omega^{\mathcal{A}\,-1\,t}_{X,Y}} \\
\omega(Z\otimes Y)\boxtimes\omega(Z\otimes Y)^\vee & \xleftarrow{\omega(f\otimes Y)\boxtimes 1} & \omega(X\otimes Y)\boxtimes\omega(Z\otimes Y)^\vee & \xrightarrow{1\boxtimes\omega(f\otimes Y)^t} & \omega(X\otimes Y)\boxtimes\omega(X\otimes Y)^\vee \\
& \searrow{\scriptstyle i_{Z\otimes Y}} & & \swarrow{\scriptstyle i_{X\otimes Y}} & \\
& & C_{\mathcal{A}} & &
\end{array}$$

Similarly,

$$\big(C^{\mathcal{O}}\bar{\otimes}J_1 \hookrightarrow C^{\mathcal{O}}\bar{\otimes}C^{\mathcal{O}} \xrightarrow{m^{\mathcal{O}}} C^{\mathcal{O}} \to C_{\mathcal{A}}\big) = 0. \quad \square$$

3.3.14. Proposition. *The multiplication $m : C_{\mathcal{A}}\bar{\otimes}C_{\mathcal{A}} \to C_{\mathcal{A}}$ induced by the multiplication $m^{\mathcal{O}}$ via Lemma 3.3.13 is associative.*

PROOF. Consider diagram (1.2.4) with $F = \omega_{\mathcal{A}}$ and $\phi = \omega^{\mathcal{A}}$. It expresses the relationship between $\omega^{\mathcal{A}}$ and the associativity constraint. Take a transposed copy of this diagram with all arrows inverted, and $\circledast$-multiply both diagrams. The obtained equation asserts that m is associative. $\square$

3.3.15. Lemma. *Let (A, m) be an associative algebra in a monoidal category $(\mathcal{W}, \otimes)$. Then the standard morphism*

$$A \to T(A)/(\mathrm{Im}(\mathrm{id}_{A\otimes A} - m))$$

is an isomorphism of algebras.

PROOF. For any $k > 0$ all morphisms $A^{\otimes k} \to A$ obtained by repeated multiplication of two neighbour factors are equal. $\square$

3.3.16. Theorem. *The standard morphism*

$$C_{\mathcal{A}} \to T(C_{\mathcal{A}})/(\mathrm{Im}(\mathrm{id}_{C_{\mathcal{A}}\otimes C_{\mathcal{A}}} - m)) = B_p \tag{3.3.2}$$

is an isomorphism of algebras and squared coalgebras. It makes $C_{\mathcal{A}}$ into a bicoalgebra. The monoidal functor $\omega_{\mathcal{A}}$ admits the factorisation

$$(\omega_{\mathcal{A}}, \omega^{\mathcal{A}}) = \big(\mathcal{A} \xrightarrow{(F_{\mathcal{A}}, f^{\mathcal{A}})} {}^{C_{\mathcal{A}}}\mathcal{V} \xrightarrow{(\mathcal{U}, \mathrm{id})} \mathcal{V}\big),$$

where $F_{\mathcal{A}}$ is the equivalence from Theorem 2.6.3 and $f^{\mathcal{A}}_{X,Y} = \omega^{\mathcal{A}}_{X,Y}$.

PROOF. Applying Lemma 3.3.15 to the algebra $(C_{\mathcal{A}}, m)$ in the category $(\widehat{\mathcal{V}\boxtimes\mathcal{V}}, \bar{\otimes})$ we deduce that $C_{\mathcal{A}} \to B_p$ is an algebra isomorphism. The algebra $B_p = T(C^{\mathcal{O}})/(J)$ is a bicoalgebra by Proposition 3.3.7. Since (3.3.2) is a coalgebra morphism as well, it makes $C_{\mathcal{A}}$ into a bicoalgebra. By Corollary 3.3.10 there exists a premonoidal prefunctor $F_p : \mathcal{P} \to {}^{B_p}\mathcal{V}$ such that

$$F_p(h_{X,Y}) = (\omega^{\mathcal{A}}_{X,Y} : \omega_{\mathcal{A}} X \otimes \omega_{\mathcal{A}} Y \to \omega_{\mathcal{A}}(X \otimes Y)).$$

In particular, $f^{\mathcal{A}}_{X,Y} = \omega^{\mathcal{A}}_{X,Y}$ is an isomorphism of B_p-comodules for $X, Y \in \mathcal{O}$. This implies that $f^{\mathcal{A}}_{X,Y}$ is an isomorphism of $C_{\mathcal{A}}$-comodules for all $X, Y \in \operatorname{Ob}\mathcal{A}$. □

Notice that F_p is the restriction of $\overline{(F_{\mathcal{A}}, f^{\mathcal{A}})} : \overline{\mathcal{A}} \to \overline{{}^{B_p}\mathcal{V}}$ to $\mathcal{P}$.

3.3.17. Proposition. *The correspondence $(\omega_{\mathcal{A}}, \omega^{\mathcal{A}}) \mapsto C_{\mathcal{A}}$ extends to a functor*

$$\Psi : \mathfrak{M}(\mathcal{V}) \to \operatorname{Bicoalg}(\widehat{\mathcal{V}}).$$

PROOF. Let $(\omega_{\mathcal{A}}, \omega^{\mathcal{A}})$ and $(\omega_{\mathcal{B}}, \omega^{\mathcal{B}})$ be monoidal fibre functors and let (G, g, γ) represent a morphism $(\omega_{\mathcal{A}}, \omega^{\mathcal{A}}) \to (\omega_{\mathcal{B}}, \omega^{\mathcal{B}})$. Proposition 2.6.4 gives us a coalgebra morphism $\boldsymbol{g} = \Psi(G, \gamma) : C_{\mathcal{A}} = \Psi(\omega_{\mathcal{A}}) \to C_{\mathcal{B}} = \Psi(\omega_{\mathcal{B}})$. Let us prove that $\boldsymbol{g}$ is an algebra morphism as well.

Recall that $\boldsymbol{g}$ is determined by the morphisms $\boldsymbol{g}_X = \gamma_X \boxtimes \gamma_X^{-1\,t} : C^X \to C^{GX}$ in diagram (2.6.3). We claim that

$$\begin{array}{ccccc}
C^X \bar{\otimes} C^Y & & \xrightarrow{\quad m^{\mathcal{O}} \quad} & & C^{X\otimes Y} \\
\Big\downarrow {\scriptstyle \boldsymbol{g}_X \bar{\otimes} \boldsymbol{g}_Y} & & & & \Big\downarrow {\scriptstyle \boldsymbol{g}_{X\otimes Y}} \\
C^{GX} \bar{\otimes} C^{GY} & \xrightarrow{m^{\mathcal{O}}} & C^{GX \otimes GY} & \xrightarrow{\chi} & C^{G(X\otimes Y)}
\end{array} \qquad (3.3.3)$$

commutes, where χ comes from the isomorphism $\omega_{\mathcal{B}} g_{X,Y} : \omega_{\mathcal{B}}(GX \otimes GY) \to \omega_{\mathcal{B}} G(X \otimes Y)$. Indeed, the expanded form of (3.3.3) is

$$\begin{array}{ccc}
\omega_{\mathcal{A}} X \otimes \omega_{\mathcal{A}} Y \boxtimes \omega_{\mathcal{A}} Y^{\vee} \otimes \omega_{\mathcal{A}} X^{\vee} & \xrightarrow{\omega^{\mathcal{A}}_{X,Y} \boxtimes \omega^{\mathcal{A}\,-1\,t}_{X,Y}} & \omega_{\mathcal{A}}(X\otimes Y) \boxtimes \omega_{\mathcal{A}}(X \otimes Y)^{\vee} \\
\Big\downarrow {\scriptstyle \gamma_X \otimes \gamma_Y \boxtimes \gamma_Y^{-1\,t} \otimes \gamma_X^{-1\,t}} & & \\
\omega_{\mathcal{B}} GX \otimes \omega_{\mathcal{B}} GY \boxtimes \omega_{\mathcal{B}} GY^{\vee} \otimes \omega_{\mathcal{B}} GX^{\vee} & & \Big\downarrow {\scriptstyle \gamma_{X\otimes Y} \boxtimes \gamma_{X\otimes Y}^{-1\,t}} \\
\Big\downarrow {\scriptstyle \omega^{\mathcal{B}}_{GX,GY} \boxtimes \omega^{\mathcal{B}\,-1\,t}_{GX,GY}} & & \\
\omega_{\mathcal{B}}(GX \otimes GY) \boxtimes \omega_{\mathcal{B}}(GX \otimes GY)^{\vee} & \xrightarrow{\omega_{\mathcal{B}} g_{X,Y} \boxtimes \omega_{\mathcal{B}} g_{X\otimes Y}^{-1\,t}} & \omega_{\mathcal{B}} G(X \otimes Y) \boxtimes \omega_{\mathcal{B}} G(X \otimes Y)^{\vee}
\end{array}$$

which is the $\boxtimes$-product of two commutative diagrams. Diagram (3.3.3) implies that $\boldsymbol{g}$ is an algebra homomorphism. □

3.3.18. Corollary. *The monoidal functor $(F_{\mathcal{A}}, f^{\mathcal{A}}) : \mathcal{A} \to \Phi(\Psi(\mathcal{A})) = {}^{C_{\mathcal{A}}}\mathcal{V}$ constructed in Theorem 3.3.16 together with* $\mathrm{id} : (\omega_{\mathcal{A}}, \omega^{\mathcal{A}}) \to (\mathcal{U}, \mathrm{id}) \circ (F_{\mathcal{A}}, f^{\mathcal{A}})$ *gives an isomorphism of functors*

$$F : \mathrm{Id}_{\mathfrak{M}(\mathcal{V})} \xrightarrow{\sim} \Phi\Psi.$$

PROOF. We already know by Proposition 2.6.6 that F is an isomorphism of the forgetful functor $\mathfrak{M}(\mathcal{V}) \to \mathfrak{A}(\mathcal{V})$ with $\Phi\Psi : \mathfrak{M}(\mathcal{V}) \to \mathfrak{M}(\mathcal{V}) \to \mathfrak{A}(\mathcal{V})$. Theorem 3.3.16 implies that this isomorphism is in $\mathfrak{M}(\mathcal{V})$. □

3.3.19. Theorem. *The functors* $\Phi : \operatorname{Bicoalg}(\widehat{\mathcal{V}}) \to \mathfrak{M}(\mathcal{V})$, $B \mapsto {}^B\mathcal{V}$ *and* $\Psi : \mathfrak{M}(\mathcal{V}) \to \operatorname{Bicoalg}(\widehat{\mathcal{V}})$, $(\omega_{\mathcal{A}}, \omega^{\mathcal{A}}) \mapsto C_{\mathcal{A}}$ *are equivalences, quasiinverse to each other.*

PROOF. By Theorem 3.3.16 Φ is essentially surjective on objects. Corollary 3.3.18 implies that Φ is full. By Theorem 2.6.7 Φ is faithful, hence, Φ is an equivalence. Ψ is quasi-inverse to Φ by Corollary 3.3.18. □

In the case $\mathcal{V} = \Bbbk$-vect this theorem was proved by Schauenburg [**27**].

3.3.20. Example (Bicoalgebras obtained from ordinary Hopf algebras). Suppose that $\mathcal{V} = \mathcal{C} = C$-comod, where C is a Hopf $\Bbbk$-algebra. How to describe the squared Hopf coalgebra $H \in \widehat{\mathcal{C} \boxtimes \mathcal{C}} = C \otimes_{\Bbbk} C$-Comod reconstructed from the identity functor $\mathrm{Id}_{\mathcal{C}} : \mathcal{C} \to \mathcal{C}$? Note that this would give a solution also to a more general problem — reconstruct a squared Hopf coalgebra H' from a functor $F : \mathcal{C} \to \mathcal{V}$, where $\mathcal{C} = C$-comod, $\mathcal{V} = V$-comod, C, V are Hopf $\Bbbk$-algebras, $F = f$-comod, $f : C \to V$ is a homomorphism of Hopf algebras. Indeed, $H' = (F \boxtimes F)H \in \widehat{\mathcal{V} \boxtimes \mathcal{V}} = V \otimes_{\Bbbk} V$-Comod, $F \boxtimes F = f \otimes_{\Bbbk} f$-Comod, since F is an exact monoidal functor. As a $\Bbbk$-vector space H' coincides with H; its coend structure $i_X : X \otimes_{\Bbbk} X^{\vee} \to H = H'$ coincides with that of H; multiplication

$$m' = (F \boxtimes F)m : H'_{1'2''} \otimes H'_{1''2'} \to H'_{12}$$

coincides with m as a linear map; the unit $\eta' = (F \boxtimes F)\eta : \Bbbk_1 \otimes_{\Bbbk} \Bbbk_2 \to H'_{12}$ coincides with η as a linear map; the comultiplication

$$\Delta' = (F \boxtimes F \boxtimes F)\Delta : H'_{13} \otimes \Bbbk_2 \to H'_{12'} \otimes H'_{2''3}$$

coincides with Δ as a linear map; the counit $\varepsilon' = F\varepsilon : \Bbbk \to H'_{1'1''}$ coincides with ε as a linear map. Considering a particular case $V = \Bbbk$, $f = \varepsilon : C \to \Bbbk$ we find by usual reconstruction theorems that $H' = C$ as a Hopf algebra. We conclude that $H = C$ as a $\Bbbk$-vector space and the squared bicoalgebra structure morphisms $m, \eta, \Delta, \varepsilon$ of H coincide with those of C as linear maps.

The maps $i_X : X_1 \otimes_{\Bbbk} X_2^{\vee} \to C_{12}$ induce on $H = C$ the left $C \otimes_{\Bbbk} C$-comodule structure

$$\delta : C \to C \otimes C \otimes C, \quad c \mapsto c_{(1)} \otimes \gamma(c_{(3)}) \otimes c_{(2)}.$$

Indeed, the $C \otimes C$-coaction on $X \otimes X^{\vee}$ is

$$\delta(x_a \otimes x^b) = t_{X\,a}{}^{c} \otimes \gamma(t_{X\,d}{}^{b}) \otimes x_c \otimes x^d \in C \otimes C \otimes X \otimes X^{\vee}$$

in notations

$$\delta(x_a) = t_{X\,a}{}^{c} \otimes x_c \in C \otimes X,$$
$$\delta(x^b) = \gamma(t_{X\,d}{}^{b}) \otimes x^d \in C \otimes X^{\vee}$$

using dual bases $(x_a) \subset X$, $(x^b) \subset X^{\vee}$.

As a corollary we get the following interpretation of structure morphisms of H:

$\eta : \Bbbk \to C = H$, $1 \mapsto 1$ is in $C^{\otimes 2}$-Comod;

$m : C_{1'2''} \otimes C_{1''2'} \to C \in C^{\otimes 2}$-Comod, where the first comodule is equipped with the coaction

$$\delta(c \otimes d) = c_{(1)}d_{(1)} \otimes \gamma(d_{(3)})\gamma(c_{(3)}) \otimes c_{(2)} \otimes d_{(2)} \in C^{\otimes 2} \otimes C \otimes C;$$

$\Delta : C_{13} \otimes \Bbbk_2 \to C_{12'} \otimes C_{2''3} \in C^{\otimes 3}$-Comod, where the first comodule is equipped with the coaction

$$\delta(c) = c_{(1)} \otimes 1 \otimes \gamma(c_{(3)}) \otimes c_{(2)} \in C^{\otimes 3} \otimes C$$

and the second with

$$\delta(c \otimes d) = c_{(1)} \otimes \gamma(c_{(3)}) d_{(1)} \otimes \gamma(d_{(3)}) \otimes c_{(2)} \otimes d_{(2)} \in C^{\otimes 3} \otimes C \otimes C;$$

$\varepsilon : C_{1'1''} \to \Bbbk \in C$-Comod, where the first comodule carries the coaction

$$\delta(c) = c_{(1)} \gamma(c_{(3)}) \otimes c_{(2)} \in C \otimes C.$$

To construct explicitly the isomorphism $\Psi\Phi \to \mathrm{Id}_{\mathfrak{M}(\mathcal{V})}$ let us use the results for $\mathfrak{A}(\mathcal{V})$. Let B be a bicoalgebra in $\widehat{\mathcal{V}}$, let $\mathcal{A} = {}^B\mathcal{V}$, let $\omega_{\mathcal{A}} = \mathcal{U} : {}^B\mathcal{V} \to \mathcal{V}$ and $\omega^{\mathcal{A}} = \mathrm{id}$. It was shown in Propositions 2.6.9, 2.6.10 that there is a unique coalgebra isomorphism $h_B : C_{\mathcal{A}} = \Psi\Phi(B) \to B$ such that for any $X \in {}^B\mathcal{V}$ the composite $X \boxtimes X^\vee \xrightarrow{i_X} C_{\mathcal{A}} \xrightarrow{h_B} B$ is the canonical coalgebra morphism (2.6.4).

3.3.21. Proposition (bicoalgebra reconstruction). *The morphism $h_B : C_{\mathcal{A}} \to B$ is an isomorphism of bicoalgebras giving the functorial isomorphism*

$$h : \Psi\Phi \to \mathrm{Id}_{\mathrm{Bicoalg}(\widehat{\mathcal{V}})} .$$

PROOF. Since $\omega^{\mathcal{A}} = \mathrm{id}$, the multiplication $m^{\mathcal{O}}$ is simply the identification morphism $m^{\mathcal{O}} = 1 \boxtimes j_+ : C^X \bar{\otimes} C^Y = X \otimes Y \boxtimes Y^\vee \otimes X^\vee \to X \otimes Y \boxtimes (X \otimes Y)^\vee = C^{X \otimes Y}$. Let us prove that

$$\begin{array}{ccc} C^X \bar{\otimes} C^Y & \xrightarrow{m^{\mathcal{O}}} & C^{X \otimes Y} \\ {\scriptstyle \ddot{\imath}_X \bar{\otimes} \ddot{\imath}_Y} \downarrow & & \downarrow {\scriptstyle \ddot{\imath}_{X \otimes Y}} \\ B \bar{\otimes} B & \xrightarrow{m} & B \end{array} \tag{3.3.4}$$

is commutative. It would imply that h_B is an algebra homomorphism. The definition (2.6.4) of $\ddot{\imath}_X$ plugged into diagram (3.3.4) gives

$$\begin{array}{ccc} X_{1'} \otimes Y_{1''} \boxtimes \mathbb{1}_{2'} \otimes Y^\vee_{2''} \otimes \mathbb{1}_{2'''} \boxtimes X^\vee_{2^4} & \xrightarrow{\sim} & X_{1'} \otimes Y_{1''} \boxtimes \mathbb{1}_{2'} \otimes (X \otimes Y)^\vee_{2''} \\ {\scriptstyle \delta_X \otimes \delta_Y \otimes Y^\vee \otimes X^\vee} \downarrow & & \downarrow {\scriptstyle \delta_{X \otimes Y}} \\ B_{1'2^4} \otimes B_{1''2'} \otimes Y_{2''} \otimes Y^\vee_{2'''} \otimes X_{2^5} \otimes X^\vee_{2^6} & & B_{12'} \otimes (X \otimes Y)_{2''} \otimes (X \otimes Y)^\vee_{2'''} \\ {\scriptstyle B \otimes B \otimes \mathrm{ev} \otimes \mathrm{ev}} \downarrow & & \downarrow {\scriptstyle B \otimes \mathrm{ev}} \\ B_{1'2''} \otimes B_{1''2'} & \xrightarrow{m} & B_{12} \end{array}$$

This diagram commutes by definition of $\delta_{X \otimes Y}$ (3.2.3). Hence, h_B is a bicoalgebra isomorphism. The isomorphism h is functorial by Proposition 2.6.9. □

3.4. Tensor generators and relations

One of the ways to construct monoidal categories is to use monoidal construction data. Here we do it directly, which allows to drop the assumption of rigidity of $\mathcal{V}$. So, let $\mathcal{V}$ be a monoidal abelian category with $\Bbbk$-linear structure induced by $\Bbbk = \mathrm{End}\, \mathbb{1}$. Similarly to morphisms in the category $\mathfrak{M}(\mathcal{V})$ one can consider morphisms from a premonoidal prefunctor $p : \mathcal{P} \to \overline{\mathcal{V}}$ to a monoidal fibre functor $(B, b) : (\mathcal{B}, \otimes) \to (\mathcal{V}, \otimes)$. They are defined as equivalence classes of pairs (F, ϕ), where $F : \mathcal{P} \to \overline{\mathcal{B}}$ is a premonoidal prefunctor and $\phi : p \to \bar{B}F$ is an isomorphism of premonoidal prefunctors from $\mathcal{P}$ to $\overline{\mathcal{V}}$. The equivalence relation is the same as in Definition 3.3.11: two pairs (F, ϕ) and (G, γ) are equivalent if there is an isomorphism of premonoidal prefunctors $\zeta : F \to G$ such that

$$\gamma = (p \xrightarrow{\phi} \bar{B}F \xrightarrow{\bar{B}\zeta} \bar{B}G).$$

3.4.1. Theorem. *Let $\mathcal{P}$ be a small premonoidal category and let $p : \mathcal{P} \to \overline{\mathcal{V}}$ be a premonoidal prefunctor. There is a monoidal fibre functor $(A, a) : (\mathcal{A}, \otimes) \to (\mathcal{V}, \otimes)$ and a morphism $(r, \mathrm{id}) : p \to (A, a)$, where $r : \mathcal{P} \to \overline{\mathcal{A}}$ is a premonoidal prefunctor, such that the map induced by the composition*

$$\mathrm{Hom}_{\mathfrak{M}(\mathcal{V})}\big((A, a), (B, b)\big) \longrightarrow \mathrm{Hom}(p, \bar{B})$$

$$(F, f, \phi) \longmapsto \big(p \xrightarrow{r} \overline{\mathcal{A}} \xrightarrow{\bar{F}} \overline{\mathcal{B}}, \phi r\big)$$

is bijective.

PROOF. Let us extend step by step $p : \mathcal{P} \to \overline{\mathcal{V}}$ to $p' : \mathcal{P}' \to \overline{\mathcal{V}}$, checking that the restriction map $\mathrm{Hom}(p', \bar{B}) \to \mathrm{Hom}(p, \bar{B})$ is bijective. Due to faithfulness of B we may assume that the initial p and all p' are faithful. The resulting p' is denoted again by p and used for the next step. In the limit we will get a monoidal category.

In order to construct the tensor product let us divide $\mathrm{Ob}\,\mathcal{P} = \mathcal{O}$ at each step into two disjoint parts: $\mathcal{D}$, where the tensor product is already defined, and $\mathcal{U}$, where it is undefined yet, $\mathcal{O} = \mathcal{D} \sqcup \mathcal{U}$. For the initial precategory $\mathcal{P}$ we set $\mathcal{D} = \varnothing$ and $\mathcal{U} = \mathcal{O}$.

Step 0. Add an object $\mathbb{1}$ to $\mathcal{U} \subset \mathrm{Ob}\,\mathcal{P}$ getting $\mathrm{Ob}\,\mathcal{P}'$, add a morphism $i : \mathbb{1} \to \varnothing$ to $\mathrm{Mor}\,\mathcal{P}$ getting $\mathrm{Mor}\,\mathcal{P}'$. The source of i is the one-letter word $\mathbb{1}$, the target is the empty word. Set $p'\mathbb{1} = \mathbb{1}$, $p'i = \mathrm{id}_{\mathbb{1}} : \mathbb{1} \to \mathbb{1}$. Let us give an inverse map for $\mathrm{Hom}(p', \bar{B}) \to \mathrm{Hom}(p, \bar{B})$. That is, given a morphism $[(F, \phi)] : p \to \bar{B}$ we construct its unique extension $[(F', \phi')] : p' \to \bar{B}$. The isomorphism $F'i : F'\mathbb{1} \to \mathbb{1}$ can be chosen arbitrarily and it determines uniquely the isomorphism $\phi'_{\mathbb{1}} : \mathbb{1} \to BF'\mathbb{1}$ such that

$$\mathrm{Id}_{\mathbb{1}} = \big(\mathbb{1} \xrightarrow{\phi'_{\mathbb{1}}} BF'\mathbb{1} \xrightarrow{BF'i} B\mathbb{1} \xrightarrow{b} \mathbb{1}\big).$$

Any other choice $F''i : F''\mathbb{1} \xrightarrow{\sim} \mathbb{1}$ yields an isomorphism $\zeta_{\mathbb{1}} = (F''i)^{-1} \circ (F'i) : F'\mathbb{1} \to F''\mathbb{1}$. Setting $\zeta_X = \mathrm{id} : FX \to FX$ for $X \in \mathcal{O}$ we get an isomorphism $\zeta : F' \to F''$. It makes pairs (F', ϕ') and (F'', ϕ'') equivalent.

Steps 1–6. Coincide with Steps 1–6 in the proof of Theorem 2.8.1. New objects are added to $\mathcal{U}$ at Steps 1, 5, 6, so that $\mathcal{D}' = \mathcal{D}$, $\mathcal{U}' \supset \mathcal{U}$.

Step 7. We set $\mathcal{D}' = \mathcal{O} = \mathcal{D} \sqcup \mathcal{U}$ and $\mathcal{U}' = \mathcal{D} \times \mathcal{U} \sqcup \mathcal{U} \times \mathcal{D} \sqcup \mathcal{U} \times \mathcal{U}$. By induction, the map $\otimes : \mathcal{D} \times \mathcal{D} \to \mathcal{O}$ is already defined. Extend it to $\otimes' : \mathcal{D}' \times \mathcal{D}' \to \mathcal{O}'$ setting $\otimes' = \mathrm{id}_{\mathcal{U}'} : \mathcal{D} \times \mathcal{U} \sqcup \mathcal{U} \times \mathcal{D} \sqcup \mathcal{U} \times \mathcal{U} \to \mathcal{U}'$. The result of applying $\otimes'$ to (X, Y) is denoted $X \otimes Y \in \mathcal{O}'$. We add a morphism $h_{X,Y} : XY \to X \otimes Y$ to $\mathrm{Mor}\,\mathcal{P}$ for each pair $X, Y \in \mathcal{D}'$, getting $\mathrm{Mor}\,\mathcal{P}'$. Here $\mathrm{source}(h_{X,Y})$ is a two-letter word, and $\mathrm{target}(h_{X,Y})$ is a one-letter word. Set $p'(X \otimes Y) = pX \otimes pY$ and $p'(h_{X,Y}) = \mathrm{id} : pX \otimes pY \to pX \otimes pY$.

Let us show the bijectivity of the map $\mathrm{Hom}(p', \bar{B}) \to \mathrm{Hom}(p, \bar{B})$. Given a morphism $[(F, \phi)] : p \to \bar{B}$ we construct its unique extension $[(F', \phi')] : p' \to \bar{B}$. We may choose arbitrarily the isomorphism $f_{X,Y} = F'(h_{X,Y}) : FX \otimes FY \to F'(X \otimes Y)$.

It must be an isomorphism because $Bf_{X,Y}$ satisfies the commutative diagram:

$$\begin{array}{ccc} pX \otimes pY & \xrightarrow{\phi_X \otimes \phi_Y} & BFX \otimes BFY \\ {\scriptstyle \phi'_{X\otimes Y}}\downarrow & & \downarrow{\scriptstyle b_{FX,FY}} \\ BF'(X \otimes Y) & \xleftarrow{Bf_{X,Y}} & B(FX \otimes FY) \end{array}$$

So, $f_{X,Y}$ determines $\phi'_{X\otimes Y}$ uniquely. Any other choice $F''(h_{X,Y}) : FX \otimes FY \to F''(X \otimes Y)$ yields an isomorphism

$$\zeta_{X\otimes Y} = F''(h_{X,Y}) \circ F'(h_{X,Y})^{-1} : F'(X \otimes Y) \to F''(X \otimes Y).$$

Hence the isomorphism $\zeta : F' \to F''$ and equivalence of (F', ϕ') and (F'', ϕ'').

Step 8. For any $X \in \mathcal{D}$ add to $\operatorname{Mor}\mathcal{P}$ morphisms $r : X\mathbb{1} \to X$, $l : \mathbb{1}X \to X$ getting $\operatorname{Mor}\mathcal{P}'$. Set $p'(r) = r_{pX} : pX \otimes \mathbb{1} \to pX$, $p'(l) = l_{pX} : \mathbb{1} \otimes pX \to pX$. Let $(F, \phi) : p \to \bar{B}$ represent a morphism. Clearly, there might be but one $Fr : FX \otimes F\mathbb{1} \to FX$ and $Fl : F\mathbb{1} \otimes FX \to FX$, which gives the extended morphism $(F', \phi) : p' \to \bar{B}$. Indeed,

$$Fr = \left(FX \otimes F\mathbb{1} \xrightarrow{1\otimes Fi} FX \otimes \mathbb{1} \xrightarrow{r_{FX}} FX\right), \tag{3.4.1}$$

$$Fl = \left(F\mathbb{1} \otimes FX \xrightarrow{Fi\otimes 1} \mathbb{1} \otimes FX \xrightarrow{l_{FX}} FX\right)$$

determines (F', ϕ). This follows from the commutative diagram

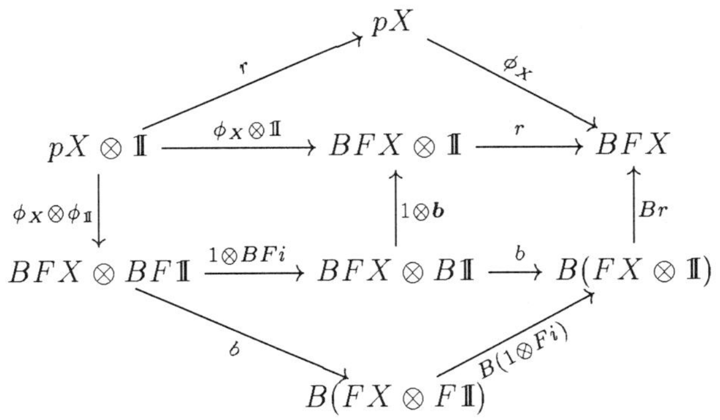

Similarly for Fl.

Step 9. For $X, Y, Z, X \otimes Y, Y \otimes Z \in \mathcal{D}$ add morphisms $a : X(YZ) \to (XY)Z$ to $\operatorname{Mor}\mathcal{P}$ getting $\operatorname{Mor}\mathcal{P}'$. Set

$$p'a = a_{pX,pY,pZ} : pX \otimes (pY \otimes pZ) \to (pX \otimes pY) \otimes pZ.$$

If $[(F, \phi)] : p \to \bar{B}$ is a morphism, there is only one way to extend F on a, namely:

$$Fa = a_{FX,FY,FZ} : FX \otimes (FY \otimes FZ) \to (FX \otimes FY) \otimes FZ, \tag{3.4.2}$$

for (B, b) is monoidal.

Step 10. Let $X, Y, V, W \in \operatorname{Words}(\mathcal{O})$ and $g : X \to Y$, $h : V \to W \in \operatorname{Mor}\mathcal{P}$, then add morphism $g{\cdot}h : XV \to YW$. Recall that $pX = pX_1 \otimes \cdots \otimes pX_n$ if $X = X_1 \dots X_n$ and the brackets are the same. Set $p'(g \cdot h) = pg \otimes ph : pX \otimes pV \to pY \otimes pW$. If (F, ϕ) represents a morphism $p \to \bar{B}$, there is only one way to extend it to $(F', \phi) : p' \to \bar{B}$, namely,

$$F(g \cdot h) = Fg \otimes Fh : FX \otimes FV \to FY \otimes FW.$$

Repeat Steps 1–10 cyclically. Denote $p_\infty : \mathcal{P}_\infty \to \overline{\mathcal{V}}$ the inductive limit of the obtained system of premonoidal functors $\mathcal{P} \to \mathcal{P}_1 \to \mathcal{P}_2 \to \dots$, commuting with $p_i : \mathcal{P}_i \to \overline{\mathcal{V}}$.

Introduce a precategory $\mathcal{A}$ with $\operatorname{Ob}\mathcal{A} = \operatorname{Ob}\mathcal{P}_\infty$ and those morphisms of $\mathcal{P}_\infty$, whose source and target are one-letter words. Steps 1–6 make sure that $\mathcal{A}$ is a $\Bbbk$-linear abelian category. There is a faithful functor $A : \mathcal{A} \to \mathcal{V}$, which is the prefunctor p_∞ restricted to $\mathcal{A}$. The partially defined at Step 7 tensor product on objects induces the map $\otimes : \operatorname{Ob}\mathcal{A} \times \operatorname{Ob}\mathcal{A} \to \operatorname{Ob}\mathcal{A}$. The tensor product of morphisms $g : X \to Y$, $h : V \to W \in \operatorname{Mor}\mathcal{A}$ is defined as the composite

$$g \otimes h = \big(X \otimes V \xrightarrow{h^{-1}_{X,V}} XV \xrightarrow{g\cdot h} YW \xrightarrow{h_{Y,W}} Y \otimes W\big)$$

by Step 10. This gives a functor $\otimes : \mathcal{A} \times \mathcal{A} \to \mathcal{A}$. An associativity constraint is given by

$$a_{X,Y,Z} = \big(X \otimes (Y \otimes Z) \xrightarrow{h^{-1}_{X,Y\otimes Z}} X(Y \otimes Z) \xrightarrow{1_X\cdot h^{-1}_{Y,Z}} X(YZ)$$
$$\xrightarrow{a} (XY)Z \xrightarrow{h_{X,Y}\cdot 1_Z} (X \otimes Y)Z \xrightarrow{h_{X\otimes Y,Z}} (X \otimes Y) \otimes Z\big). \tag{3.4.3}$$

It is mapped to $a_{pX,pY,pZ}$ by p_∞ (Step 9). Functoriality and the pentagon equation follows from that of $a_{pX,pY,pZ}$ by faithfulness of p_∞. Step 0 gives a unit object $\mathbb{1}$ and Step 8 gives the isomorphisms

$$r_X : X \otimes \mathbb{1} \xrightarrow{h^{-1}_{X,\mathbb{1}}} X\mathbb{1} \xrightarrow{r} X, \tag{3.4.4}$$

$$l_X : \mathbb{1} \otimes X \xrightarrow{h^{-1}_{\mathbb{1},X}} \mathbb{1}X \xrightarrow{l} X, \tag{3.4.5}$$

which are mapped to r_{pX}, l_{pX} by p_∞. Therefore, $(\mathcal{A}, \otimes, \mathbb{1}, r, l)$ is a monoidal category. By construction there is a monoidal functor $(A, \mathrm{id}, \mathrm{id}) : (\mathcal{A}, \otimes, \mathbb{1}) \to (\mathcal{V}, \otimes, \mathbb{1})$, where A is the restriction of p_∞.

The premonoidal precategory $\mathcal{P}_\infty$ is isomorphic to $\overline{\mathcal{A}}$. Indeed, for $X = X_1 \dots X_n$, $Y = Y_1 \dots Y_k$ there are unique isomorphisms $h_X : X_1 \dots X_n \to X_1 \otimes \dots \otimes X_n$, $h_Y : Y_1 \dots Y_k \to Y_1 \otimes \dots \otimes Y_k$ composed with $1 \cdot h_{M,N} \cdot 1$. They induce bijections

$$\operatorname{Hom}_{\mathcal{P}}(X, Y) \to \operatorname{Hom}_{\mathcal{P}}(X_1 \otimes \dots \otimes X_n, Y_1 \otimes \dots \otimes Y_k).$$

The functor $\overline{\mathcal{A}} \simeq \mathcal{P}_\infty \xrightarrow{F} \overline{\mathcal{B}}$ induces a monoidal functor $(F, f, \boldsymbol{f}) : (\mathcal{A}, \otimes, \mathbb{1}) \to (\mathcal{B}, \otimes, \mathbb{1})$. Indeed, take $f_{X,Y} = F(h_{X,Y})$ constructed at Step 7 and $\boldsymbol{f} = Fi : F\mathbb{1} \to \mathbb{1}$ constructed at Step 0. Equation (3.4.3) implies that associativity is coherent with (F, f). Indeed, by Step 10

$$F(h_{X,Y} \cdot 1_Z) = f_{X,Y} \otimes 1_{FZ} : (FX \otimes FY) \otimes FZ \to F(X \otimes Y) \otimes FZ,$$

and, similarly, $F(1_X \cdot h^{-1}_{Y,Z}) = 1_{FX} \otimes f^{-1}_{Y,Z}$. Also $Fa = a_{FX,FY,FZ}$ by (3.4.2).

Equations (3.4.4), (3.4.5) imply that $(F, f, \boldsymbol{f})$ is coherent with r and l. Indeed, applying F to (3.4.4) we get

$$Fr = \big(FX \otimes F\mathbb{1} \xrightarrow{f_{X,\mathbb{1}}} F(X \otimes \mathbb{1}) \xrightarrow{Fr_X} FX\big).$$

Together with definition (3.4.1) of Fr

$$Fr = \big(FX \otimes F\mathbb{1} \xrightarrow{1\otimes \boldsymbol{f}} FX \otimes \mathbb{1} \xrightarrow{r} FX\big),$$

it gives coherence of r and $(F, f, \boldsymbol{f})$. Similarly for l.

Clearly, the functor $\overline{\mathcal{A}} \to \overline{\mathcal{B}}$ is nothing else but $\overline{(F, f, \boldsymbol{f})}$. The isomorphism $\phi : \overline{(A, a, \boldsymbol{a})} \to \overline{(B, b, \boldsymbol{b})} \circ \overline{(F, f, \boldsymbol{f})}$ is an isomorphism of monoidal functors ϕ : $(A, a, \boldsymbol{a}) \to (B, b, \boldsymbol{b}) \circ (F, f, \boldsymbol{f})$ as well. The theorem is proved. □

3.4.2. Theorem. *In addition, let $\mathcal{V}$ be rigid, so it satisfies the assumptions of Section 1.3. Let $\mathcal{P}$ be a small premonoidal category and let $p : \mathcal{P} \to \bar{\mathcal{V}}$ be a premonoidal prefunctor. Let B_p be the bicoalgebra associated with p in Proposition 3.3.7. Then the category $(\mathcal{A}, \otimes, \mathbb{1}) = ({}^{B_p}\mathcal{V}, \otimes, \Bbbk)$, the underlying functor $\overline{(A, a, \boldsymbol{a})} = (\mathcal{U}, \mathrm{id}, \mathrm{id})$ and the premonoidal functor $F_p : \mathcal{P} \to \overline{{}^{B_p}\mathcal{V}}$ are the canonical ones constructed in Theorem 3.4.1.*

PROOF. Let us prove that each step at the proof of Theorem 3.4.1 does not change the associated category of comodules and the induced morphism of bicoalgebras $B_p \to B_{p'}$ is an isomorphism. Then the bicoalgebra $B_\infty = B_{p_\infty}$ corresponding to the limit precategory $\mathcal{P}_\infty$ is isomorphic to the initial bicoalgebra B_p. According to the proof of Theorem 3.4.1 we may assume that $\mathcal{P} = \overline{\mathcal{A}}$ and $p = \overline{(A, a, \boldsymbol{a})}$ for some small monoidal category $(\mathcal{A}, \otimes, \mathbb{1})$ and some faithful $\Bbbk$-linear exact monoidal functor $(A, a, \boldsymbol{a}) : (\mathcal{A}, \otimes, \mathbb{1}) \to (\mathcal{V}, \otimes, \mathbb{1})$. The category of comodules ${}^{B_p}\mathcal{V}$ associated with $\overline{(A, a, \boldsymbol{a})}$ is monoidally equivalent to $\mathcal{A}$ by Theorem 3.3.16. Finally, the map

$$\mathrm{Hom}_{\mathfrak{M}(\mathcal{V})}\big((A, a, \boldsymbol{a}), (B, b, \boldsymbol{b})\big) \to \mathrm{Hom}\big(\overline{(A, a, \boldsymbol{a})}, \overline{(B, b, \boldsymbol{b})}\big)$$

is a bijection, whence the theorem will follow.

We have to prove that $B_p \to B_{p'}$ is an isomorphism at each step.

Step 0. We add one generator, thus $C^{\mathcal{O}'} = C^{\mathcal{O}} \oplus \mathbb{1} \boxtimes \mathbb{1}^\vee$ with the embedding $i_{\mathbb{1}} : \mathbb{1} \boxtimes \mathbb{1}^\vee \to C^{\mathcal{O}'}$. We add one relation, namely,

$$\mathrm{Im}(\eta - i_{\mathbb{1}} : \mathbb{1} \boxtimes \mathbb{1} \to T(C^{\mathcal{O}'})).$$

Hence, the quotient $B_{p'} = T(C^{\mathcal{O}'})/\sim$ remains isomorphic to B_p.

Steps 1–6 are considered in Theorem 2.8.3.

Step 7. We add objects $X \otimes Y$ for $(X, Y) \in \mathcal{D} \times \mathcal{U}$, $\mathcal{U} \times \mathcal{D}$ and $\mathcal{U} \times \mathcal{U}$. Thus $C^{\mathcal{O}}$ increases by $(X \otimes Y) \boxtimes (X \otimes Y)^\vee$. We add also morphisms $h_{X,Y} : XY \to X \otimes Y$, hence, new relations identify $(X \otimes Y) \boxtimes (X \otimes Y)^\vee \subset C^{\mathcal{O}'}$ with $(X \boxtimes X^\vee) \bar{\otimes} (Y \boxtimes Y^\vee) \subset T^2(C^{\mathcal{O}})$ and the bicoalgebras $B_{p'}$ and B_p are isomorphic.

Step 8. The relations coming from the morphisms $r : X\mathbb{1} \to X$, $l : \mathbb{1}X \to X$ identify $(X \boxtimes X^\vee) \bar{\otimes} (\mathbb{1} \boxtimes \mathbb{1}^\vee)$ and $(\mathbb{1} \boxtimes \mathbb{1}^\vee) \bar{\otimes} (X \boxtimes X^\vee)$ with $X \boxtimes X^\vee$, so that $i_{\mathbb{1}} : \mathbb{1} \boxtimes \mathbb{1}^\vee \to B_{p'}$ is the unit of the algebra. However, this was already postulated at Step 0, thus the algebra B_p does not change.

Step 9. The isomorphisms $a : X(YZ) \to (XY)Z$ identify $(X \boxtimes X^\vee) \bar{\otimes} [(Y \boxtimes Y^\vee) \bar{\otimes} (Z \boxtimes Z^\vee)]$ with $[(X \boxtimes X^\vee) \bar{\otimes} (Y \boxtimes Y^\vee)] \bar{\otimes} (Z \boxtimes Z^\vee)$. The identification isomorphism is nothing else but the associativity isomorphism. However, the algebra B_p is associative by definition, so it does not change.

Step 10. Given $g : X \to Y$, $h : V \to W \in \mathrm{Mor}\,\mathcal{P}$ we get $\mathrm{Rel}\,g$, $\mathrm{Rel}\,h$. We want to show that the image of the difference

$$\begin{aligned} &(X \boxtimes Y^\vee) \bar{\otimes} (V \boxtimes W^\vee) \xrightarrow{X \boxtimes g^t \bar{\otimes} V \boxtimes h^t} (X \boxtimes X^\vee) \bar{\otimes} (V \boxtimes V^\vee) \\ -&(X \boxtimes Y^\vee) \bar{\otimes} (V \boxtimes W^\vee) \xrightarrow{g \boxtimes Y^\vee \bar{\otimes} h \boxtimes W^\vee} (Y \boxtimes Y^\vee) \bar{\otimes} (W \boxtimes W^\vee) \end{aligned}$$

lies in the ideal generated by Rel g, Rel h. This is proved by adding and subtracting the morphism

$$g \boxtimes Y^{\vee} \bar{\otimes} V \boxtimes h^{t} : (X \boxtimes Y^{\vee}) \bar{\otimes} (V \boxtimes W^{\vee}) \to (Y \boxtimes Y^{\vee}) \bar{\otimes} (V \boxtimes V^{\vee}).$$

Therefore, the theorem is proved. $\square$

3.5. Relationship with braided bialgebras

Let us consider the case of braided $\mathcal{V}$. Then it makes sense to consider braided (quasiclassical) bialgebras.

There is a unique functorial isomorphism $\phi : (\circledast X) \otimes (\circledast Y) \to \circledast (X \bar{\otimes} Y)$, $X, Y \in \mathcal{V} \boxtimes \mathcal{V}$, such that for $X = A \boxtimes B$, $Y = C \boxtimes D$ it equals

$$\phi = (432)^{\sim}_{+} : (A \otimes B) \otimes (C \otimes D) \to (A \otimes C) \otimes (D \otimes B),$$

$\phi =$ 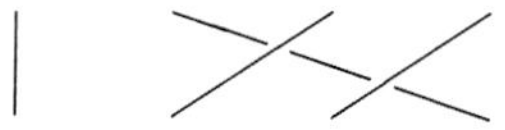 .

Indeed, it extends uniquely to arbitrary $X, Y \in \mathcal{V} \boxtimes \mathcal{V}$ via resolutions (1.6.2).

3.5.1. Proposition. *When $\mathcal{V}$ is braided, there is a monoidal functor*

$$(\circledast, \phi, r_{\mathbb{I}}^{-1}) : (\mathcal{V} \boxtimes \mathcal{V}, \bar{\otimes}, \mathbb{I} \boxtimes \mathbb{I}) \to (\mathcal{V}, \otimes, \mathbb{I}).$$

PROOF. It suffices to check coherence of ϕ with associativity on objects of the form $X = A \boxtimes B$, $Y = C \boxtimes D$, $Z = E \boxtimes F$. That is, to prove commutativity of the diagram

$$\begin{array}{ccc} A \otimes B \otimes C \otimes D \otimes E \otimes F & \xrightarrow{A \otimes B \otimes \phi_{C,D,E,F}} & A \otimes B \otimes C \otimes E \otimes F \otimes D \\ {\scriptstyle \phi_{A,B,C,D \otimes E \otimes F}} \downarrow & & \downarrow {\scriptstyle \phi_{A,B,C \otimes E, F \otimes D}} \\ A \otimes C \otimes D \otimes B \otimes E \otimes F & \xrightarrow{\phi_{A \otimes C, D \otimes B, E, F}} & A \otimes C \otimes E \otimes F \otimes D \otimes B \end{array}$$

This follows from the equation

$$\begin{pmatrix} 123456 \\ 125634 \end{pmatrix}^{\sim}_{+} \circ (432)^{\sim}_{+} = (65432)^{\sim}_{+} \circ (654)^{\sim}_{+}. \quad \square$$

3.5.2. Proposition. *Let $B = (B, \Delta, \varepsilon, m, \eta)$ be a bicoalgebra in $\widehat{\mathcal{V}}$. Denote*

$$\bar{B} = \circledast B = B^{12},$$

$$\bar{\Delta} = \left(B^{12} \simeq B^{13} \otimes \mathbb{I}^{2} \xrightarrow{\Delta^{123}} B^{12} \otimes B^{34}\right),$$

$$\bar{\eta} = \left(\mathbb{I} \xrightarrow{r_{\mathbb{I}}^{-1}} \mathbb{I} \otimes \mathbb{I} \xrightarrow{\circledast \eta} \circledast B\right),$$

$$\bar{m} = \left(B^{12} \otimes B^{34} \xrightarrow{\phi} B^{14} \otimes B^{23} \xrightarrow{\circledast m} B^{12}\right).$$

Then $(\bar{B}, \bar{\Delta}, \varepsilon, \bar{m}, \bar{\eta})$ is a braided bialgebra in $\widehat{\mathcal{V}}$.

PROOF. We already noticed that $(\bar{B}, \bar{\Delta}, \varepsilon)$ is a coalgebra. Since $(\circledast, \phi, r_{\mathbb{I}}^{-1})$ is a monoidal functor, it transfers the algebra (B, m, η) to an algebra $(\bar{B}, \bar{m}, \bar{\eta})$. It remains to prove the bialgebra property of $\bar{B}$. This is precisely the exterior of the diagram

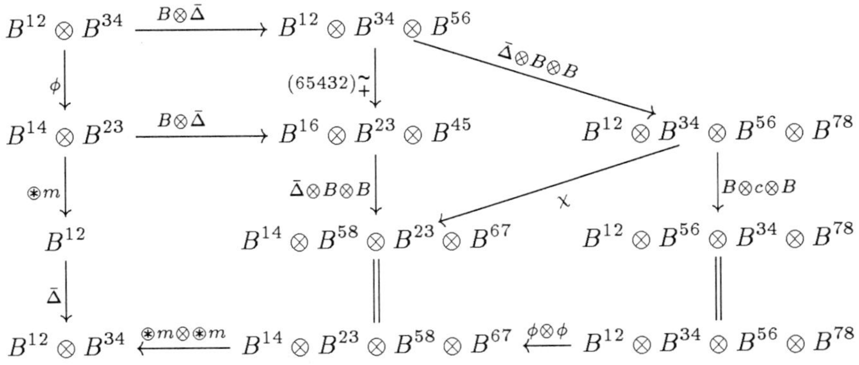

whose commutativity we are going to prove. Here

$$\chi = \begin{pmatrix} 12345678 \\ 14582367 \end{pmatrix}^{\sim}_{+}.$$

The hexagon follows from the bicoalgebra axiom (3.2.1). The pentagon is the equation

$$\chi = (432)^{\sim}_{+} \circ (876)^{\sim}_{+} \circ \begin{pmatrix} 12345678 \\ 12563478 \end{pmatrix}^{\sim}_{+}.$$

The square follows from the square

$$\begin{array}{ccc} X^1 \otimes Y^2 \otimes C^{34} & \xrightarrow{X\otimes Y\otimes\bar\Delta} & X^1 \otimes Y^2 \otimes C^{34} \otimes C^{56} \\ \phi\big\downarrow & & \big\downarrow(65432)^{\sim}_{+} \\ X^1 \otimes C^{23} \otimes Y^4 & \xrightarrow{X\otimes\bar\Delta\otimes Y} & X^1 \otimes C^{23} \otimes C^{45} \otimes Y^6 \end{array} \qquad (3.5.1)$$

valid for any squared coalgebra C in $\widehat{\mathcal{V}}$ and objects $X, Y \in \mathcal{V}$. Indeed, C can be covered by coalgebras $Z \boxtimes Z^\vee$ and for such coalgebra the diagram (3.5.1) takes the form

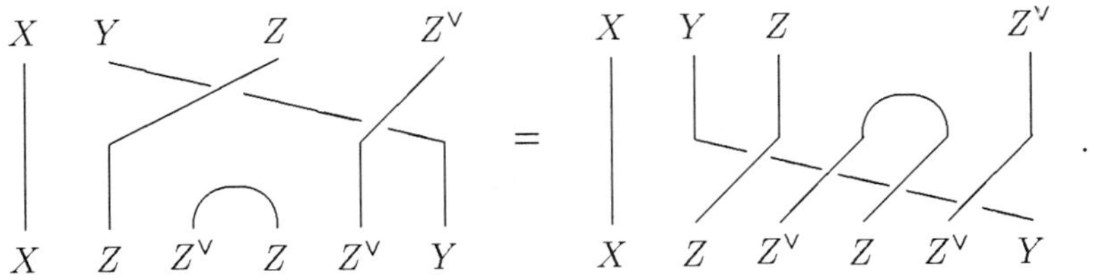

Similarly, the quadrilateral follows from the diagram

$$\begin{array}{ccc} C^{12} \otimes X^3 \otimes Y^4 & \xrightarrow{\bar\Delta\otimes X\otimes Y} & C^{12} \otimes C^{34} \otimes X^5 \otimes Y^6 \\ (432)^{\sim}_{+}\big\downarrow & & \big\downarrow\begin{pmatrix} 123456 \\ 134625 \end{pmatrix}^{\sim}_{+} \\ C^{14} \otimes X^2 \otimes Y^3 = C^{15} \otimes X^2 \otimes \mathbb{1}^3 \otimes Y^4 & \xrightarrow{\Delta^{135}} & C^{13} \otimes X^2 \otimes C^{46} \otimes Y^5 \end{array}$$

which follows from the particular case $C = Z \boxtimes Z^\vee$:

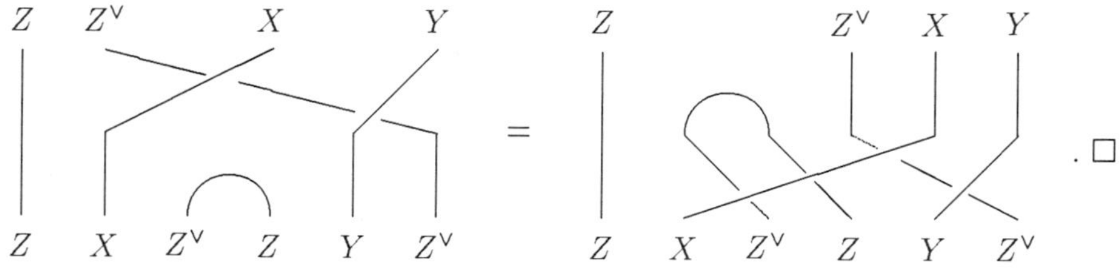

3.5.3. Proposition. *There exists a* $\Bbbk$*-linear exact faithful monoidal functor*

$$(F, \mathrm{id}, \mathrm{id}) : {}^B\mathcal{V} \to {}^{\bar{B}}\mathcal{V}, \quad F(X, \delta_X) = (X, \bar{\delta}_X), \quad F(f) = f,$$

$$\bar{\delta}_X = \big(X \xrightarrow{r_X^{-1}} X \otimes \mathbb{1} \xrightarrow{\circledast\delta_X} \bar{B} \otimes X\big),$$

commuting with the underlying functor.

PROOF. We already know that F is a functor. It remains to check that two structures of $\bar{B}$-comodule on the tensor product of B-comodules coincide. So, let (X, δ_X), $(Y, \delta_Y) \in {}^B\mathcal{V}$ be B-comodules. One $\bar{B}$-comodule structure on $X \otimes Y$ is the product of $\bar{B}$-comodule structures on X and Y:

$$X \otimes Y \xrightarrow{\bar{\delta}_X \otimes \bar{\delta}_Y} \bar{B} \otimes X \otimes \bar{B} \otimes Y \xrightarrow{\bar{B} \otimes c \otimes Y} \bar{B} \otimes \bar{B} \otimes X \otimes Y$$
$$\xrightarrow{\phi \otimes X \otimes Y} B^{14} \otimes B^{23} \otimes X^5 \otimes Y^6 \xrightarrow{\circledast m \otimes X \otimes Y} \bar{B} \otimes X \otimes Y.$$

Another comes from the product of B-comodules:

$$X \otimes Y \xrightarrow{X \otimes \bar{\delta}_Y} X \otimes \bar{B} \otimes Y \simeq X^1 \otimes B^{23} \otimes \mathbb{1}^4 \otimes Y^5$$
$$\xrightarrow{\delta^{14} \otimes B \otimes Y} B^{14} \otimes B^{23} \otimes X^5 \otimes Y^6 \xrightarrow{\circledast m \otimes X \otimes Y} \bar{B} \otimes X \otimes Y.$$

Coincidence of these two follows from a commutative diagram which one can generalise to the case of two squared coalgebras C, D and two comodules $X \in {}^C\mathcal{V}$, $Y \in {}^D\mathcal{V}$:

$$\begin{array}{ccccc} X \otimes Y & \xrightarrow{\bar{\delta}_X \otimes \bar{\delta}_Y} & \bar{C} \otimes X \otimes \bar{D} \otimes Y & \xrightarrow{1 \otimes c \otimes 1} & \bar{C} \otimes \bar{D} \otimes X \otimes Y \\ {\scriptstyle X \otimes \bar{\delta}_Y}\big\downarrow & & & & \big\downarrow{\scriptstyle \phi \otimes X \otimes Y} \\ X \otimes \bar{D} \otimes Y & \xrightarrow{\sim} & X^1 \otimes D^{23} \otimes \mathbb{1}^4 \otimes Y^5 & \xrightarrow{\delta_X^{14} \otimes \bar{D} \otimes Y} & C^{14} \otimes D^{23} \otimes X^5 \otimes Y^6 \end{array}$$

Commutativity of this diagram follows from the particular case of the canonical comodules X, Y over the coalgebras $C = X \boxtimes X^\vee$, $D = Y \boxtimes Y^\vee$, where it reduces to

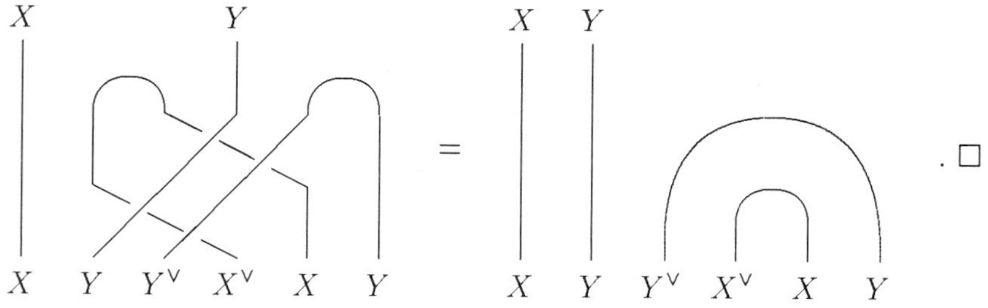

3.6. Commutative algebras and braiding

Let $\mathcal{C}$ be a $\Bbbk$-linear abelian braided rigid monoidal category, and let $(\omega, \omega^{\mathcal{C}}) : \mathcal{C} \to \mathcal{V}$ be a $\Bbbk$-linear exact faithful monoidal functor. Then the coend C reconstructed from ω as in (2.5.3) is an algebra, commutative in a certain sense. Define a morphism $\chi : C\bar{\otimes}C \to C\bar{\otimes}C \in \widehat{\mathcal{V}\boxtimes\mathcal{V}}$ via commutative diagram

$$\begin{array}{ccccc}
\omega(X\otimes Y)\boxtimes\omega(Y^\vee\otimes X^\vee) & \xleftarrow{\omega^{\mathcal{C}}_{X,Y}\boxtimes\omega^{\mathcal{C}}_{Y^\vee,X^\vee}} & (\omega X\otimes\omega Y)\boxtimes(\omega Y^\vee\otimes\omega X^\vee) & \xrightarrow{i_X\bar{\otimes}i_Y} & C\bar{\otimes}C \\
{\scriptstyle \omega c\boxtimes\omega c^{-1}}\downarrow & & {\scriptstyle f\boxtimes g}\downarrow & & \downarrow{\scriptstyle \chi} \\
\omega(Y\otimes X)\boxtimes\omega(X^\vee\otimes Y^\vee) & \xrightarrow{\omega^{\mathcal{C}-1}_{Y,X}\boxtimes\omega^{\mathcal{C}-1}_{X^\vee,Y^\vee}} & (\omega Y\otimes\omega X)\boxtimes(\omega X^\vee\otimes\omega Y^\vee) & \xrightarrow{i_Y\bar{\otimes}i_X} & C\bar{\otimes}C
\end{array}$$

where f and g denote functorial isomorphisms

$$f_{X,Y} = \omega^{\mathcal{C}-1}_{Y,X} \circ \omega c \circ \omega^{\mathcal{C}}_{X,Y}, \qquad g_{X,Y} = \omega^{\mathcal{C}-1}_{X^\vee,Y^\vee} \circ \omega c^{-1} \circ \omega^{\mathcal{C}}_{Y^\vee,X^\vee}.$$

Abusing the notation we write $\chi = f \boxtimes g = \overline{\omega c} \boxtimes \overline{\omega c^{-1}}$.

3.6.1. Proposition. *The multiplication in C is commutative in the following sense*

$$m = (C\bar{\otimes}C \xrightarrow{\chi} C\bar{\otimes}C \xrightarrow{m} C).$$

PROOF. In the following diagram top and bottom quadrangles and both hexagons commute.

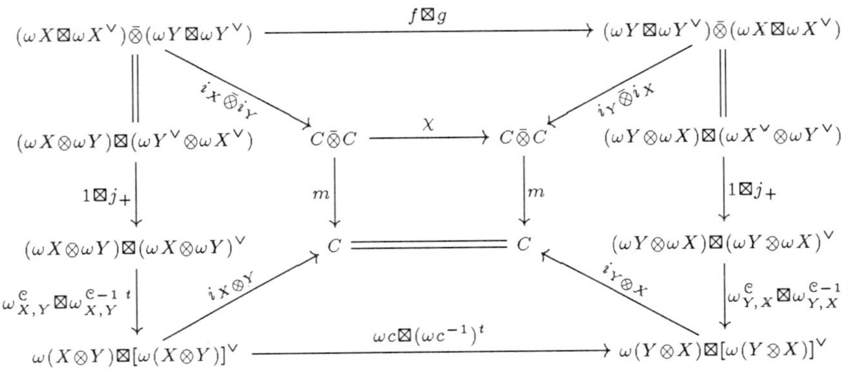

The exterior commutes due to the computation

$$\begin{aligned}
g &= \omega^{\mathcal{C}-1}_{X^\vee,Y^\vee} \circ \omega c^{-1} \circ \omega^{\mathcal{C}}_{Y^\vee,X^\vee} \\
&= \omega^{\mathcal{C}-1}_{X^\vee,Y^\vee} \circ \omega j_+^{-1} \circ \omega(c^{-1\,t}) \circ \omega j_+ \circ \omega^{\mathcal{C}}_{Y^\vee,X^\vee} \\
&= j_+^{-1} \circ \omega^{\mathcal{C}\,t}_{Y,X} \circ (\omega c^{-1})^t \circ \omega^{\mathcal{C}-1\,t}_{X,Y} \circ j_+.
\end{aligned}$$

Hence, the central square commutes. □

3.6.2. Corollary. *If, moreover, $\mathcal{V}$ is braided, then the algebra $\bar{C} \in \widehat{\mathcal{V}}$ defined in Proposition 3.5.2 is commutative in the following sense*

$$\begin{aligned}
\bar{m} &= (\bar{C}\otimes\bar{C} \xrightarrow{\xi} \bar{C}\otimes\bar{C} \xrightarrow{\bar{m}} \bar{C}), \\
\xi &= (\bar{C}\otimes\bar{C} \xrightarrow{\phi} \circledast(C\bar{\otimes}C) \xrightarrow{\circledast\chi} \circledast(C\bar{\otimes}C) \xrightarrow{\phi^{-1}} \bar{C}\otimes\bar{C}), \\
\phi &= (432)^{\sim}_{c}.
\end{aligned}$$

PROOF. The object PC_p together with $Pi_X : X^\vee \boxtimes X \to PC_p$ is the colimit of the diagram

$$pX^\vee \boxtimes pX \xleftarrow{pf^t \boxtimes pX} pY^\vee \boxtimes pX \xrightarrow{pY^\vee \boxtimes pf} pY^\vee \boxtimes pY,$$

where $f : X \to Y$ runs over $\operatorname{Mor} \mathcal{P}$. The object $C_{p^\vee}$ is the colimit of the diagram

$$pX^\vee \boxtimes pX^{\vee\vee} \xleftarrow{pf^t \boxtimes pX^{\vee\vee}} pY^\vee \boxtimes pX^{\vee\vee} \xrightarrow{pY^\vee \boxtimes pf^{tt}} pY^\vee \boxtimes pY^{\vee\vee},$$

which is isomorphic to the previous one via $1 \otimes \zeta$. Therefore, there is an isomorphism of colimits $z : PC_p \to C_{p^\vee}$, which makes diagram (4.1.1) commutative. The canonical coalgebra structure of $pX^\vee \boxtimes pX^{\vee\vee}$ pulled along $pX^\vee \boxtimes \zeta_{pX}$ onto $pX^\vee \boxtimes pX$ is given by (4.1.2), (4.1.3). □

4.1.2. Definition. Let C be a squared coalgebra in $\widehat{\mathcal{V}}$. The opposite coalgebra $C_{\text{op}} = (PC, \Delta^{\text{op}}, \varepsilon^{\text{op}})$ is the unique coalgebra structure on PC such that $P\ddot{\imath}_M : M^\vee \boxtimes M \to PC$ is a squared coalgebra homomorphism for any C-comodule $M \in {}^C\mathcal{V}$, where $M^\vee \boxtimes M$ is equipped with coalgebra structure (4.1.2), (4.1.3).

To check the existence of the opposite coalgebra notice that any coalgebra has the form C_p for some $\mathcal{P} \subset {}^C\mathcal{V}$ and apply Proposition 4.1.1.

4.1.3. Remark. The duality yields an equivalence of categories $({}^C\mathcal{V})^{\text{op}} \to {}^{C_{\text{op}}}\mathcal{V}$, $(M, \delta_M) \mapsto (M^\vee, \delta'_{M^\vee})$, where δ' is given by

$$\delta'_{M^\vee} = \big(M^\vee \boxtimes \mathbb{1} \xrightarrow{M^\vee \boxtimes \text{coev}} M^\vee \boxtimes M^{\vee\vee} \otimes M^\vee \xrightarrow{M^\vee \boxtimes \zeta^{-1} \otimes M^\vee}$$
$$M^\vee \boxtimes M \otimes M^\vee \xrightarrow{P\ddot{\imath}_M \otimes M^\vee} (PC)_{12'} \otimes M^\vee_{2''}\big). \quad (4.1.4)$$

This follows by Proposition 4.1.1.

Clearly, another choice of ζ gives an isomorphic coalgebra structure in PC. There is also a coaction in ${}^\vee M$

$${}'\delta_{{}^\vee M} = \big({}^\vee M \boxtimes \mathbb{1} \xrightarrow{{}^t\zeta^{-1} \boxtimes \text{coev}} M^\vee \boxtimes M \otimes {}^\vee M \xrightarrow{P\ddot{\imath}_M \otimes {}^\vee M} (PC)_{12'} \otimes {}^\vee M_{2''}\big).$$

In fact, ${}^t\zeta^{-1} : ({}^\vee M, {}'\delta_{{}^\vee M}) \to (M^\vee, \delta'_{M^\vee})$ is an isomorphism of C_{op}-comodules. Indeed,

$$\begin{array}{ccc} {}^\vee X \boxtimes \mathbb{1} & \xrightarrow{{}^t\zeta^{-1} \boxtimes \text{coev}} & X^\vee \boxtimes X \boxtimes {}^\vee X \\ {\scriptstyle {}^t\zeta^{-1} \boxtimes \text{coev}} \downarrow & & \downarrow {\scriptstyle X^\vee \boxtimes X \otimes {}^t\zeta^{-1}} \\ X^\vee \boxtimes X^{\vee\vee} \otimes X^\vee & \xrightarrow{X^\vee \boxtimes \zeta^{-1} \otimes X^\vee} & X^\vee \boxtimes X \otimes X^\vee \end{array}$$

commutes. Therefore, $({}^C\mathcal{V})^{\text{op}} \to {}^{C_{\text{op}}}\mathcal{V}$, $(M, \delta_M) \mapsto ({}^\vee M, {}'\delta_{{}^\vee M})$ is also an equivalence.

4.1.4. Proposition. *Each functorial isomorphism $\zeta : X \to X^{\vee\vee}$ determines a functor*

$$P_\zeta : \text{Coalgsq}(\widehat{\mathcal{V}}) \to \text{Coalgsq}(\widehat{\mathcal{V}}), \qquad h \mapsto Ph,$$
$$P_\zeta(C) = C_{\text{op}} = (PC, \Delta^{\text{op}}, \varepsilon^{\text{op}}) = (PC, \Delta^\zeta, \varepsilon^\zeta).$$

All such functors are isomorphic.

PROOF. If $h : C \to D$ is a squared coalgebra homomorphism, then each C-comodule X is a D-comodule with

$$\ddot{i}_X^D = \left(X \boxtimes X^\vee \xrightarrow{\ddot{i}_X^C} C \xrightarrow{h} D\right).$$

The morphism $Ph : C_{\text{op}} \to D_{\text{op}}$ is coherent with the counit. For,

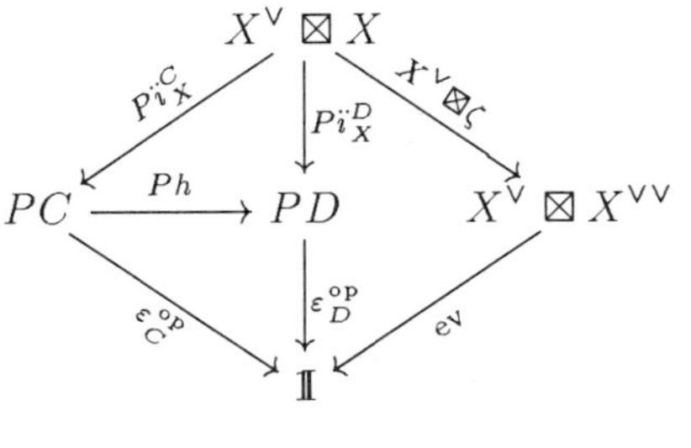

commutes.

The morphism $Ph : C_{\text{op}} \to D_{\text{op}}$ is coherent with comultiplication. Indeed, denoting

$$\text{coev}_\zeta = \left(\mathbb{1} \xrightarrow{\text{coev}} X^{\vee\vee} \otimes X^\vee \xrightarrow{\zeta^{-1} \otimes X^\vee} X \otimes X^\vee\right),$$

we get

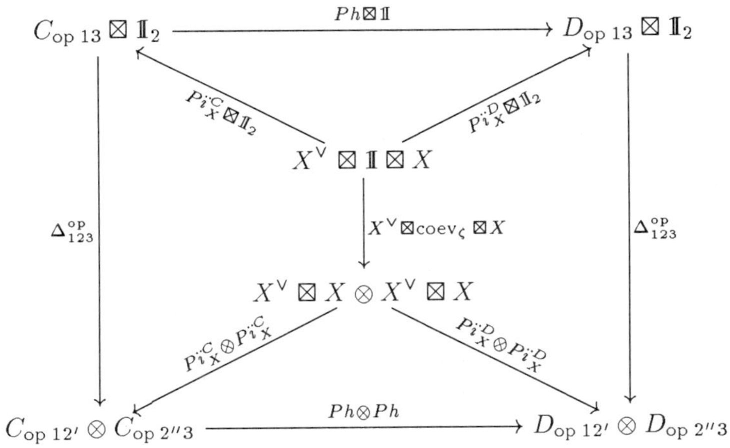

Therefore, P_ζ is a functor.

If $\zeta'_X : X \to X^{\vee\vee}$ is another functorial isomorphism, then there is a functorial isomorphism $\lambda_X : X \to X$ such that $\zeta' = \zeta \circ \lambda : X \to X^{\vee\vee}$. Clearly, $1 \boxtimes \lambda : X^\vee \boxtimes X \to X^\vee \boxtimes X$ induces an isomorphism of coalgebras $1 \boxtimes \lambda : P_\zeta C \to P_{\zeta'} C$. $\square$

4.1.5. Example. Assume that we are given a homomorphism of Hopf $\Bbbk$-algebras $f : C \to V$, which produces $F = f\text{-comod} : \mathcal{C} = C\text{-comod} \to \mathcal{V} = V\text{-comod}$. Suppose $\mathcal{V}$ is rigid-involutive, that is, there exists a functorial isomorphism $\zeta_X : X \to X^{\vee\vee}$, $X \in \mathcal{V}$. Then it is determined by an invertible element $\zeta \in V^*$ via

$$\zeta_X(x_a) = \zeta(t_{X\,a}{}^b) x_b^{\vee\vee}.$$

Here $(x_a) \subset X$, $(x^a) \subset X^\vee$, $(x_a^{\vee\vee}) \subset X^{\vee\vee}$ are dual bases. To yield a comodule morphism ζ must satisfy the condition

$$v_{(1)}\zeta(v_{(2)}) = \zeta(v_{(1)})\gamma_V^2(v_{(2)}), \quad \text{for } v \in V,$$

or $a\zeta = \zeta \cdot {}^t\gamma_V^2(a)$ for $a \in V^*$. The inverse morphism $\zeta_X^{-1} : X^{\vee\vee} \to X$ is the map

$$\zeta_X^{-1}(x_c^{\vee\vee}) = \zeta^{-1}(t_{X\,c}{}^d)x_d,$$

where $\zeta^{-1} \in V^*$ is the inverse of $\zeta \in V^*$.

Choose the functor $P : V \otimes_{\Bbbk} V\text{-Comod} \to V \otimes_{\Bbbk} V\text{-Comod}$ as follows

$$(PM, \delta^{(12)}) = \bigl(M, M \xrightarrow{\delta} V \otimes V \otimes M \xrightarrow{\tau(12)} V \otimes V \otimes M\bigr),$$

where τ is the usual action of permutations in tensor products. The symmetry $\tau(12) : M \otimes N \to N \otimes M$,

$$\begin{array}{ccc} V\text{-Comod} \times V\text{-Comod} & \xrightarrow{\boxtimes} & V \otimes_{\Bbbk} V\text{-Comod} \\ {\scriptstyle (12)}\downarrow & \overset{\tau(12)}{\swarrow\!\!\!\swarrow} & \downarrow{\scriptstyle P} \\ V\text{-Comod} \times V\text{-Comod} & \xrightarrow{\boxtimes} & V \otimes_{\Bbbk} V\text{-Comod} \end{array}$$

is an isomorphism of $V \otimes_{\Bbbk} V$-comodules

$$\bigl(M \otimes N, M \otimes N \xrightarrow{\delta\times\delta} V \otimes M \otimes V \otimes N \xrightarrow{\tau(123)} V \otimes V \otimes M \otimes N\bigr)$$
$$\text{and } \bigl(N \otimes M, N \otimes M \xrightarrow{\delta\times\delta} V \otimes N \otimes V \otimes M \xrightarrow{\tau(23)} V \otimes V \otimes N \otimes M\bigr).$$

The opposite coalgebra $H_{\mathrm{op}} \in \widehat{\mathcal{V} \boxtimes \mathcal{V}}$ to the coalgebra $H \in \widehat{\mathcal{V} \boxtimes \mathcal{V}}$ reconstructed from F is by Definition 4.1.2 the following. As a vector space $H_{\mathrm{op}} = H = C$. The left $V^{\otimes 2}$-coaction on H_{op} is

$$\delta(h) = f(\gamma(h_{(3)})) \otimes f(h_{(1)}) \otimes h_{(2)} \in V^{\otimes 2} \otimes H_{\mathrm{op}}.$$

The opposite comultiplication in H_{op} with respect to ζ is

$$\Delta_\zeta^{\mathrm{op}}(h) = h_{(3)} \otimes \zeta^{-1}(f(h_{(2)}))h_{(1)} \in H_{\mathrm{op}} \otimes H_{\mathrm{op}}.$$

It is a morphism of $V^{\otimes 3}$-comodules. The "opposite" counit is

$$\varepsilon_\zeta^{\mathrm{op}}(h) = \zeta(f(h)), \qquad \varepsilon_\zeta^{\mathrm{op}} \in V\text{-Comod}.$$

On the set of functorial isomorphisms $\zeta : X \to X^{\vee\vee}$ there is an involution $\zeta \mapsto \zeta^{-1\,t}$. Namely, $(\zeta^{-1\,t})_X = (\zeta_{{}^\vee X})^{-1\,t} : X \to X^{\vee\vee}$, where $\zeta_{{}^\vee X} : {}^\vee X \to X^\vee$. Indeed, this is an involution since

$$\bigl(X \xrightarrow{\zeta_X} X^{\vee\vee} \xrightarrow{\zeta_X^{tt}} X^{(4\vee)}\bigr) = \bigl(X \xrightarrow{\zeta_X} X^{\vee\vee} \xrightarrow{\zeta_{X^{\vee\vee}}} X^{(4\vee)}\bigr)$$

implies $\zeta_{X^{\vee\vee}} = \zeta_X^{tt} : X^{\vee\vee} \to X^{(4\vee)}$. Also $\zeta^{-1\,t} = {}^t\zeta^{-1}$. Note that fixed points of this involution, i.e. rigid categories $\mathcal{V}$ with $\zeta : X \to X^{\vee\vee}$ satisfying the equation $\zeta = \zeta^{-1\,t}$, are precisely *pivotal categories*. This equation can be rewritten as follows

$$\bigl(X^\vee \xrightarrow{\zeta_{X^\vee}} X^{\vee\vee\vee} \xrightarrow{\zeta_X^t} X^\vee\bigr) = \mathrm{id}_{X^\vee}.$$

Let functorial isomorphism $S_{\mathcal{A}\mathcal{B}} : \mathrm{Id}_{\mathcal{A}\boxtimes\mathcal{B}} \xrightarrow{\sim} P^2 : \mathcal{A} \boxtimes \mathcal{B} \to \mathcal{A} \boxtimes \mathcal{B}$ be a syllepsis of symmetric monoidal 2-category $\mathfrak{A} = \mathfrak{Ab}_{\Bbbk}^l$ (see Day and Street [6] for definition, this notion appeared first time in Breen [5]). In notations of Appendix A.6 the inverse syllepsis is

$S^{-1} = [Y]_3 : P^2 = [$ 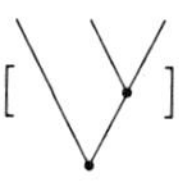 $]_2 \to \mathrm{Id} = [t^3]_2 : [$ 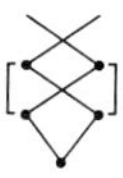 $]_1 \to \boxtimes = [t^2] : \mathfrak{A}^2 \to \mathfrak{A}.$

Denote by S also the canonical extension of $S_{\mathcal{V}\boxtimes\mathcal{V}}$ to $\widehat{\mathcal{V} \boxtimes \mathcal{V}}$.

4.1.6. Proposition. *The above S gives functorial coalgebra isomorphisms*

$$S_C : C \to P_\zeta(P_{{}^t\zeta^{-1}}C), \qquad S_C : C \to P_{{}^t\zeta^{-1}}(P_\zeta C).$$

PROOF. We have to prove that $\Delta^{\zeta, {}^t\zeta^{-1}} = \Delta$. The comultiplication Δ^ζ in $P_\zeta C$ is determined by the coalgebra homomorphisms $P\ddot{\imath}_X : X^\vee \boxtimes X \to P_\zeta C$, where X is an arbitrary C-comodule and the comultiplication in $X^\vee \boxtimes X$ is given by

$$X^\vee \boxtimes \mathbb{1} \boxtimes X \xrightarrow{X^\vee \boxtimes \mathrm{coev} \boxtimes X} X^\vee \boxtimes X^{\vee\vee} \otimes X^\vee \boxtimes X \xrightarrow{X^\vee \boxtimes \zeta^{-1} \otimes X^\vee \boxtimes X} X^\vee \boxtimes X \otimes X^\vee \boxtimes X.$$

Then $X^\vee$ is a $P_\zeta C$-comodule and by (4.1.4)

$$\ddot{\imath}_{X^\vee} = \big(X^\vee \boxtimes X^{\vee\vee} \xrightarrow{X^\vee \boxtimes \zeta^{-1}} X^\vee \boxtimes X \xrightarrow{P\ddot{\imath}_X} PC\big).$$

The coalgebra homomorphisms

$$P\ddot{\imath}_{X^\vee} = \big(X^{\vee\vee} \boxtimes X^\vee \xrightarrow{\zeta^{-1} \boxtimes X^\vee} X \boxtimes X^\vee \xrightarrow{\ddot{\imath}_X} C\big)$$

determine the comultiplication in $P_{{}^t\zeta^{-1}} P_\zeta C$. The comultiplication in $X^{\vee\vee} \boxtimes X^\vee$ is

$$X^{\vee\vee} \boxtimes \mathbb{1} \boxtimes X^\vee \xrightarrow{X^{\vee\vee} \boxtimes \mathrm{coev} \boxtimes X^\vee} X^{\vee\vee} \boxtimes X^{\vee\vee\vee} \otimes X^{\vee\vee} \boxtimes X^\vee \xrightarrow{X^{\vee\vee} \boxtimes \zeta^t \otimes X^{\vee\vee} \boxtimes X^\vee} X^{\vee\vee} \boxtimes X^\vee \otimes X^{\vee\vee} \boxtimes X^\vee.$$

Therefore, the exterior of the following diagram is commutative,

$$\begin{array}{ccccc}
X^{\vee\vee}\boxtimes\mathbb{1}\boxtimes X^\vee & \xrightarrow{1\boxtimes\mathrm{coev}\boxtimes 1} & X^{\vee\vee}\boxtimes X^{\vee\vee\vee}\otimes X^{\vee\vee}\boxtimes X^\vee & \xrightarrow{1\boxtimes\zeta^t_X\otimes 1\boxtimes 1} & X^{\vee\vee}\boxtimes X^\vee\otimes X^{\vee\vee}\boxtimes X^\vee \\
\downarrow{\scriptstyle \zeta^{-1}\boxtimes\mathbb{1}\boxtimes 1} & & \downarrow{\scriptstyle \zeta^{-1}\boxtimes 1\otimes 1\boxtimes 1} & & \downarrow{\scriptstyle \zeta^{-1}\boxtimes 1\otimes\zeta^{-1}\boxtimes 1} \\
X\boxtimes\mathbb{1}\boxtimes X^\vee & \xrightarrow{1\boxtimes\mathrm{coev}\boxtimes 1} & X\boxtimes X^{\vee\vee\vee}\otimes X^{\vee\vee}\boxtimes X^\vee & \xrightarrow{1\boxtimes\zeta^t_X\otimes\zeta^{-1}\boxtimes 1} & X\boxtimes X^\vee\otimes X\boxtimes X^\vee \\
\downarrow{\scriptstyle \ddot{\imath}_{X\,13}\boxtimes\mathbb{1}} & & & & \downarrow{\scriptstyle \ddot{\imath}_X\otimes\ddot{\imath}_X} \\
C_{13}\boxtimes\mathbb{1}_2 & & \xrightarrow{\qquad\Delta^{\zeta,{}^t\zeta^{-1}}_{123}\qquad} & & C_{12'}\otimes C_{2''3}
\end{array}$$

hence, the whole diagram is commutative. The middle row can be replaced by

$$X \boxtimes \mathrm{coev} \boxtimes X^\vee : X \boxtimes \mathbb{1} \boxtimes X^\vee \to X \boxtimes X^\vee \otimes X^{\vee\vee} \boxtimes X^\vee,$$

hence, the lower row can be replaced by Δ_{123} without breaking the commutativity of the diagram. This implies $\Delta^{\zeta, {}^t\zeta^{-1}} = \Delta$. □

4.2. Comparison with opposite coalgebra in braided case

If $\mathcal{V}$ is braided, we have the usual notion of an opposite coalgebra. The following proposition shows that the new notion of opposite coincides with traditional one at the quasiclassical level.

4.2.1. Proposition. *Let C be a squared coalgebra in $\widehat{\mathcal{V}}$, let $\zeta = u_1^2 : X \to X^{\vee\vee}$ and let $\bar{C}_{\mathrm{op}} = (\bar{C}, \bar{\Delta}^{\mathrm{op}})$ be the quasiclassical opposite to $\bar{C}$:*

$$\bar{\Delta}^{\mathrm{op}} = \big(\bar{C} \xrightarrow{\bar{\Delta}} \bar{C} \otimes \bar{C} \xrightarrow{c} \bar{C} \otimes \bar{C}\big).$$

Then the isomorphism, induced by the braiding

$$c : \bar{C}_{\mathrm{op}} = (C^{12}, \bar{\Delta}^{\mathrm{op}}, \varepsilon) \to (C^{21}, \overline{\Delta^{\mathrm{op}}}, \varepsilon^{\mathrm{op}}) = \overline{C_{\mathrm{op}}}$$

is a coalgebra isomorphism.

PROOF. It suffices to prove for $C = X \boxtimes X^\vee$. Coherence with the comultiplication is the equation

$$
\begin{aligned}
&\big(X \otimes X^\vee \xrightarrow{c} X^\vee \otimes X \xrightarrow{X^\vee \otimes \mathrm{coev} \otimes X} X^\vee \otimes X^{\vee\vee} \otimes X^\vee \otimes X \\
&\qquad \xrightarrow{X^\vee \otimes u_{-1}^{-2} \otimes X^\vee \otimes X} X^\vee \otimes X \otimes X^\vee \otimes X\big) \\
&= \big(X \otimes X^\vee \xrightarrow{X \otimes \mathrm{coev} \otimes X^\vee} (X \otimes X^\vee) \otimes (X \otimes X^\vee) \xrightarrow{c} \\
&\qquad (X \otimes X^\vee) \otimes (X \otimes X^\vee) \xrightarrow{c \otimes c} X^\vee \otimes X \otimes X^\vee \otimes X\big).
\end{aligned}
$$

It is obvious in graphical notation

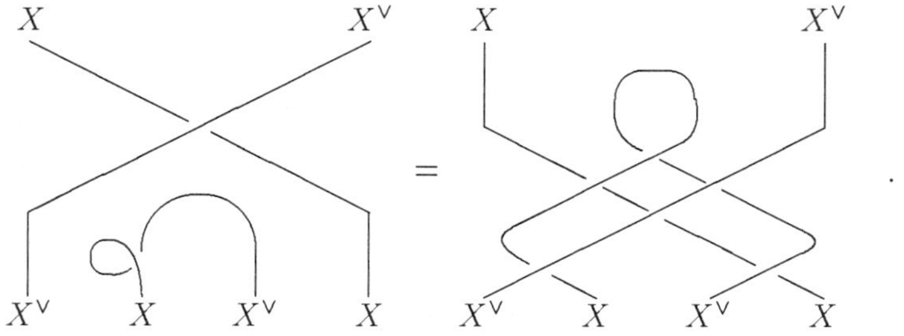

Coherence with the counit is similar. □

4.3. The antipode

Hopf coalgebras are bicoalgebras, whose categories of comodules are rigid. However, at the moment we use another definition: Hopf coalgebras are bicoalgebras with an antipode. The first definition will be made a result.

4.3.1. Definition. Let H be a bicoalgebra in $\widehat{\mathcal{V}}$. A *right antipode* in H (with respect to ζ) is a morphism $\gamma' = \gamma_\zeta : H_{\mathrm{op}} \to H \in \widehat{\mathcal{V} \boxtimes \mathcal{V}}$ such that

$$
\begin{aligned}
&\big(H_{\mathrm{op}1''1'} \boxtimes \mathbb{I}_2 \xrightarrow{\Delta^{\mathrm{op}}_{1''21'}} H_{\mathrm{op}1''2'} \otimes H_{\mathrm{op}2''1'} =\!=\!= H_{1'2''} \otimes H_{\mathrm{op}1''2'} \\
&\qquad \xrightarrow{H \otimes \gamma'_{1''2'}} H_{1'2''} \otimes H_{1''2'} \xrightarrow{m} H_{12}\big) \\
&= \big(H_{1'1''} \boxtimes \mathbb{I}_2 \xrightarrow{\varepsilon \boxtimes \mathbb{I}} \mathbb{I}_1 \boxtimes \mathbb{I}_2 \xrightarrow{\eta} H_{12}\big),
\end{aligned}
\tag{4.3.1a}
$$

$$
\begin{aligned}
&\big(\mathbb{I}_1 \boxtimes H_{2''2'} \xrightarrow{\Delta_{2''12'}} H_{2''1'} \otimes H_{1''2'} =\!=\!= H_{\mathrm{op}1'2''} \otimes H_{1''2'} \\
&\qquad \xrightarrow{\gamma'_{1'2''} \otimes H} H_{1'2''} \otimes H_{1''2'} \xrightarrow{m} H_{12}\big) \\
&= \big(\mathbb{I}_1 \boxtimes H_{\mathrm{op}2'2''} \xrightarrow{\mathbb{I} \boxtimes \varepsilon^{\mathrm{op}}} \mathbb{I}_1 \boxtimes \mathbb{I}_2 \xrightarrow{\eta} H_{12}\big).
\end{aligned}
\tag{4.3.1b}
$$

A *left antipode* in H (with respect to ζ) is a morphism ${}'\gamma = {}_\zeta\gamma : H_{\mathrm{op}} \to H \in \widehat{\mathcal{V} \boxtimes \mathcal{V}}$ such that

$$\begin{aligned} \big(H_{1''1'} \boxtimes \mathbb{1}_2 \xrightarrow{\Delta_{1''21'}} H_{1''2'} \otimes H_{2''1'} &=\!=\!= H_{\mathrm{op}1'2''} \otimes H_{1''2'} \\ &\xrightarrow{{}'\gamma_{1'2''} \otimes H} H_{1'2''} \otimes H_{1''2'} \xrightarrow{m} H_{12}\big) \end{aligned} \tag{4.3.2a}$$

$$= \big(H_{\mathrm{op}1'1''} \boxtimes \mathbb{1}_2 \xrightarrow{\varepsilon^{\mathrm{op}} \boxtimes \mathbb{1}_2} \mathbb{1}_1 \boxtimes \mathbb{1}_2 \xrightarrow{\eta} H_{12}\big),$$

$$\begin{aligned} \big(\mathbb{1}_1 \boxtimes H_{\mathrm{op}2''2'} \xrightarrow{\Delta^{\mathrm{op}}_{2''12'}} H_{\mathrm{op}2''1'} \otimes H_{\mathrm{op}1''2'} &=\!=\!= H_{1'2''} \otimes H_{\mathrm{op}1''2'} \\ &\xrightarrow{H \otimes {}'\gamma_{1''2'}} H_{1'2''} \otimes H_{1''2'} \xrightarrow{m} H_{12}\big) \end{aligned} \tag{4.3.2b}$$

$$= \big(\mathbb{1}_1 \boxtimes H_{2'2''} \xrightarrow{\mathbb{1} \boxtimes \varepsilon} \mathbb{1}_1 \boxtimes \mathbb{1}_2 \xrightarrow{\eta} H_{12}\big).$$

A (*squared*) *Hopf coalgebra* is a bicoalgebra which has a right and a left antipode.

Graphical expression of these equations is the following. Here X is an H-comodule and $\ddot{\imath}_X : X \boxtimes X^\vee \to H$ is implicit.

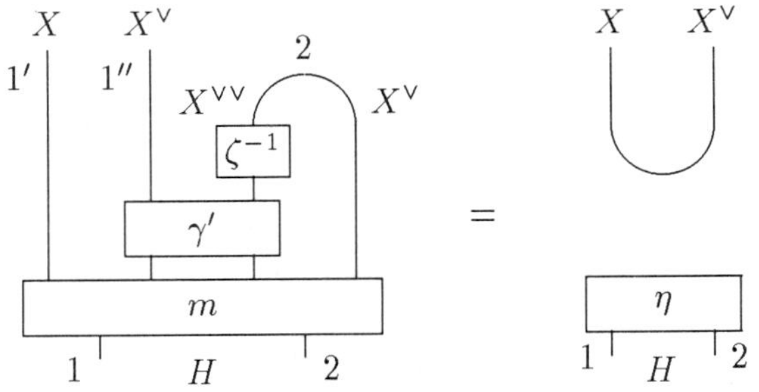

(4.3.3a)

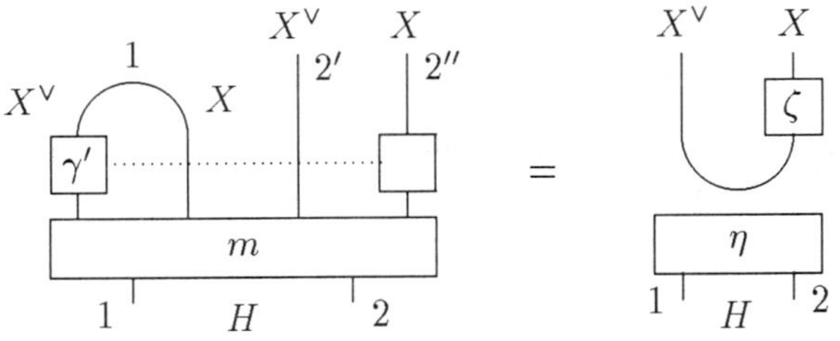

(4.3.3b)

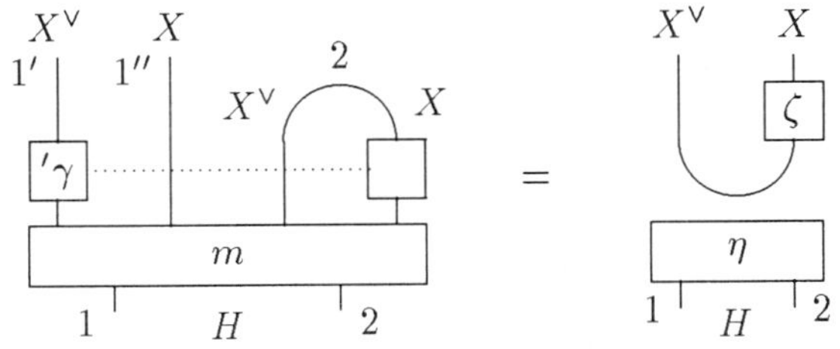

(4.3.4a)

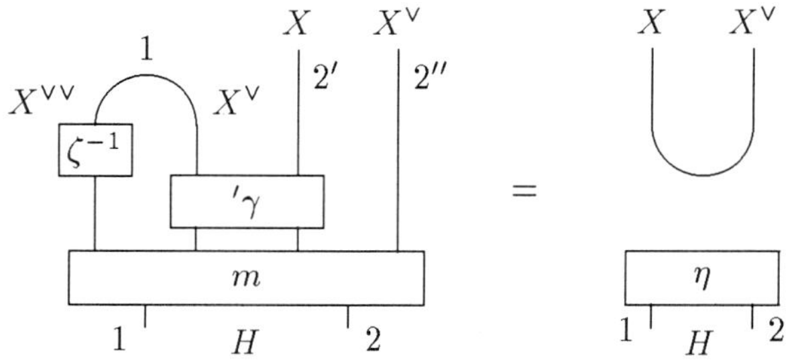

(4.3.4b)

The dual notion is called squared Hopf algebra.

4.3.2. Example. Assume that we are given a homomorphism of Hopf $\Bbbk$-algebras $f : C \to V$, which produces $F = f\text{-comod} : \mathcal{C} = C\text{-comod} \to \mathcal{V} = V\text{-comod}$. Suppose $\mathcal{V}$ is rigid-involutive, that is, there exists a functorial isomorphism $\zeta_X : X \to X^{\vee\vee}$, $X \in \mathcal{V}$. It is shown in Example 4.1.5 that it is determined by an invertible element $\zeta \in V^*$ and the opposite operations in $H_{\mathrm{op}} = C$ are

$$\Delta_\zeta^{\mathrm{op}}(h) = h_{(3)} \otimes \zeta^{-1}(f(h_{(2)}))h_{(1)} \in H_{\mathrm{op}} \otimes H_{\mathrm{op}},$$
$$\varepsilon_\zeta^{\mathrm{op}}(h) = \zeta(f(h)).$$

As will be shown in Proposition 4.6.3 rigidity implies that the bicoalgebra H has right and left antipodes, hence, it is a Hopf coalgebra. We shall verify this directly. By Definition 4.3.1 the right antipode γ' in H satisfies equations (4.3.1a), (4.3.1b):

$$(H \xrightarrow{\Delta_\zeta^{\mathrm{op}}} H \otimes H \xrightarrow{P} H \otimes H \xrightarrow{1\otimes\gamma'} H \otimes H \xrightarrow{m} H) = (H \xrightarrow{\varepsilon} \Bbbk \xrightarrow{\eta} H),$$

$$(H \xrightarrow{\Delta} H \otimes H \xrightarrow{\gamma'\otimes 1} H \otimes H \xrightarrow{m} H) = (H \xrightarrow{\varepsilon_\zeta^{\mathrm{op}}} \Bbbk \xrightarrow{\eta} H),$$

or

$$h_{(1)}\,\zeta^{-1}(f(h_{(2)}))\,\gamma'(h_{(3)}) = \varepsilon(h),$$

$$\gamma'(h_{(1)})\,h_{(2)} = \zeta(f(h))$$

for any $h \in H$. Since the ordinary antipode $\gamma : C \to C$ is the inverse to id_C with respect to convolution product, the second equation admits a unique solution

$$\gamma'(h) = \zeta(f(h_{(1)}))\,\gamma(h_{(2)}).$$

One can check directly that it satisfies also the first equation and the requirement that $\gamma' : H_{21} \to H_{12}$ is a morphism of $V \otimes_\Bbbk V$-comodules.

The left antipode $'\gamma$ in H satisfies equations (4.3.2a), (4.3.2b)

$$(H \xrightarrow{\Delta} H \otimes H \xrightarrow{P} H \otimes H \xrightarrow{'\gamma\otimes 1} H \otimes H \xrightarrow{m} H) = (H \xrightarrow{\varepsilon_\zeta^{\mathrm{op}}} \Bbbk \xrightarrow{\eta} H),$$

$$(H \xrightarrow{\Delta_\zeta^{\mathrm{op}}} H \otimes H \xrightarrow{1\otimes '\gamma} H \otimes H \xrightarrow{m} H) = (H \xrightarrow{\varepsilon} \Bbbk \xrightarrow{\eta} H),$$

or

$$'\gamma(h_{(2)})\,h_{(1)} = \zeta(f(h)),$$

$$h_{(3)}\,\zeta^{-1}(f(h_{(2)}))\,'\gamma(h_{(1)}) = \varepsilon(h).$$

The first equation has a unique solution

$$'\gamma(h) = \zeta(f(h_{(2)}))\,\gamma^{-1}(h_{(1)}).$$

One can check directly that this map satisfies also the second equation, and $'\gamma : H_{21} \to H_{12}$ is a homomorphism of $V \otimes_{\Bbbk} V$-comodules.

4.3.3. Proposition. *Let* $(H, \gamma', '\gamma)$ *be a Hopf coalgebra. Then* $\gamma', '\gamma : H_{\mathrm{op}} \to H$ *are homomorphisms of squared coalgebras.*

PROOF. Let us prove for γ', the case of $'\gamma$ being similar. The scheme of the proof is the same as for ordinary Hopf algebras. We recall it:

$$\begin{aligned} \gamma(x_{(1)})x_{(2)} = \varepsilon(x) &\implies \gamma(x_{(1)})_{(1)}x_{(2)} \otimes \gamma(x_{(1)})_{(2)}x_{(3)} = 1 \otimes \varepsilon(x) \\ &\implies \gamma(x_{(1)})_{(1)}x_{(2)} \otimes \gamma(x_{(1)})_{(2)}x_{(3)}\gamma(x_{(4)}) = 1 \otimes \gamma(x) \\ &\implies \gamma(x_{(1)})_{(1)}x_{(2)} \otimes \gamma(x_{(1)})_{(2)} = 1 \otimes \gamma(x) \\ &\implies \gamma(x_{(1)})_{(1)}x_{(2)}\gamma(x_{(3)}) \otimes \gamma(x_{(1)})_{(2)} = \gamma(x_{(2)}) \otimes \gamma(x_{(1)}) \\ &\implies \gamma(x)_{(1)} \otimes \gamma(x)_{(2)} = \gamma(x_{(2)}) \otimes \gamma(x_{(1)}). \end{aligned}$$

However, the same reasoning for squared Hopf coalgebras is much more technical and involved, so we illustrate it with graphical notations.

Equation (4.3.1b) composed with Δ gives

$$\begin{aligned} &\Big(\mathbb{1}_1 \boxtimes \mathbb{1}_2 \boxtimes H_{3''3'} \xrightarrow{\Delta_{3''13'} \boxtimes \mathbb{1}_2} H_{3''1'} \otimes H_{1''3'} \boxtimes \mathbb{1}_2 \xrightarrow{\gamma'_{1'3''} \otimes H \boxtimes \mathbb{1}} \\ &\qquad H_{1'3''} \otimes H_{1''3'} \boxtimes \mathbb{1}_2 \xrightarrow{m \boxtimes \mathbb{1}} H_{13} \boxtimes \mathbb{1}_2 \xrightarrow{\Delta_{123}} H_{12'} \otimes H_{2''3}\Big) \\ &= \Big(\mathbb{1}_1 \boxtimes \mathbb{1}_2 \boxtimes H_{\mathrm{op}3'3''} \xrightarrow{\mathbb{1} \boxtimes \mathbb{1} \boxtimes \varepsilon^{\mathrm{op}}} \mathbb{1}_1 \boxtimes \mathbb{1}_2 \boxtimes \mathbb{1}_3 \xrightarrow{\eta_{13} \boxtimes \mathbb{1}_2} \\ &\qquad H_{13} \boxtimes \mathbb{1}_2 \xrightarrow{\Delta_{123}} H_{12'} \otimes H_{2''3}\Big). \end{aligned}$$

From the definition of bicoalgebra we get

$$\begin{aligned} &\Big(\mathbb{1}_1 \boxtimes \mathbb{1}_2 \boxtimes H_{3''3'} \xrightarrow{\Delta_{3''13'} \boxtimes \mathbb{1}_2} H_{3''1'} \otimes H_{1''3'} \boxtimes \mathbb{1}_2 \xrightarrow{\gamma'_{1'3''} \otimes H \boxtimes \mathbb{1}} H_{1'3''} \otimes H_{1''3'} \boxtimes \mathbb{1}_2 \\ &\qquad \xrightarrow{H \otimes \Delta_{1''23'}} H_{1'3''} \otimes H_{1''2'} \otimes H_{2''3} \simeq H_{1'3''} \otimes H_{1''2'} \otimes \mathbb{1}_{2''} \otimes H_{2'''3'} \\ &\qquad \xrightarrow{\Delta_{1'2''3''} \otimes H \otimes H} H_{1'2''} \otimes H_{1''2'} \otimes H_{2'''3''} \otimes H_{2^43'} \xrightarrow{m \otimes m} H_{12'} \otimes H_{2''3}\Big) \\ &= \Big(\mathbb{1}_1 \boxtimes \mathbb{1}_2 \boxtimes H_{\mathrm{op}3'3''} \xrightarrow{\mathbb{1} \boxtimes r_{\mathbb{1}}^{-1} \boxtimes \varepsilon^{\mathrm{op}}} \mathbb{1}_1 \boxtimes \mathbb{1}_{2'} \otimes \mathbb{1}_{2''} \boxtimes \mathbb{1}_3 \xrightarrow{\eta \boxtimes \eta} H_{12'} \otimes H_{2''3}\Big), \end{aligned}$$

or

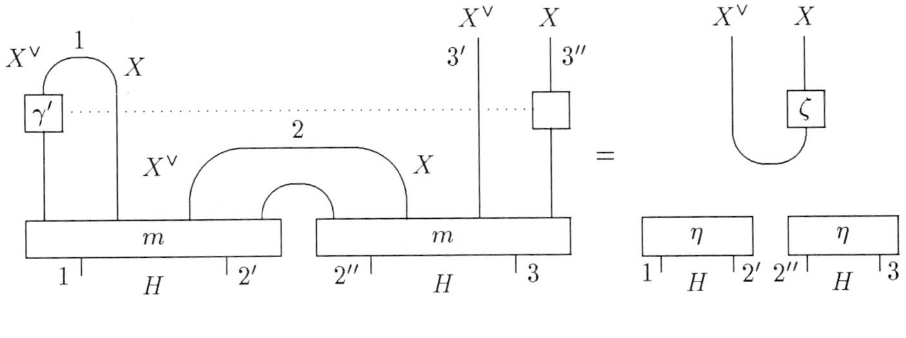

Multiplying it with an expression taken from (4.3.1a) we get

$$
\begin{aligned}
&\Big(\mathbb{I}_1 \boxtimes \mathbb{I}_{2'} \otimes H_{\mathrm{op}2''3''} \otimes \mathbb{I}_{3'} \xrightarrow{\mathbb{I}\boxtimes\mathbb{I}\boxtimes\Delta^{\mathrm{op}}_{2''3'3''}} \mathbb{I}_1 \boxtimes \mathbb{I}_{2'} \otimes H_{\mathrm{op}2''3'} \otimes H_{3'''3''} \\
&\qquad \xrightarrow{\mathbb{I}_{2'}\otimes\gamma'_{2''3'}\otimes\Delta_{3'''13''}} \mathbb{I}_{2'} \otimes H_{2''3'} \otimes H_{3'''1'} \otimes H_{1''3''} \\
&\qquad \xrightarrow{H\otimes\gamma'_{1'3'''}\otimes\Delta_{1''2'3''}} H_{2'''3'} \otimes H_{1'3'''} \otimes H_{1''2'} \otimes H_{2''3'''} \\
&\qquad \simeq H_{1'3'''} \otimes H_{1''2'} \otimes \mathbb{I}_{2''} \otimes H_{2'''3''} \otimes H_{2^43'} \xrightarrow{\Delta_{1'2''3'''}\otimes H\otimes m} \\
&\qquad H_{1'2''} \otimes H_{1''2'} \otimes H_{2'''3''} \otimes H_{2^43'} \xrightarrow{m\otimes m} H_{12'} \otimes H_{2''3}\Big) \\
&= \Big(\mathbb{I}_1 \boxtimes \mathbb{I}_{2'} \otimes H_{\mathrm{op}2''3''} \otimes \mathbb{I}_{3'} \xrightarrow{\mathbb{I}\boxtimes\mathbb{I}\otimes\Delta^{\mathrm{op}}_{2''3'3''}} \mathbb{I}_1 \boxtimes \mathbb{I}_{2'} \otimes H_{\mathrm{op}2''3'} \otimes H_{\mathrm{op}3''3'''} \\
&\qquad \xrightarrow{\mathbb{I}\boxtimes r_{\mathbb{I}}^{-1}\otimes\gamma'_{2''3'}\otimes\varepsilon^{\mathrm{op}}_{3''}} \mathbb{I}_1 \boxtimes \mathbb{I}_{2'} \otimes \mathbb{I}_{2''} \otimes H_{2'''3'} \otimes \mathbb{I}_{3''} \\
&\qquad \xrightarrow{\eta\otimes\eta\otimes H} H_{12'} \otimes H_{2''3''} \otimes H_{2'''3'} \xrightarrow{H\otimes m} H_{12'} \otimes H_{2''3}\Big),
\end{aligned}
$$

or

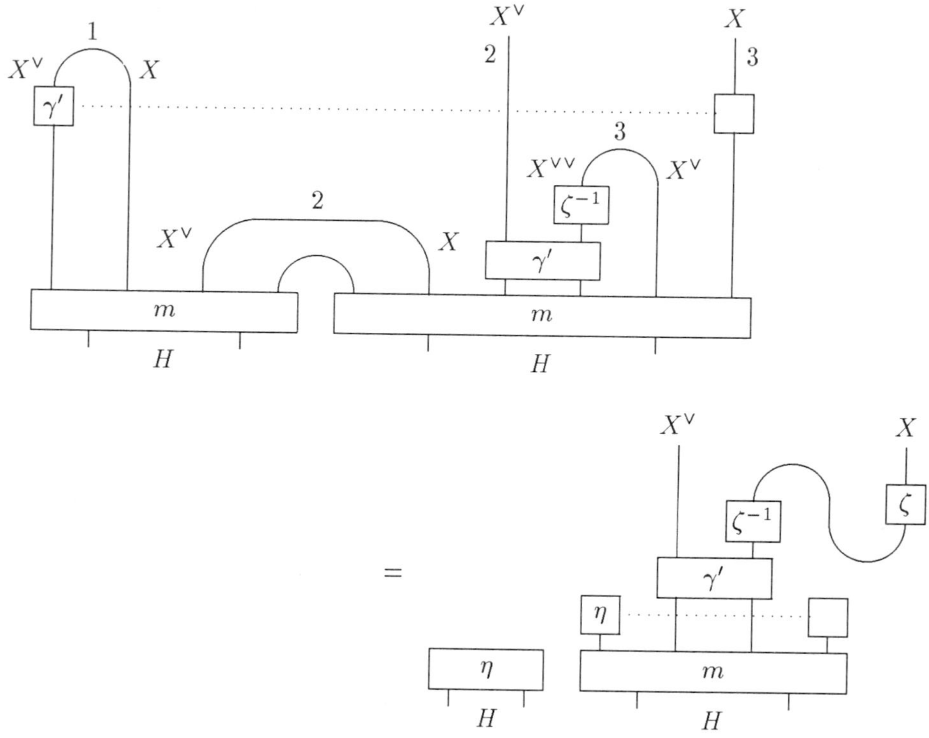

Simplifying this with the help of (4.3.1a) one gets

$$
\begin{aligned}
&\Big(\mathbb{I}_1 \boxtimes H_{32} \xrightarrow{\Delta_{312}} H_{31'} \otimes H_{1''2} \xrightarrow{\gamma'_{1'3}\otimes r_H^{-1}} H_{1'3} \otimes H_{1''2'} \otimes \mathbb{I}_{2''} \\
&\qquad \xrightarrow{\Delta_{1'2''3}\otimes H} H_{1'2''} \otimes H_{1''2'} \otimes H_{2'''3} \xrightarrow{m\otimes H} H_{12'} \otimes H_{2''3}\Big) \\
&= \Big(\mathbb{I}_1 \boxtimes H_{\mathrm{op}23} \simeq \mathbb{I}_1 \boxtimes \mathbb{I}_{2'} \otimes H_{\mathrm{op}2''3} \xrightarrow{\eta\otimes\gamma'_{2''3}} H_{12'} \otimes H_{2''3}\Big),
\end{aligned}
$$

or

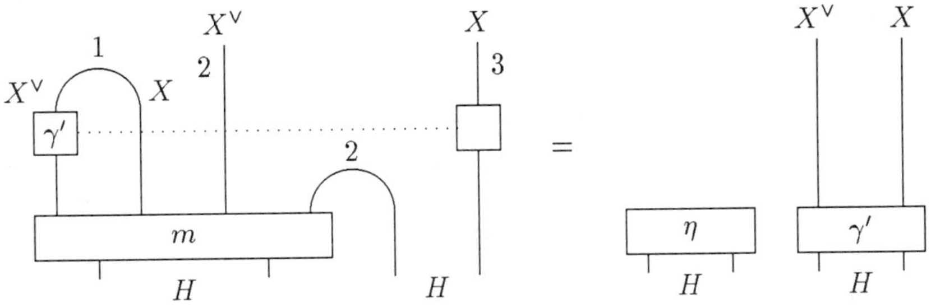

Repeating the same trick one gets

$$\begin{aligned}
&\big(H_{\mathrm{op}13} \boxtimes \mathbb{1}_2 \xrightarrow{\Delta^{\mathrm{op}}_{123}} H_{\mathrm{op}12'} \otimes H_{\mathrm{op}2''3} \simeq \mathbb{1}_{1'} \otimes H_{\mathrm{op}1''2'} \otimes H_{32''} \xrightarrow{H_{\mathrm{op}} \otimes \Delta_{31'2''}} \\
&\qquad H_{31'} \otimes H_{\mathrm{op}1'''2'} \otimes H_{1''2''} \xrightarrow{\gamma'_{1'3} \otimes \gamma'_{1'''2'} \otimes r_H^{-1}} H_{1'3} \otimes H_{1'''2'} \otimes H_{1''2''} \otimes \mathbb{1}_{2'''} \\
&\qquad \xrightarrow{\Delta_{1'2'''3} \otimes H \otimes H} H_{1'2'''} \otimes H_{1''2''} \otimes H_{1'''2'} \otimes H_{2^43} \xrightarrow{m^{(2)} \otimes H} H_{12'} \otimes H_{2''3}\big) \\
&= \big(H_{\mathrm{op}13} \boxtimes \mathbb{1}_2 \xrightarrow{\Delta^{\mathrm{op}}_{123}} H_{\mathrm{op}12'} \otimes H_{\mathrm{op}2''3} \xrightarrow{\gamma'_{12'} \otimes \gamma'_{2''3}} H_{12'} \otimes H_{2''3} \\
&\qquad \simeq \mathbb{1}_{1'} \otimes H_{1''2'} \otimes \mathbb{1}_{2''} \otimes H_{2'''3} \xrightarrow{\eta_{1'2''} \otimes H \otimes H} H_{1'2''} \otimes H_{1''2'} \otimes H_{2'''3} \\
&\qquad \xrightarrow{m \otimes H} H_{12'} \otimes H_{2''3}\big)
\end{aligned}$$

or

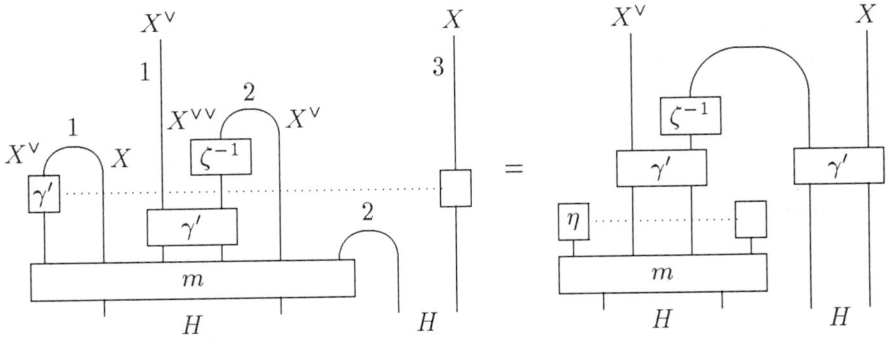

Applying (4.3.1a) one gets

$$\begin{aligned}
&\big(H_{\mathrm{op}13} \boxtimes \mathbb{1}_2 \xrightarrow{\gamma'_{13} \boxtimes \mathbb{1}} H_{13} \boxtimes \mathbb{1}_2 \xrightarrow{\Delta_{123}} H_{12'} \otimes H_{2''3}\big) \\
&\qquad = \big(H_{\mathrm{op}13} \boxtimes \mathbb{1}_2 \xrightarrow{\Delta^{\mathrm{op}}_{123}} H_{\mathrm{op}12'} \otimes H_{\mathrm{op}2''3} \xrightarrow{\gamma'_{12'} \otimes \gamma'_{2''3}} H_{12'} \otimes H_{2''3}\big),
\end{aligned}$$

or

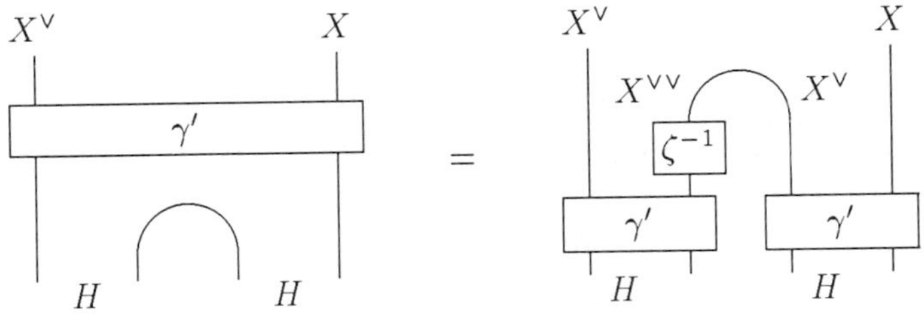

Therefore, γ' is coherent with the comultiplication.

Let us prove that $\gamma' : H_{\mathrm{op}} \to H$ preserves the counit. Applying $\circledast$ and ε to (4.3.1b) one gets

$$\big(\mathbb{I}^1 \otimes H^{32} \xrightarrow{\Delta^{312}} H^{41} \otimes H^{23} \xrightarrow{\circledast\gamma'\otimes H} H^{14} \otimes H^{23} \xrightarrow{\circledast m} H^{12} \xrightarrow{\varepsilon} \mathbb{I}^1\big)$$
$$= \big(\mathbb{I}^1 \otimes H^{23}_{\mathrm{op}} \xrightarrow{\mathbb{I}\otimes\varepsilon^{\mathrm{op}}} \mathbb{I}^1 \otimes \mathbb{I}^2 \xrightarrow{\circledast\eta} H^{12} \xrightarrow{\varepsilon} \mathbb{I}^1\big).$$

By (3.2.2) this implies

$$\big(\mathbb{I}^1 \otimes H^{32} \xrightarrow{\Delta^{312}} H^{41} \otimes H^{23} \xrightarrow{\circledast\gamma'\otimes\varepsilon} H^{13} \otimes \mathbb{I}^2 \simeq H^{12} \xrightarrow{\varepsilon} \mathbb{I}^1\big)$$
$$=\big(\mathbb{I}^1 \otimes H^{23}_{\mathrm{op}} \xrightarrow{\mathbb{I}\otimes\varepsilon^{\mathrm{op}}} \mathbb{I}^1 \otimes \mathbb{I}^2 \xrightarrow{l_{\mathbb{I}}} \mathbb{I}^1\big).$$

This can be rewritten as

$$\big(H^{21} \simeq \mathbb{I}^1 \otimes H^{32} \xrightarrow{\Delta^{312}} H^{41} \otimes H^{23} \xrightarrow{H\otimes\varepsilon} H^{31} \otimes \mathbb{I}^2 \simeq H^{21} \xrightarrow{\circledast\gamma'} H^{12} \xrightarrow{\varepsilon} \mathbb{I}^1\big)$$
$$=\big(H^{12}_{\mathrm{op}} \xrightarrow{l_H^{-1}} \mathbb{I}^1 \otimes H^{23}_{\mathrm{op}} \xrightarrow{\mathbb{I}\otimes\varepsilon^{\mathrm{op}}} \mathbb{I}^1 \otimes \mathbb{I}^2 \xrightarrow{l_{\mathbb{I}}} \mathbb{I}^1\big),$$

or as

$$\big(H^{12}_{\mathrm{op}} \xrightarrow{\circledast\gamma'} H^{12} \xrightarrow{\varepsilon} \mathbb{I}^1\big) = \big(H^{12}_{\mathrm{op}} \xrightarrow{\varepsilon^{\mathrm{op}}} \mathbb{I}^1\big).$$

Therefore, γ' is a squared coalgebra homomorphism. □

4.3.4. Theorem. *The category ${}^H\mathcal{V}$ of comodules over a Hopf coalgebra H is rigid. For the right dual of X one can take*

$$\big(X^\vee, \delta_{X^\vee} : X^\vee_1 \boxtimes \mathbb{I}_2 \xrightarrow{\delta'_{X^\vee}} H_{\mathrm{op}12'} \otimes X^\vee_{2''} \xrightarrow{\gamma'_{12'}\otimes X^\vee_{2''}} H_{12'} \otimes X^\vee_{2''}\big),$$

for the left dual –

$$\big({}^\vee X, \delta_{{}^\vee X} : {}^\vee X_1 \boxtimes \mathbb{I}_2 \xrightarrow{{}'\delta_{{}^\vee X}} H_{\mathrm{op}12'} \otimes {}^\vee X_{2''} \xrightarrow{{}'\gamma_{12'}\otimes {}^\vee X_{2''}} H_{12'} \otimes {}^\vee X_{2''}\big).$$

The evaluation and coevaluation morphisms are the same as in $\mathcal{V}$.

Proof. By Remark 4.1.3 $(X^\vee, \delta'_{X^\vee})$ and $({}^\vee X, {}'\delta_{{}^\vee X})$ are H_{op}-comodules. Therefore, $(X^\vee, \delta_{X^\vee})$ and $({}^\vee X, \delta_{{}^\vee X})$ are H-comodules by Proposition 4.3.3, where

$$\delta_{X^\vee} = \big(X^\vee \boxtimes \mathbb{I} \xrightarrow{X^\vee\boxtimes\mathrm{coev}} X^\vee \boxtimes X^{\vee\vee} \otimes X^\vee \xrightarrow{X^\vee\otimes\zeta^{-1}\otimes X^\vee} X^\vee_1 \boxtimes X_{2'} \otimes X^\vee_{2''}$$
$$\xrightarrow{\ddot{\imath}_{2'1}\otimes X^\vee} H_{\mathrm{op}12'} \otimes X^\vee_{2''} \xrightarrow{\gamma'\otimes X^\vee} H_{12'} \otimes X^\vee_{2''}\big), \quad (4.3.5)$$

$$\delta_{{}^\vee X} = \big({}^\vee X \boxtimes \mathbb{I} \xrightarrow{{}^\vee X\boxtimes\mathrm{coev}} {}^\vee X \boxtimes X \otimes {}^\vee X \xrightarrow{{}^t\zeta^{-1}\boxtimes X\otimes {}^\vee X} X^\vee_1 \boxtimes X_{2'} \otimes {}^\vee X_{2''}$$
$$\xrightarrow{\ddot{\imath}_{2'1}\otimes {}^\vee X} H_{\mathrm{op}12'} \otimes {}^\vee X_{2''} \xrightarrow{{}'\gamma\otimes {}^\vee X} H_{12'} \otimes {}^\vee X_{2''}\big). \quad (4.3.6)$$

It remains to prove that evaluation and coevaluation are homomorphisms of comodules. This is deduced from (4.3.1a)–(4.3.2b).

The evaluation $\mathrm{ev} : X \otimes X^\vee \to \mathbb{I}$ is a comodule morphism iff

$$\begin{aligned}
\Big(X \otimes X^\vee \boxtimes \mathbb{I} &\xrightarrow{X\otimes X^\vee \boxtimes \mathrm{coev}} X \otimes X^\vee \boxtimes X^{\vee\vee} \otimes X^\vee \\
&\xrightarrow{X\otimes X^\vee\boxtimes \zeta^{-1}\otimes l^{-1}_{X^\vee}} X \otimes X^\vee \boxtimes X \otimes \mathbb{I} \otimes X^\vee \\
&\xrightarrow{X\otimes P\ddot{\imath}_X \otimes \mathrm{coev} \otimes X^\vee} X_{1'} \otimes H_{\mathrm{op}1''2'} \otimes X^\vee_{2''} \otimes X_{2'''} \otimes X^\vee_{2^4} \\
&\xrightarrow{\ddot{\imath}_{X\,1'2''} \otimes \gamma_{1''2'} \otimes \mathrm{ev}} H_{1'2''} \otimes H_{1''2'} \otimes \mathbb{I}_{2'''} \xrightarrow{m\otimes \mathbb{I}} H_{12'} \otimes \mathbb{I}_{2''}\Big) \\
= \Big(X \otimes X^\vee \boxtimes \mathbb{I} &\xrightarrow{\mathrm{ev}\boxtimes\mathbb{I}} \mathbb{I} \boxtimes \mathbb{I} \xrightarrow{\eta} H_{12} \simeq H_{12'} \otimes \mathbb{I}_{2''}\Big),
\end{aligned}$$

or

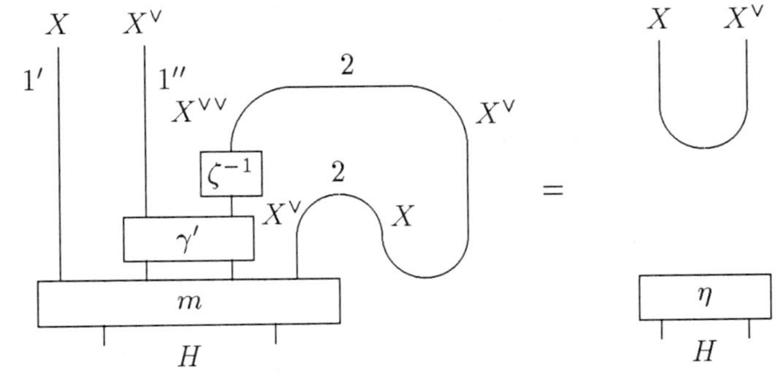

This is equivalent to

$$\begin{aligned}
\Big(X \otimes X^\vee \boxtimes \mathbb{I} &\xrightarrow{\ddot{\imath}_X \boxtimes \mathbb{I}} H_{\mathrm{op}1''1'} \boxtimes \mathbb{I}_2 \xrightarrow{\Delta^{\mathrm{op}}_{1''21'}} H_{\mathrm{op}1''2'} \otimes H_{\mathrm{op}2''1'} \\
&\xrightarrow{\gamma'_{1''2'}\otimes H} H_{1''2'} \otimes H_{1'2''} \xrightarrow{m} H_{12}\Big) \\
= \Big(X \otimes X^\vee \boxtimes \mathbb{I} &\xrightarrow{\ddot{\imath}_X \boxtimes \mathbb{I}} H_{1'1''} \boxtimes \mathbb{I}_2 \xrightarrow{\varepsilon\otimes\mathbb{I}} \mathbb{I}_1 \boxtimes \mathbb{I}_2 \xrightarrow{\eta} H_{12}\Big),
\end{aligned}$$

or (4.3.3a). Thus, it is equivalent to (4.3.1a).

The coevaluation $\mathrm{coev} : \mathbb{I} \to X^\vee \otimes X$ is a comodule morphism iff

$$\begin{aligned}
\Big(\mathbb{I}_1 \boxtimes \mathbb{I}_2 &\xrightarrow{\mathrm{coev}\otimes\mathrm{coev}} X^\vee \otimes X \boxtimes X^\vee \otimes X \xrightarrow{X^\vee \otimes \ddot{\imath}_X \otimes l^{-1}_X} X^\vee_{1'} \otimes H_{1''2'} \otimes \mathbb{I}_{2''} \otimes X_{2'''} \\
&\xrightarrow{X^\vee \otimes H \otimes \mathrm{coev} \otimes X} X^\vee_{1'} \otimes H_{1''2'} \otimes X^{\vee\vee}_{2''} \otimes X^\vee_{2'''} \otimes X_{2^4} \xrightarrow{X^\vee \otimes H \otimes \zeta^{-1} \otimes X^\vee \otimes X} \\
&X^\vee_{1'} \otimes H_{1''2'} \otimes X_{2''} \otimes X^\vee_{2'''} \otimes X_{2^4} \xrightarrow{\ddot{\imath}_{X\,2''1'} \otimes H \otimes X^\vee \otimes X} \\
&H_{\mathrm{op}1'2''} \otimes H_{1''2'} \otimes X^\vee_{2'''} \otimes X_{2^4} \xrightarrow{\gamma'_{1'2''} \otimes H \otimes X^\vee \otimes X} \\
&H_{1'2''} \otimes H_{1''2'} \otimes X^\vee_{2'''} \otimes X_{2^4} \xrightarrow{m \otimes X^\vee \otimes X} H_{12'} \otimes X^\vee_{2''} \otimes X_{2'''}\Big) \\
= \Big(\mathbb{I}_1 \boxtimes \mathbb{I}_2 &\xrightarrow{\eta} H_{12} \simeq H_{12'} \otimes \mathbb{I}_{2''} \xrightarrow{H\otimes \mathrm{coev}} H_{12'} \otimes X^\vee_{2''} \otimes X_{2'''}\Big),
\end{aligned}$$

or

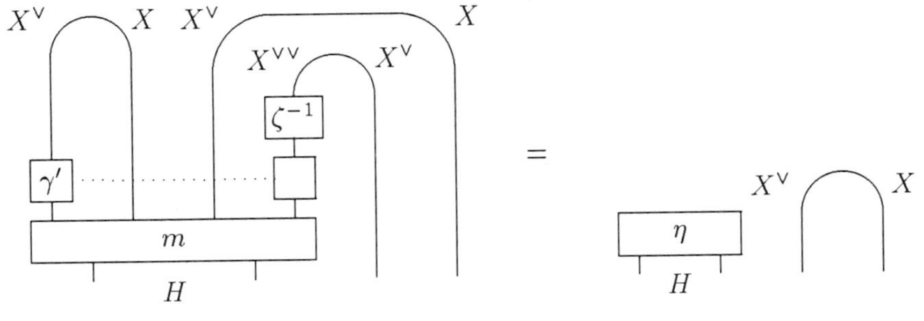

This is equivalent to

$$
\begin{aligned}
&\big(\mathbb{1} \boxtimes X^\vee \otimes X^{\vee\vee} \xrightarrow{\operatorname{coev} \otimes X^\vee \otimes \zeta^{-1}} X^\vee_{1'} \otimes X_{1''} \boxtimes X^\vee_{2'} \otimes X_{2''} \\
&\qquad \xrightarrow{\ddot{\imath}_{2''1'} \otimes \ddot{\imath}_{1''2'}} H_{\mathrm{op}1'2''} \otimes H_{1''2'} \xrightarrow{\gamma'_{1'2''} \otimes H} H_{1'2''} \otimes H_{1''2'} \xrightarrow{m} H_{12}\big) \\
&= \big(\mathbb{1} \boxtimes X^\vee \otimes X^{\vee\vee} \xrightarrow{\mathbb{1} \boxtimes \operatorname{ev}} \mathbb{1} \boxtimes \mathbb{1} \xrightarrow{\eta} H_{12}\big),
\end{aligned}
$$

or (4.3.3b).

Thus, it can be rewritten as

$$
\begin{aligned}
&\big(\mathbb{1}_1 \boxtimes X^\vee_{2'} \otimes X_{2''} \xrightarrow{\mathbb{1} \boxtimes \ddot{\imath}_{X\,2''2'}} \mathbb{1}_1 \boxtimes H_{2''2'} \xrightarrow{\Delta_{2''12'}} H_{2''1'} \otimes H_{1''2'} \\
&\qquad \xrightarrow{\gamma'_{1'2''} \otimes H} H_{1'2''} \otimes H_{1''2'} \xrightarrow{m} H_{12}\big) \\
&= \big(\mathbb{1}_1 \boxtimes X^\vee_{2'} \otimes X_{2''} \xrightarrow{\mathbb{1} \boxtimes \ddot{\imath}_{X\,2''2'}} \mathbb{1}_1 \boxtimes H_{\mathrm{op}2'2''} \xrightarrow{\mathbb{1} \boxtimes \varepsilon^{\mathrm{op}}} \mathbb{1}_1 \boxtimes \mathbb{1}_2 \xrightarrow{\eta} H_{12}\big),
\end{aligned}
$$

and it is equivalent to (4.3.1b).

The evaluation $\operatorname{ev} : {}^\vee X \otimes X \to \mathbb{1}$ is a comodule morphism iff

$$
\begin{aligned}
&\big({}^\vee X \otimes X \boxtimes \mathbb{1} \xrightarrow{{}^\vee X \otimes X \boxtimes \operatorname{coev}} {}^\vee X \otimes X \boxtimes X^\vee \otimes X \xrightarrow{{}^\vee X \otimes \ddot{\imath}_X \otimes l_X^{-1}} \\
&\qquad {}^\vee X_{1'} \otimes H_{1''2'} \otimes \mathbb{1}_{2''} \otimes X_{2'''} \xrightarrow{{}^t\zeta^{-1} \otimes H \otimes \operatorname{coev} \otimes X} \\
&\qquad X^\vee_{1'} \otimes H_{1''2'} \otimes X_{2''} \otimes {}^\vee X_{2'''} \otimes X_{2^4} \xrightarrow{\ddot{\imath}_{X\,2''1'} \otimes H \otimes \operatorname{ev}} \\
&\qquad H_{\mathrm{op}1'2''} \otimes H_{1''2'} \otimes \mathbb{1} \xrightarrow{{}'\gamma_{1'2''} \otimes r_H} H_{1'2''} \otimes H_{1''2'} \xrightarrow{m} H_{12}\big) \\
&= \big({}^\vee X \otimes X \boxtimes \mathbb{1} \xrightarrow{\operatorname{ev} \boxtimes \mathbb{1}} \mathbb{1}_1 \boxtimes \mathbb{1}_2 \xrightarrow{\eta} H_{12}\big),
\end{aligned}
$$

or

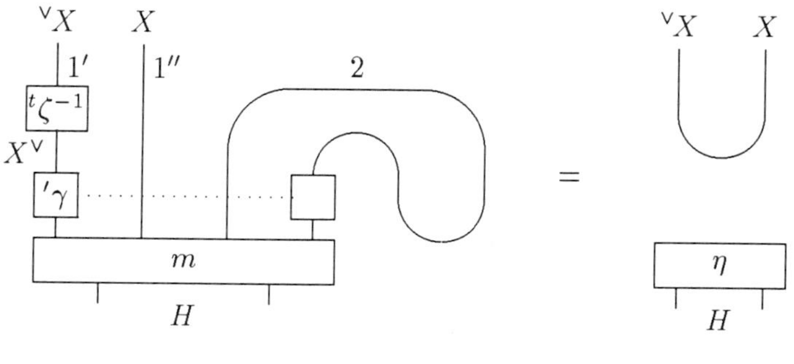

This is equivalent to

$$
\begin{aligned}
&\big(X^\vee_{1'} \otimes X_{1''} \boxtimes \mathbb{1}_2 \xrightarrow{\ddot{\imath}_{X\,1''1'} \boxtimes \mathbb{1}} H_{1''1'} \boxtimes \mathbb{1}_2 \xrightarrow{\Delta_{1''21'}} H_{1''2'} \otimes H_{2''1'} \\
&\qquad \xrightarrow{H \otimes {}'\gamma_{1'2''}} H_{1''2'} \otimes H_{1'2''} \xrightarrow{m} H_{12}\big) \\
&= \big(X^\vee_{1'} \otimes X_{1''} \boxtimes \mathbb{1}_2 \xrightarrow{\ddot{\imath}_{X\,1''1'} \boxtimes \mathbb{1}} H_{\mathrm{op}1'1''} \boxtimes \mathbb{1}_2 \xrightarrow{\varepsilon^{\mathrm{op}} \boxtimes \mathbb{1}} \mathbb{1}_1 \boxtimes \mathbb{1}_2 \xrightarrow{\eta} H_{12}\big)
\end{aligned}
$$

or to (4.3.4a). Thus, it is equivalent to (4.3.2a).

The coevaluation $\mathrm{coev} : \mathbb{I} \to X \otimes {}^\vee X$ is a comodule morphism iff

$$
\begin{aligned}
\big(\mathbb{I} \boxtimes \mathbb{I} &\xrightarrow{\mathrm{coev}\otimes\mathrm{coev}} X \otimes {}^\vee X \boxtimes X \otimes {}^\vee X \xrightarrow{X \otimes {}^t\zeta^{-1} \boxtimes r_X^{-1} \otimes {}^\vee X} \\
& X_{1'} \otimes X^\vee_{1''} \boxtimes X_{2'} \otimes \mathbb{I}_{2''} \otimes {}^\vee X_{2'''} \xrightarrow{X \otimes \ddot{\imath}_{X\,2'1''} \otimes \mathrm{coev} \otimes {}^\vee X} \\
& X_{1'} \otimes H_{\mathrm{op}1''2'} \otimes X^\vee_{2''} \otimes X_{2'''} \otimes {}^\vee X_{2^4} \xrightarrow{\ddot{\imath}_{X\,1'2''} \otimes {}'\gamma_{1''2'} \otimes X \otimes {}^\vee X} \\
& H_{1'2''} \otimes H_{1''2'} \otimes X_{2'''} \otimes {}^\vee X_{2^4} \xrightarrow{m \otimes X \otimes {}^\vee X} H_{12'} \otimes X_{2''} \otimes {}^\vee X_{2'''}\big) \\
= \big(\mathbb{I}_1 \boxtimes \mathbb{I}_2 &\xrightarrow{\eta} H_{12} \simeq H_{12'} \otimes \mathbb{I}_{2''} \xrightarrow{H \otimes \mathrm{coev}} H_{12'} \otimes X_{2''} \otimes {}^\vee X_{2'''}\big),
\end{aligned}
$$

or

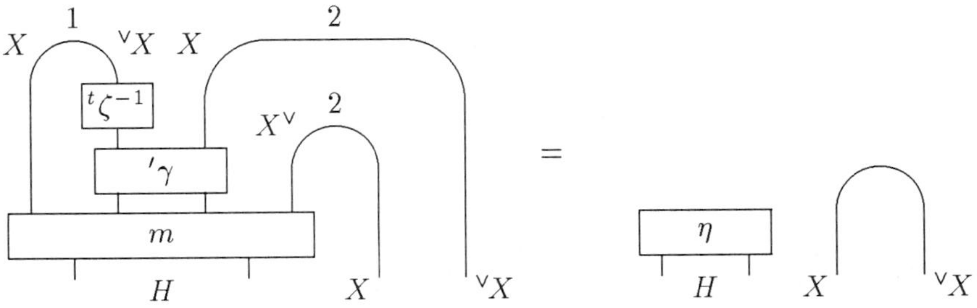

This is equivalent to

$$
\begin{aligned}
\big(\mathbb{I} \boxtimes X \otimes X^\vee &\xrightarrow{\mathrm{coev} \boxtimes X \otimes X^\vee} X^{\vee\vee}_{1'} \otimes X^\vee_{1''} \boxtimes X_{2'} \otimes X^\vee_{2''} \xrightarrow{\zeta^{-1} \otimes \ddot{\imath}_{X\,2'1''} \otimes X^\vee} \\
& X_{1'} \otimes H_{\mathrm{op}1''2'} \otimes X^\vee_{2''} \xrightarrow{\ddot{\imath}_{X\,1'2''} \otimes {}'\gamma_{1''2'}} H_{1'2''} \otimes H_{1''2'} \xrightarrow{m} H_{12}\big) \\
= \big(\mathbb{I} \boxtimes X \otimes X^\vee &\xrightarrow{\mathbb{I} \boxtimes \mathrm{ev}} \mathbb{I} \boxtimes \mathbb{I} \xrightarrow{\eta} H_{12}\big),
\end{aligned}
$$

or (4.3.4b). Thus, it can be rewritten as

$$
\begin{aligned}
\big(\mathbb{I}_1 \boxtimes X_{2'} \otimes X^\vee_{2''} &\xrightarrow{\mathbb{I} \boxtimes \ddot{\imath}_X} \mathbb{I}_1 \boxtimes H_{\mathrm{op}2''2'} \xrightarrow{\Delta^{\mathrm{op}}_{2''12'}} H_{\mathrm{op}2''1'} \otimes H_{\mathrm{op}1''2'} \\
& \xrightarrow{H \otimes {}'\gamma} H_{1'2''} \otimes H_{1''2'} \xrightarrow{m} H_{12}\big) \\
= \big(\mathbb{I}_1 \boxtimes X_{2'} \otimes X^\vee_{2''} &\xrightarrow{\mathbb{I} \boxtimes \ddot{\imath}_X} \mathbb{I}_1 \boxtimes H_{2'2''} \xrightarrow{\mathbb{I} \boxtimes \varepsilon} \mathbb{I}_1 \boxtimes \mathbb{I}_2 \xrightarrow{\eta} H_{12}\big),
\end{aligned}
$$

and it is equivalent to (4.3.2b). □

The formulae (4.3.5), (4.3.6) together with definition (2.6.4) of $\ddot{\imath}_X$ imply

$$
\begin{array}{ccc}
X_1^\vee \boxtimes X_2 & \xrightarrow{X^\vee \boxtimes \zeta} & X_1^\vee \boxtimes X_2^{\vee\vee} \\
{\scriptstyle P\ddot{\imath}_X}\downarrow & & \downarrow{\scriptstyle \ddot{\imath}_{X^\vee}} \\
H_{\mathrm{op}12} & \xrightarrow{\gamma'} & H_{12}
\end{array}, \qquad (4.3.7)
$$

$$
\begin{array}{ccc}
X_1^\vee \boxtimes X_2 & \xrightarrow{{}^t\zeta \boxtimes X} & {}^\vee X_1 \boxtimes X_2 \\
{\scriptstyle P\ddot{\imath}_X}\downarrow & & \downarrow{\scriptstyle \ddot{\imath}_{{}^\vee X}} \\
H_{\mathrm{op}12} & \xrightarrow{{}'\gamma} & H_{12}
\end{array}. \qquad (4.3.8)
$$

4.3.5. Corollary. *The right and the left antipodes $\gamma', {}'\gamma : H_{\mathrm{op}} \to H$ of a Hopf coalgebra are unique and invertible.*

Uniqueness follows from the fundamental theorem on coalgebras (Corollary 2.6.11) and diagrams (4.3.7), (4.3.8). Invertibility follows from the fact that $X \mapsto X^\vee$, $X \mapsto {}^\vee X$ are equivalences. If only right antipode exists for H, it need not be invertible.

4.3.6. Remark. If $\tilde\zeta_X : X \to X^{\vee\vee}$ is another functorial isomorphism, there exists a functorial automorphism $\lambda : X \to X$ such that $\tilde\zeta = \zeta \circ \lambda : X \to X^{\vee\vee}$. Let a bialgebra H have right and left antipodes $\gamma_\zeta, {}_\zeta\gamma : H_{\text{op}} \to H$ with respect to ζ. Then H has also right and left antipodes $\gamma_{\tilde\zeta}, {}_{\tilde\zeta}\gamma : H_{\text{op}} \to H$ with respect to $\tilde\zeta$. Namely, there is a functorial automorphism $1 \boxtimes \lambda : \mathrm{Id}_{\widehat{\mathcal{V}\boxtimes\mathcal{V}}} \to \mathrm{Id}_{\widehat{\mathcal{V}\boxtimes\mathcal{V}}}$ and

$$\gamma_{\tilde\zeta} = \bigl(H_{\text{op}} \xrightarrow{1\boxtimes\lambda} H_{\text{op}} \xrightarrow{\gamma_\zeta} H\bigr),$$
$${}_{\tilde\zeta}\gamma = \bigl(H_{\text{op}} \xrightarrow{1\boxtimes\lambda} H_{\text{op}} \xrightarrow{{}_\zeta\gamma} H\bigr).$$

The proof is left to the reader. Therefore, the property of being a Hopf coalgebra does not depend on the choice of ζ.

4.3.7. Proposition. *Let H be a bicoalgebra in $\widehat{\mathcal{V}}$.*

(a) *If H has a right antipode $\gamma_\zeta : H_{\text{op}} \to H$ with respect to ζ and it is invertible, then ${}_{{}^t\zeta^{-1}}\gamma = P\gamma_\zeta^{-1}$ is a left antipode for H with respect to ${}^t\zeta^{-1}$.*

(b) *If H has a left antipode ${}_\zeta\gamma : H_{\text{op}} \to H$ with respect to ζ and it is invertible, then $\gamma_{{}^t\zeta^{-1}} = P{}_\zeta\gamma^{-1}$ is a right antipode for H with respect to ${}^t\zeta^{-1}$.*

In both cases H is a Hopf coalgebra.

Proof. (a) Equation (4.3.1a) implies that

$$\begin{aligned}\bigl(H_{1''1'} \boxtimes \mathbb{1}_2 &\xrightarrow{\circledast\gamma_\zeta^{-1}\boxtimes\mathbb{1}} H_{\text{op}1''1'} \boxtimes \mathbb{1}_2 \xrightarrow{\Delta^\zeta_{1''21'}} H_{\text{op}1''2'} \otimes H_{\text{op}2''1'} \\ &\xrightarrow{\gamma_\zeta\otimes H} H_{1''2'} \otimes H_{1'2''} \xrightarrow{m} H_{12}\bigr) \qquad (4.3.9)\\ = \bigl(H_{1''1'} \boxtimes \mathbb{1}_2 &\xrightarrow{\circledast\gamma_\zeta^{-1}\boxtimes\mathbb{1}} H_{\text{op}1''1'} \boxtimes \mathbb{1}_2 \xrightarrow{\varepsilon\boxtimes\mathbb{1}} \mathbb{1}_1 \boxtimes \mathbb{1}_2 \xrightarrow{\eta} H_{12}\bigr).\end{aligned}$$

Diagram (4.3.7) implies the diagram

$$\begin{array}{ccccc} X_{1'}^{\vee\vee} \otimes X_{1''}^{\vee} & \xrightarrow{\zeta^{-1}\otimes X^\vee} & X_{1'} \otimes X_{1''}^{\vee} & \xrightarrow{\text{ev}} & \mathbb{1} \\ {\scriptstyle \circledast P\ddot\imath_{X^\vee}}\downarrow & & \downarrow{\scriptstyle\circledast\ddot\imath_X} & & \| \\ H_{1''1'} & \xrightarrow{\circledast\gamma_\zeta^{-1}} & H_{\text{op}1''1'} & \xrightarrow{\varepsilon} & \mathbb{1} \end{array} .$$

The upper row can be replaced by

$$X^{\vee\vee} \boxtimes X^\vee \xrightarrow{X^{\vee\vee}\boxtimes\zeta_X^{-1\,t}} X^{\vee\vee} \boxtimes X^{\vee\vee\vee} \xrightarrow{\text{ev}} \mathbb{1},$$

hence the lower row can be replaced by

$$\varepsilon^{\zeta^{-1\,t}} : (PH)_{1'1''} \to \mathbb{1}.$$

Plugging this into (4.3.9) and noticing that γ_ζ^{-1} is a coalgebra morphism we get

$$\begin{aligned}(H_{1''1'} \boxtimes \mathbb{I}_2 &\xrightarrow{\Delta_{1''21'}} H_{1''2'} \otimes H_{2''1'} \xrightarrow{\gamma_\zeta^{-1}\otimes\gamma_\zeta^{-1}} H_{\mathrm{op}1''2'} \otimes H_{\mathrm{op}2''1'} \\ &\xrightarrow{\gamma_\zeta\otimes H} H_{1''2'} \otimes H_{1'2''} \xrightarrow{m} H_{12}) \\ = (H_{\mathrm{op}1'1''} \boxtimes \mathbb{I}_2 &\xrightarrow{\varepsilon^{\zeta^{-1\,t}}\boxtimes\mathbb{I}} \mathbb{I}_1 \boxtimes \mathbb{I}_2 \xrightarrow{\eta} H_{12}).\end{aligned}$$

This is equivalent to equation (4.3.2a) for $'\gamma = P\gamma_\zeta^{-1}$ with respect to $\zeta^{-1\,t}$.

Equation (4.3.1b) implies that

$$\begin{aligned}(\mathbb{I}_1 \boxtimes H_{\mathrm{op}2''2'} &\xrightarrow{\mathbb{I}\boxtimes P\gamma_\zeta^{-1}} \mathbb{I}_1 \boxtimes H_{2''2'} \xrightarrow{\Delta_{2''12'}} H_{2''1'} \otimes H_{1''2'} \\ &== H_{\mathrm{op}1'2''} \otimes H_{1''2'} \xrightarrow{\gamma_{\zeta 1'2''}\otimes H} H_{1'2''} \otimes H_{1''2'} \xrightarrow{m} H_{12}) \qquad (4.3.10) \\ = (\mathbb{I}_1 \boxtimes H_{\mathrm{op}2''2'} &\xrightarrow{\mathbb{I}\boxtimes P\gamma_\zeta^{-1}} \mathbb{I}_1 \boxtimes H_{2''2'} \xrightarrow{\mathbb{I}\boxtimes\varepsilon^\zeta} \mathbb{I}_1 \boxtimes \mathbb{I}_2 \xrightarrow{\eta} H_{12}).\end{aligned}$$

Diagram (4.3.7) implies commutativity of

$$\begin{array}{ccccc} X^\vee \otimes X^{\vee\vee} & \xrightarrow{X^\vee\otimes\zeta^{-1}} & X^\vee \otimes X & \xrightarrow{X^\vee\otimes\zeta} & X^\vee \otimes X^{\vee\vee} \\ {\scriptstyle \circledast\ddot{\imath}_{X^\vee}}\downarrow & & \downarrow{\scriptstyle \circledast P\ddot{\imath}_X} & & \downarrow{\scriptstyle \mathrm{ev}} \\ H^{21}_{\mathrm{op}} & \xrightarrow{\circledast P\gamma_\zeta^{-1}} & H^{21} & \xrightarrow{\varepsilon^\zeta} & \mathbb{I} \end{array} .$$

The upper path equals ev : $X^\vee \otimes X^{\vee\vee} \to \mathbb{I}$, hence, the lower row can be replaced by $\varepsilon : H^{12} \to \mathbb{I}$. By Proposition 4.3.3 $\gamma_\zeta^{-1} : (H, \Delta, \varepsilon) \to (PH, \Delta^\zeta, \varepsilon^\zeta)$ is a squared coalgebra homomorphism. Applying to it the functor $P_{^t\zeta^{-1}}$ (see Proposition 4.1.4), we get that

$$P\gamma_\zeta^{-1} : (PH, \Delta^{^t\zeta^{-1}}, \varepsilon^{^t\zeta^{-1}}) \to (H, \Delta, \varepsilon)$$

is a coalgebra morphism by Proposition 4.1.6.

Thus we can transform (4.3.10) to

$$\begin{aligned}(\mathbb{I}_1 \boxtimes H_{\mathrm{op}2''2'} &\xrightarrow{\Delta^{^t\zeta^{-1}}_{2''12'}} H_{\mathrm{op}2''1'} \otimes H_{\mathrm{op}1''2'} \xrightarrow{P\gamma_\zeta^{-1}\otimes P\gamma_\zeta^{-1}} H_{2''1'} \otimes H_{1''2'} \\ &\xrightarrow{P\gamma_\zeta\otimes H} H_{\mathrm{op}2''1'} \otimes H_{1''2'} == H_{1'2''} \otimes H_{1''2'} \xrightarrow{m} H_{12}) \\ = (\mathbb{I}_1 \boxtimes H_{2'2''} &\xrightarrow{\mathbb{I}\boxtimes\varepsilon} \mathbb{I}_1 \boxtimes \mathbb{I}_2 \xrightarrow{\eta} H_{12}).\end{aligned}$$

This equivalent to equation (4.3.2b) for $'\gamma = P\gamma_\zeta^{-1}$ with respect to $\zeta^{-1\,t}$.

Therefore, $P\gamma_\zeta^{-1}$ is the left antipode with respect to $\zeta^{-1\,t}$. By Remark 4.3.6 H is a Hopf coalgebra.

(b) is similar. □

4.4. Comparison with braided Hopf algebras

We recall that if $\mathcal{V}$ is braided, the tensor functor $(\circledast, \phi, r_{\mathbb{I}}^{-1}) : \mathcal{V} \boxtimes \mathcal{V} \to \mathcal{V}$ from Proposition 3.5.1 transforms squared bicoalgebras B into quasiclassical bialgebras $\bar{B}$ (see Proposition 3.5.2).

4.4.1. Proposition. *Let H be a squared Hopf algebra in $\widehat{\mathcal{V}}$. Then $\bar{H}$ is a quasi-classical Hopf algebra in $\widehat{\mathcal{V}}$. If γ' is the right antipode for H with respect to $\zeta = u_1^2$, then*

$$\gamma_{\bar{H}} = \big(H^{12} \xrightarrow{c} H^{21} \xrightarrow{\circledast\gamma'} H^{12}\big)$$

is the antipode for $\bar{H}$. If ${}'\gamma$ is the left antipode for H with respect to $\tilde{\zeta} = u_{-1}^2$, then

$$\tilde{\gamma}_{\bar{H}} = \gamma_{\bar{H}}^{-1} = \big(H^{12} \xrightarrow{c^{-1}} H^{21} \xrightarrow{\circledast'\gamma} H^{12}\big)$$

is the skew antipode for $\bar{H}$.

PROOF. For $\zeta = u_1^2$ equation (4.3.3a) projected to $\widehat{\mathcal{V}}$ by $\circledast$ takes the form

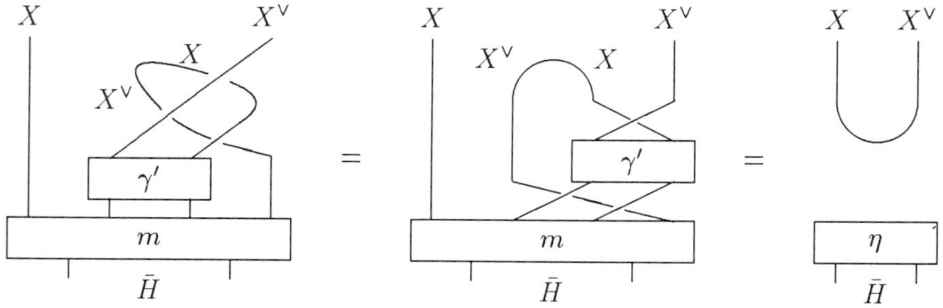

Hence,

$$\begin{aligned}
&\big(\bar{H} \xrightarrow{\bar{\Delta}} \bar{H} \otimes \bar{H} \xrightarrow{\bar{H}\otimes\gamma_{\bar{H}}} \bar{H} \otimes \bar{H} \xrightarrow{\bar{m}} \bar{H}\big) \\
&= \big(H^{12} \xrightarrow{\bar{\Delta}} H^{12} \otimes H^{34} \xrightarrow{H\otimes c} H^{12} \otimes H^{43} \xrightarrow{H\otimes\circledast\gamma'} H^{12} \otimes H^{34} \\
&\qquad \xrightarrow{\phi} H^{14} \otimes H^{23} \xrightarrow{\circledast m} H^{12}\big) \\
&= \big(H^{12} \xrightarrow{\varepsilon} \mathbb{1} \xrightarrow{\bar{\eta}} H^{12}\big).
\end{aligned}$$

When we project equation (4.3.3b) for $\zeta = u_1^2$ to $\widehat{\mathcal{V}}$, we have to assemble together both halves of γ' in (4.3.3b) along the dotted line. Choose arbitrarily the (over- or under-) crossing signs in the intersection points of the dotted line with the solid lines, and make the picture into a tangled one pushing the second half of γ' along the dotted line towards the first half. The dotted line shrinks leaving a solid line instead. This recipe would certainly work if we have $\sum_i f_i \boxtimes g_i$ instead of γ' or any other general morphism. However, the general case reduces to the special one with the help of resolutions (1.6.2).

For a particular choice of crossings we get from (4.3.3b)

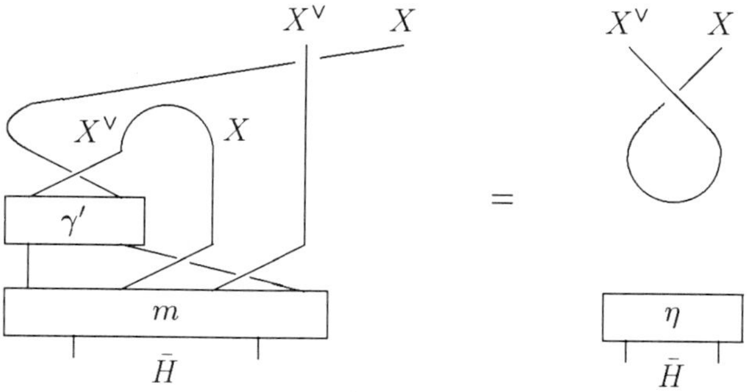

Multiplying this with $c : H^{12} \to H^{21}$ we may write it as

$$\begin{aligned}
&(\bar{H} \xrightarrow{\bar{\Delta}} \bar{H} \otimes \bar{H} \xrightarrow{\gamma_{\bar{H}} \otimes \bar{H}} \bar{H} \otimes \bar{H} \xrightarrow{\bar{m}} \bar{H}) \\
&= \big(H^{12} \xrightarrow{\bar{\Delta}} H^{12} \otimes H^{34} \xrightarrow{c \otimes H} H^{21} \otimes H^{34} \xrightarrow{\circledast\gamma' \otimes H} H^{12} \otimes H^{34} \\
&\qquad\qquad \xrightarrow{\phi} H^{14} \otimes H^{23} \xrightarrow{\circledast m} H^{12}\big) \\
&= \big(H^{12} \xrightarrow{\varepsilon} \mathbb{1} \xrightarrow{\bar{\eta}} H^{12}\big).
\end{aligned}$$

Therefore, $\gamma_{\bar{H}}$ is the antipode of $\bar{H}$.

Similarly one can prove for $\tilde{\gamma}_{\bar{H}}$. However, there is a shorter way. We already know that $\gamma_{\bar{H}}$ is invertible, thus, $\tilde{\gamma}_{\bar{H}} = \gamma_{\bar{H}}^{-1}$ is the skew antipode. Since $u^2_{-1} = (u^2_1)^{-1\,t}$, we have $'\gamma = P\gamma_\zeta^{-1}$ for $\zeta = u_1^2$ by Proposition 4.3.7. Hence,

$$\begin{aligned}
&\big(H^{12} \xrightarrow{c^{-1}} H^{21} \xrightarrow{P\gamma_\zeta^{-1}} H^{12}\big) \\
&= \big(H^{12} \xrightarrow{P\gamma_\zeta} H^{21} \xrightarrow{c} H^{12}\big)^{-1} \\
&= \big(H^{12} \xrightarrow{c} H^{21} \xrightarrow{\gamma_\zeta} H^{12}\big)^{-1} = \gamma_{\bar{H}}^{-1} = \tilde{\gamma}_{\bar{H}}. \qquad \square
\end{aligned}$$

4.5. Multiplication in the opposite bialgebra

The monoidal functor $(\text{-}^\vee, j_+, d) : (\mathcal{C}^{\text{op}}, \otimes, \mathbb{1}) \to (\mathcal{C}, \otimes_{\text{op}}, \mathbb{1})$, $X \mapsto X^\vee$, $f \mapsto f^t$, $j_+ : Y^\vee \otimes X^\vee \to (X \otimes Y)^\vee$, $d : \mathbb{1} \to \mathbb{1}^\vee$, is an equivalence. Here j_+ and d are uniquely determined by rigid monoidal structure in Section 1.2.6. The square of this functor, $(\text{-}^{\vee\vee}, j_{+2}, d_2) : (\mathcal{C}, \otimes, \mathbb{1}) \to (\mathcal{C}, \otimes, \mathbb{1})$, $X \mapsto X^{\vee\vee}$, $f \mapsto f^{tt}$

$$j_{+2} = \big(X^{\vee\vee} \otimes Y^{\vee\vee} \xrightarrow{j_+} (Y^\vee \otimes X^\vee)^\vee \xrightarrow{j_+^{-1\,t}} (X \otimes Y)^{\vee\vee}\big),$$

$$d_2 = \big(\mathbb{1} \xrightarrow{d} \mathbb{1}^\vee \xrightarrow{d^{-1\,t}} \mathbb{1}^{\vee\vee}\big),$$

is a monoidal self-equivalence of $\mathcal{C}$.

4.5.1. Definition. Let us call a functorial isomorphism $\zeta_X : X \to X^{\vee\vee}$ *normalised* if

$$\zeta_{\mathbb{1}} = d_2 = \big(\mathbb{1} \xrightarrow{d} \mathbb{1}^\vee \xrightarrow{d^{-1\,t}} \mathbb{1}^{\vee\vee}\big).$$

For any functorial isomorphism $\zeta : X \to X^{\vee\vee}$ there is a unique constant $\lambda \in \Bbbk$ such that $\lambda\zeta : X \to X^{\vee\vee}$ is normalised.

Take any normalised $\zeta_X : X \to X^{\vee\vee}$. Since the functor $\text{Id}_{\mathcal{C}}$ is isomorphic to $\text{-}^{\vee\vee}$ by ζ, it can be given a monoidal structure isomorphic to $(\text{-}^{\vee\vee}, j_{+2}, d_2)$. Namely,

$$\zeta : (\text{Id}, \psi, \text{id}_{\mathbb{1}}) \to (\text{-}^{\vee\vee}, j_{+2}, d_2)$$

is an isomorphism of monoidal functors, where

$$\psi_{X,Y} = \big(X \otimes Y \xrightarrow{\zeta \otimes \zeta} X^{\vee\vee} \otimes Y^{\vee\vee} \xrightarrow{j_{+2}} (X \otimes Y)^{\vee\vee} \xrightarrow{\zeta^{-1}} X \otimes Y\big).$$

If ζ were not normalised $\mathbb{1} \xrightarrow{d_2} \mathbb{1}^{\vee\vee} \xrightarrow{\zeta_{\mathbb{1}}^{-1}} \mathbb{1}$ would appear instead of $\text{id}_{\mathbb{1}}$.

We would like to introduce an algebra structure in H_{op} so that $\gamma_\zeta : H_{\text{op}} \to H$ were a bicoalgebra isomorphism. Choose the functors $\text{-}^\vee$, ${}^\vee\text{-}$ and a monoidally closed set $\mathcal{O} \subset \text{Ob}\,{}^H\mathcal{V}$ such that each H-comodule is isomorphic to an object from $\mathcal{O}$. Let $\mathcal{C}$ be the full subcategory of ${}^H\mathcal{V}$ with $\text{Ob}\,\mathcal{C} = \mathcal{O}$, and let $p = \mathcal{U} : \mathcal{P} = \overline{\mathcal{C}} \to \overline{\mathcal{V}}$. We may choose H as the bicoalgebra reconstructed from p by Theorem 3.3.19. The bicoalgebra $H_{p^\vee}$ is reconstructed from the premonoidal prefunctor $p^\vee : \mathcal{P}^{\text{op}} \to \overline{\mathcal{V}}$

associated to the monoidal functor $\flat = \left(\mathcal{C}^{\mathrm{op}} \xrightarrow{\mathcal{U}} \mathcal{V}^{\mathrm{op}} \xrightarrow{(\text{-}^\vee, j_+, d)} \mathcal{V}\right)$. As shown in Proposition 4.1.1, there is a coalgebra isomorphism $z : H^{\zeta}_{\mathrm{op}} \to H_{p^\vee}$. The functor $(\text{-}^\vee, j_+, d) : (\mathcal{C}^{\mathrm{op}}, \otimes, \mathbb{1}) \to (\mathcal{C}, \otimes_{\mathrm{op}}, \mathbb{1})$ and the identity

$$\flat = \left(\mathcal{C}^{\mathrm{op}} \xrightarrow{(\text{-}^\vee, j_+, d)} \mathcal{C} \xrightarrow{\mathcal{U}} \mathcal{V}\right)$$

determine an isomorphism $p^\vee \to p$ of monoidal construction data, thus, an isomorphism $\kappa : H_{p^\vee} \to H$ of bicoalgebras. If $X \in {}^H\mathcal{V}$, the comodule structure of $X^\vee$ can be found from the following diagram

$$\begin{array}{ccccccc}
X^\vee \boxtimes \mathbb{1} & \xrightarrow{X^\vee \boxtimes \mathrm{coev}} & X^\vee \boxtimes X^{\vee\vee} \otimes X^\vee & \xrightarrow{i_{X^\vee} \otimes X^\vee} & H_{p^\vee 12'} \otimes X^\vee_{2''} & & \\
 & & \Big\downarrow {\scriptstyle X^\vee \boxtimes \zeta^{-1} \otimes X^\vee} & & \Big\uparrow {\scriptstyle z \otimes X^\vee} & \searrow^{\kappa \otimes X^\vee} & \\
 & & X^\vee \boxtimes X \otimes X^\vee & \xrightarrow[P\ddot{\imath}_X \otimes X^\vee]{} & H_{\mathrm{op}12'} \otimes X^\vee_{2''} & \xrightarrow[\gamma_\zeta \otimes X^\vee]{} & H_{12'} \otimes X^\vee_{2''}
\end{array}$$

In particular, $\gamma_\zeta = \kappa \circ z$. Thus, if H_{op} is given such multiplication that z is an algebra isomorphism, then H_{op} is a bialgebra and z and γ_ζ are bialgebra isomorphisms.

4.5.2. Proposition. *Let H be a Hopf coalgebra, and let $\zeta : X \to X^{\vee\vee}$ be a normalised functorial isomorphism. Then the coalgebra H_{op} has a (unique) algebra structure*

$$\begin{aligned}
m^{\mathrm{op}} &= \Big(PH_{1'2''} \otimes PH_{1''2'} \xrightarrow{1_1 \boxtimes \psi_{2'2''}} PH_{1'2''} \otimes PH_{1''2'} \\
&\qquad = H_{2''1'} \otimes H_{2'1''} \xrightarrow{m_{21}} H_{21} = PH_{12}\Big), \\
\eta^{\mathrm{op}} &= P\eta : \mathbb{1} \boxtimes \mathbb{1} \to PH,
\end{aligned}$$

making it into a bicoalgebra, such that $z : H_{\mathrm{op}} \to H_{p^\vee}$ and $\gamma_\zeta : H_{\mathrm{op}} \to H$ are bicoalgebra isomorphisms.

PROOF. Consider the cubic diagram (C) on the facing page. The top commutes by definition of ψ. Two side walls commute by diagram (4.1.1). The front wall commutes by definition of j_{+2} and multiplication in $H_{p^\vee}$ (see Remark 3.1.2). The back wall commutes because

$$\begin{array}{ccccc}
X^\vee \otimes Y^\vee \boxtimes Y \otimes X & \xrightarrow{1 \boxtimes \psi_{Y,X}} & X^\vee \otimes Y^\vee \boxtimes Y \otimes X & \xrightarrow{j_+ \boxtimes 1} & (Y \otimes X)^\vee \boxtimes Y \otimes X \\
\Big\downarrow {\scriptstyle P\ddot{\imath}_X \otimes P\ddot{\imath}_Y} & & \Big\downarrow {\scriptstyle P\ddot{\imath}_X \otimes P\ddot{\imath}_Y} & & \Big\downarrow {\scriptstyle P\ddot{\imath}_{Y \otimes X}} \\
H_{\mathrm{op}1'2''} \otimes H_{\mathrm{op}1''2'} & \xrightarrow{1_1 \boxtimes \psi_{2'2''}} & H_{\mathrm{op}1'2''} \otimes H_{\mathrm{op}1''2'} & \xrightarrow{Pm} & H_{\mathrm{op}12}
\end{array}$$

commutes (see Remark 3.1.2) and the lower row here equals to m^{op}. Therefore, the bottom of diagram (C) commutes, and z is coherent with the multiplication.

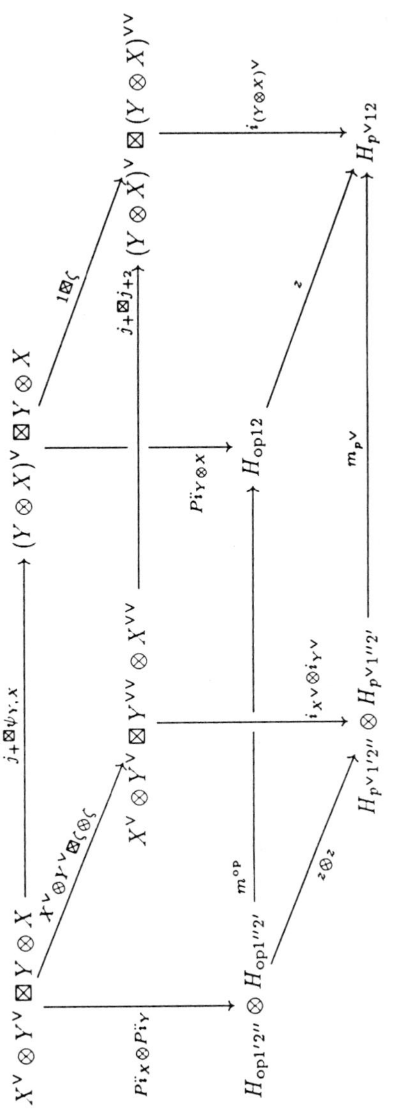

FIGURE 4.1. Diagram C

Also, as the following diagram shows, z is coherent with the unit.

$$\begin{array}{c}
\xrightarrow{P\eta} \\
\mathbb{1} \boxtimes \mathbb{1} \xrightarrow{d \boxtimes \mathbb{1}} \mathbb{1}^\vee \boxtimes \mathbb{1} = \mathbb{1}^\vee \boxtimes \mathbb{1} \xrightarrow{P\ddot{i}_\mathbb{1}} H_{\mathrm{op}} \\
\downarrow{\mathbb{1} \boxtimes d} \quad \downarrow{\mathbb{1}^\vee \boxtimes d} \quad \downarrow{\mathbb{1}^\vee \boxtimes \zeta} \quad \downarrow{z} \\
\mathbb{1} \boxtimes \mathbb{1}^\vee \xrightarrow{d \boxtimes \mathbb{1}^\vee} \mathbb{1}^\vee \boxtimes \mathbb{1}^\vee \xrightarrow{\mathbb{1}^\vee \boxtimes d^{-1\,t}} \mathbb{1}^\vee \boxtimes \mathbb{1}^{\vee\vee} \xrightarrow{i_{\mathbb{1}^\vee}} H_{p^\vee} \\
\xrightarrow{i_\mathbb{1}}
\end{array}$$

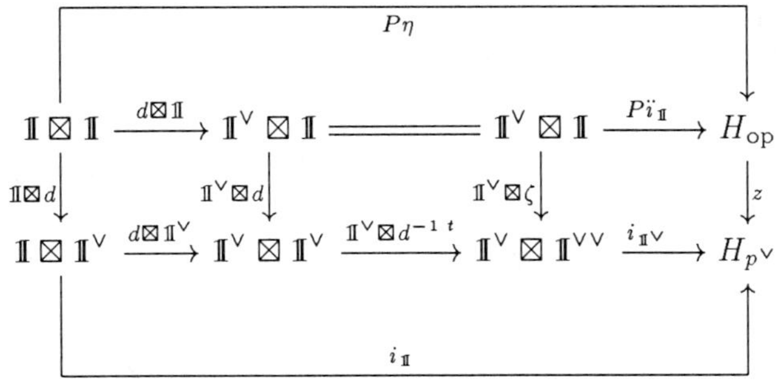

Therefore, $z : H_{\mathrm{op}} \to H$ is an isomorphism of algebras. Since it is a coalgebra isomorphism as well, the proposition is proved. □

4.5.3. Example. If $\mathcal{V}$ is braided and $\zeta = u_1^2$ we have

$$\psi_{X,Y} = c^{-2} : X \otimes Y \to X \otimes Y.$$

4.5.4. Lemma. *Let C, D be squared coalgebras. Then*

$$\tau = \big((C_{\mathrm{op}}\bar{\otimes} D_{\mathrm{op}})_{12} = C_{\mathrm{op}1'2''} \otimes D_{\mathrm{op}1''2'} \xrightarrow{1_1\boxtimes\psi_{2'2''}} C_{2''1'} \otimes D_{2'1''} = (D\bar{\otimes} C)_{21} = (D\bar{\otimes} C)_{\mathrm{op}12}\big)$$

is an isomorphism of squared coalgebras.

Proof. We have to prove that

$$\begin{aligned}
&\big(C_{\mathrm{op}1'3''} \otimes D_{\mathrm{op}1''3'} \boxtimes \mathbb{1}_2 \xrightarrow{C\otimes\Delta_D^{\mathrm{op}}} C_{\mathrm{op}1'3''} \otimes D_{\mathrm{op}1''2'} \otimes \mathbb{1}_{2''} \otimes D_{2'''3'} \\
&\xrightarrow{\Delta_C^{\mathrm{op}}\otimes D\otimes D} C_{\mathrm{op}1'2''} \otimes D_{\mathrm{op}1''2'} \otimes C_{\mathrm{op}2'''3''} \otimes D_{\mathrm{op}2^4 3'} \\
&\xrightarrow{\psi_{2'2''}\boxtimes\psi_{3'3''}} C_{2''1'} \otimes D_{2'1''} \otimes C_{3''2'''} \otimes D_{3'2^4}\big) \qquad (4.5.1)\\
&= \big(C_{3''1'} \otimes D_{3'1''} \boxtimes \mathbb{1}_2 \xrightarrow{1\boxtimes 1\boxtimes\psi_{3'3''}} (D\bar{\otimes}C)_{31} \boxtimes \mathbb{1}_2 \xrightarrow{\Delta^{\mathrm{op}}_{D\bar{\otimes}C}} \\
&(D\bar{\otimes}C)_{2'1} \otimes (D\bar{\otimes}C)_{32''} = D_{2'1''} \otimes C_{2''1'} \otimes D_{3'2^4} \otimes C_{3''2'''}\big).
\end{aligned}$$

It suffices to consider $C = X \boxtimes X^\vee$, $D = Y \boxtimes Y^\vee$. In graphical notation

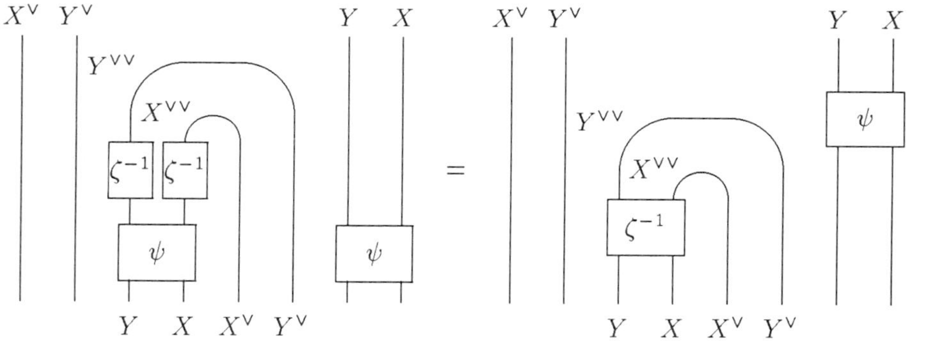

which follows from the definition of ψ. □

4.5.5. Proposition. *Let B be a bicoalgebra and let ζ be normalised. Then $B^{\mathrm{op}}_{\mathrm{op}} = (PB, \Delta^{\mathrm{op}}, \varepsilon^{\mathrm{op}}, m^{\mathrm{op}}, \eta^{\mathrm{op}})$,*

$$\begin{aligned}
m^{\mathrm{op}} &= \big(PB_{1'2''} \otimes PB_{1''2'} \xrightarrow{1_1\boxtimes\psi_{2'2''}} B_{2''1'} \otimes B_{2'1''} \xrightarrow{m_{21}} PB_{12}\big),\\
\eta^{\mathrm{op}} &= P\eta : \mathbb{1} \boxtimes \mathbb{1} \to PB,
\end{aligned}$$

is a bicoalgebra. It is called the opposite bicoalgebra *(with respect to ζ).*

Proof. Since $(\mathrm{Id}, \psi, \mathrm{id})$ is monoidal, the algebra $(PB, m^{\mathrm{op}}, \eta^{\mathrm{op}})$ is associative and unital.

To deduce the main bicoalgebra property (3.2.1), set $C = D = B$ in the previous lemma. Then

$$
\begin{aligned}
&\big(B_{\mathrm{op}1'3''} \otimes B_{\mathrm{op}1''3'} \boxtimes \mathbb{1}_2 \xrightarrow{\Delta_{B_{\mathrm{op}}\bar{\otimes} B_{\mathrm{op}}}} B_{\mathrm{op}1'2''} \otimes B_{\mathrm{op}1''2'} \otimes B_{\mathrm{op}2'''3''} \otimes B_{\mathrm{op}2^43'} \\
&\qquad \xrightarrow{1_1 \boxtimes \psi_{2'2''} \boxtimes \psi_{3'3''}} B_{2''1'} \otimes B_{2'1''} \otimes B_{3''2'''} \otimes B_{3'2^4} \xrightarrow{m_{2'1} \otimes m_{32''}} B_{2'1} \otimes B_{32''}\big) \\
&= \big(B_{3''1'} \otimes B_{3'1''} \boxtimes \mathbb{1}_2 \xrightarrow{1 \boxtimes 1 \boxtimes \psi_{3'3''}} (B \bar{\otimes} B)_{31} \boxtimes \mathbb{1}_2 \xrightarrow{\Delta^{\mathrm{op}}_{B\bar{\otimes}B}} (B\bar{\otimes}B)_{2'1} \otimes (B\bar{\otimes}B)_{32''} \\
&\qquad = B_{2'1''} \otimes B_{2''1'} \otimes B_{3'2^4} \otimes B_{3''2'''} \xrightarrow{m_{2'1} \otimes m_{32''}} B_{2'1} \otimes B_{32''}\big) \\
&= \big(B_{3''1'} \otimes B_{3'1''} \boxtimes \mathbb{1}_2 \xrightarrow{1 \boxtimes 1 \boxtimes \psi_{3'3''}} B_{3''1'} \otimes B_{3'1''} \boxtimes \mathbb{1}_2 \\
&\qquad \xrightarrow{m_{31} \boxtimes \mathbb{1}} B_{31} \boxtimes \mathbb{1}_2 \xrightarrow{\Delta^{\mathrm{op}}_{123}} B_{\mathrm{op}12'} \otimes B_{\mathrm{op}2''3}\big).
\end{aligned}
$$

The last equation expresses the fact that $Pm : (B\bar{\otimes}B)_{\mathrm{op}} \to B_{\mathrm{op}}$ is a coalgebra morphism (by Proposition 4.1.4). Thence, m^{op} is a coalgebra morphism.

Diagram (3.2.2) takes the form

$$
\begin{array}{ccccc}
B^{14}_{\mathrm{op}} \otimes B^{23}_{\mathrm{op}} & \xrightarrow{B \otimes \varepsilon^{\mathrm{op}}} & B^{13}_{\mathrm{op}} \otimes \mathbb{1}^2 & \xrightarrow{\sim} & B^{12}_{\mathrm{op}} \\
{\scriptstyle \psi^{34}} \downarrow & & & & \downarrow {\scriptstyle \varepsilon^{\mathrm{op}}} \\
B^{41} \otimes B^{32} & \xrightarrow{m^{21}} & B^{21} & \xrightarrow{\varepsilon^{\mathrm{op}}} & \mathbb{1}
\end{array}
$$

To prove it we consider two B-comodules $X, Y \in {}^B\mathcal{V}$, and rewrite graphically the diagram as

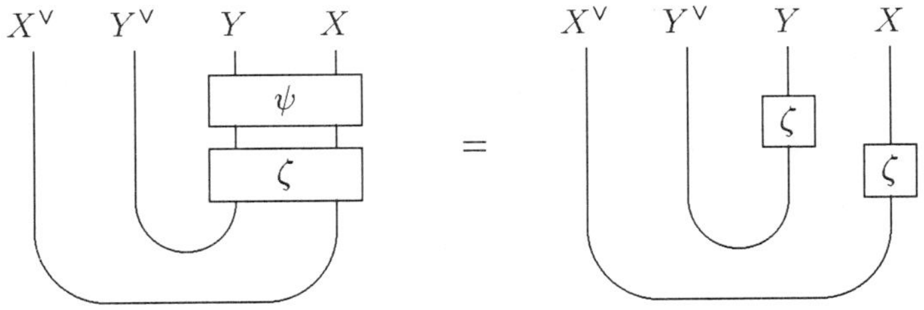

(4.5.2)

The remaining check is left to the reader. □

4.5.6. Proposition. *Any normalised functorial isomorphism $\zeta : X \to X^{\vee\vee}$ determines a functor*

$$P_\zeta : \mathrm{Bicoalg}(\widehat{\mathcal{V}}) \to \mathrm{Bicoalg}(\widehat{\mathcal{V}}), \quad B \mapsto B^{\mathrm{op}}_{\mathrm{op}}, \quad h \mapsto Ph.$$

All such functors are isomorphic.

PROOF. $1_1 \boxtimes \psi_{2'2''} : \bar{\otimes} \to \bar{\otimes}$ is a functorial isomorphism. Hence, Proposition 4.1.4 extends to bicoalgebras. All ψ are isomorphic to j_{+2} by ζ, therefore, all P_ζ are isomorphic. □

4.5.7. Lemma. *Let C be a squared coalgebra, let B be a bicoalgebra, and let $\gamma : C_{\mathrm{op}} \to B$ be a coalgebra morphism. Then there is a unique algebra morphism $\gamma' : T(C)^{\mathrm{op}}_{\mathrm{op}} \to B$ whose restriction to PC is γ. Moreover, γ' is a bicoalgebra homomorphism.*

PROOF. By Example 3.2.4 there is a unique bicoalgebra morphism $T(C_{\mathrm{op}}) \to B$ extending γ. It remains to notice that the coalgebra embedding $C_{\mathrm{op}} \hookrightarrow T(C)^{\mathrm{op}}_{\mathrm{op}}$ induces a bicoalgebra isomorphism $T(C_{\mathrm{op}}) \to T(C)^{\mathrm{op}}_{\mathrm{op}}$ by the same Example 3.2.4.
□

4.5.8. Lemma. *Let C be a squared coalgebra and let B be a bicoalgebra. Let $\gamma : C_{\mathrm{op}} \to B$, $\varsigma : C \to B$ be coalgebra homomorphisms such that*

$$\big(C_{\mathrm{op}1''1'} \boxtimes \mathbb{1}_2 \xrightarrow{\Delta^{\mathrm{op}}_{1''21'}} C_{\mathrm{op}1''2'} \otimes C_{\mathrm{op}2''1'} \xrightarrow{\gamma_{1''2'}\otimes\varsigma_{1'2''}} B_{1''2'} \otimes B_{1'2''} \xrightarrow{m} B_{12}\big)$$
$$= \big(C_{1'1''} \otimes \mathbb{1}_2 \xrightarrow{\varepsilon\boxtimes\mathbb{1}} \mathbb{1}_1 \boxtimes \mathbb{1}_2 \xrightarrow{\eta} B_{12}\big), \quad (4.5.3)$$

$$\big(\mathbb{1}_1 \boxtimes C_{2''2'} \xrightarrow{\Delta_{2''12'}} C_{2''1'} \otimes C_{1''2'} \xrightarrow{\gamma_{1'2''}\otimes\varsigma_{1''2'}} B_{1'2''} \otimes B_{1''2'} \xrightarrow{m} B_{12}\big)$$
$$= \big(\mathbb{1}_1 \boxtimes C_{\mathrm{op}2'2''} \xrightarrow{\mathbb{1}\boxtimes\varepsilon^{\mathrm{op}}} \mathbb{1}_1 \boxtimes \mathbb{1}_2 \xrightarrow{\eta} B_{12}\big). \quad (4.5.4)$$

Then the unique bicoalgebra morphisms $\gamma' : T(C)^{\mathrm{op}}_{\mathrm{op}} \to B$, $\varsigma' : T(C) \to B$ which extend γ, ς satisfy

$$\big(T(C)_{\mathrm{op}1''1'} \boxtimes \mathbb{1}_2 \xrightarrow{\Delta^{\mathrm{op}}_{1''21'}} T(C)_{\mathrm{op}1''2'} \otimes T(C)_{\mathrm{op}2''1'}$$
$$\xrightarrow{\gamma'_{1''2'}\otimes\varsigma'_{1'2''}} B_{1''2'} \otimes B_{1'2''} \xrightarrow{m} B_{12}\big) \quad (4.5.5)$$
$$= \big(T(C)_{1'1''} \otimes \mathbb{1}_2 \xrightarrow{\varepsilon\boxtimes\mathbb{1}} \mathbb{1}_1 \boxtimes \mathbb{1}_2 \xrightarrow{\eta} B_{12}\big),$$

$$\big(\mathbb{1}_1 \boxtimes T(C)_{2''2'} \xrightarrow{\Delta_{2''12'}} T(C)_{2''1'} \otimes T(C)_{1''2'} \xrightarrow{\gamma'_{1'2''}\otimes\varsigma'_{1''2'}} B_{1'2''} \otimes B_{1''2'} \xrightarrow{m} B_{12}\big)$$
$$= \big(\mathbb{1}_1 \boxtimes T(C)_{\mathrm{op}2'2''} \xrightarrow{\mathbb{1}\boxtimes\varepsilon^{\mathrm{op}}} \mathbb{1}_1 \boxtimes \mathbb{1}_2 \xrightarrow{\eta} B_{12}\big). \quad (4.5.6)$$

PROOF. The bicoalgebra morphisms γ' ς' exist by Lemma 4.5.7 and Example 3.2.4. We shall prove the result by induction. Assuming that (4.5.5), (4.5.6) hold when restricted to degree a and d, we deduce the same for $a+d$. In other words, assuming that (4.5.3), (4.5.4) hold for $C = A$, $\gamma = \gamma_A : A_{\mathrm{op}} \to B$, $\varsigma = \varsigma_A : A \to B$ and for $C = D$, $\gamma = \gamma_D : D_{\mathrm{op}} \to B$, $\varsigma = \varsigma_D : D \to B$, we shall deduce the same equations for $C = A\bar{\otimes}D$,

$$\gamma_{A\bar{\otimes}D} = \big((A\bar{\otimes}D)_{\mathrm{op}12} = D_{2''1'} \otimes A_{2'1''} \xrightarrow{1\boxtimes\psi^{-1}_{2'2''}} D_{\mathrm{op}1'2''} \otimes A_{\mathrm{op}1''2'}$$
$$\xrightarrow{\gamma_D\otimes\gamma_A} B_{1'2''} \otimes B_{1''2'} \xrightarrow{m} B_{12}\big), \quad (4.5.7)$$
$$\varsigma_{A\bar{\otimes}D} = \big(A_{1'2''} \otimes D_{1''2'} \xrightarrow{\varsigma_A\otimes\varsigma_D} B_{1'2''} \otimes B_{1''2'} \xrightarrow{m} B_{12}\big).$$

This choice of $\gamma_{A\bar{\otimes}D}$ suits well to our purposes, since in case $A = C^{\bar{\otimes}a}$, $D = C^{\bar{\otimes}d}$ the composite

$$(A\bar{\otimes}D)_{\mathrm{op}} \hookrightarrow (TC\bar{\otimes}TC)_{\mathrm{op}} \xrightarrow{Pm} TC_{\mathrm{op}} \xrightarrow{\gamma'} B$$

coincides with (4.5.7) by the homomorphism property of γ'.

Let us prove (4.5.4) for $A\bar{\otimes}D$:

$$
\begin{aligned}
\big(\mathbb{1}_1 \boxtimes (A\bar{\otimes}D)_{2''2'} &= \mathbb{1}_1 \boxtimes D_{2^4 2'} \otimes A_{2'''2''} \xrightarrow{\Delta_{2^4 1 2'}\otimes A} D_{2^4 1'} \otimes \mathbb{1}_{1''} \otimes D_{1'''2'} \otimes A_{2'''2''} \\
&\xrightarrow{D\otimes D\otimes \Delta_{2'''1''2''}} D_{2^4 1'} \otimes A_{2'''1''} \otimes D_{1^4 2'} \otimes A_{1'''2''} \\
&\xrightarrow{\psi^{-1}_{2'''2^4}} D_{\mathrm{op}1'2^4} \otimes A_{\mathrm{op}1''2'''} \otimes A_{1'''2''} \otimes D_{1^4 2'} \xrightarrow{\gamma_D\otimes\gamma_A\otimes\varsigma_A\otimes\varsigma_D} \\
&B_{1'2^4} \otimes B_{1''2'''} \otimes B_{1'''2''} \otimes B_{1^4 2'} \xrightarrow{m\otimes m} B_{1'2''} \otimes B_{1''2'} \xrightarrow{m} B_{12}\big) \\
= \big(\mathbb{1}_1 \boxtimes D_{2^4 2'} \otimes A_{2'''2''} &\xrightarrow{\psi^{-1}_{2'''2^4}} \mathbb{1}_1 \boxtimes D_{2^4 2'} \otimes A_{2'''2''} \\
&\xrightarrow{\Delta_{2^4 1 2'}\otimes A} D_{2^4 1'} \otimes \mathbb{1}_{1''} \otimes D_{1'''2'} \otimes A_{2'''2''} \\
&\xrightarrow{D\otimes D\otimes \Delta_{2'''1''2''}} D_{2^4 1'} \otimes A_{2'''1''} \otimes A_{1'''2''} \otimes D_{1^4 2'} \\
&\xrightarrow{\gamma_D\otimes\gamma_A\otimes\varsigma_A\otimes\varsigma_D} B_{1'2^4} \otimes B_{1''2'''} \otimes B_{1'''2''} \otimes B_{1^4 2'} \xrightarrow{m^{(3)}} B_{12}\big) \\
= \big(\mathbb{1}_1 \boxtimes D_{2^4 2'} \otimes A_{2'''2''} &\xrightarrow{\psi^{-1}_{2'''2^4}} \mathbb{1}_1 \boxtimes D_{2^4 2'} \otimes A_{2'''2''} \xrightarrow{\Delta_{2^4 1 2'}\otimes \varepsilon^{\mathrm{op}}} \\
&D_{2'''1'} \otimes D_{1''2'} \otimes \mathbb{1}_{2''} \xrightarrow{\sim} D_{\mathrm{op}1'2''} \otimes D_{1''2'} \xrightarrow{\gamma\otimes\varsigma} B_{1'2''} \otimes B_{1''2'} \xrightarrow{m} B_{12}\big) \\
= \big(\mathbb{1}_1 \boxtimes D_{2^4 2'} \otimes A_{2'''2''} &\xrightarrow{\psi^{-1}_{2'''2^4}} \mathbb{1}_1 \boxtimes D_{2^4 2'} \otimes A_{2'''2''} \xrightarrow{\mathbb{1}\boxtimes D\otimes\varepsilon^{\mathrm{op}}} \mathbb{1}_1 \boxtimes D_{2'''2'} \otimes \mathbb{1}_{2''} \\
&\simeq \mathbb{1}_1 \boxtimes D_{2''2'} \xrightarrow{\mathbb{1}\boxtimes\varepsilon^{\mathrm{op}}} \mathbb{1}_1 \boxtimes \mathbb{1}_2 \xrightarrow{\eta} B_{12}\big) \\
= \big(\mathbb{1}_1 \boxtimes (A\bar{\otimes}D)_{\mathrm{op}2'2''} &\xrightarrow{\mathbb{1}\boxtimes\varepsilon^{\mathrm{op}}} \mathbb{1}_1 \boxtimes \mathbb{1}_2 \xrightarrow{\eta} B_{12}\big),
\end{aligned}
$$

for

$$
\begin{aligned}
\big(D^{41} \otimes A^{32} \xrightarrow{\psi^{-1(34)}} D^{41} \otimes A^{32} \xrightarrow{D\otimes\varepsilon^{\mathrm{op}}} D^{31} \otimes \mathbb{1}^2 \xrightarrow{\sim} D^{21} \xrightarrow{\varepsilon^{\mathrm{op}}} \mathbb{1}^1\big) \\
= \big((A\bar{\otimes}D)^{21} \xrightarrow{\varepsilon^{\mathrm{op}}} \mathbb{1}^1\big)
\end{aligned}
$$

by (4.5.2).

Now let us prove (4.5.3) for $A\bar{\otimes}D$. The opposite comultiplication on $A\bar{\otimes}D$ is found from (4.5.1):

$$
\begin{aligned}
\Delta^{\mathrm{op}}_{A\bar{\otimes}D} = \big((A\bar{\otimes}D)_{\mathrm{op}13} \boxtimes \mathbb{1}_2 &= D_{\mathrm{op}1'3''} \otimes A_{\mathrm{op}1''3'} \boxtimes \mathbb{1}_2 \\
&\xrightarrow{D\otimes\Delta^{\mathrm{op}}_A} D_{\mathrm{op}1'3''} \otimes A_{\mathrm{op}1''2'} \otimes \mathbb{1}_{2''} \otimes A_{2'''3'} \\
&\xrightarrow{\Delta^{\mathrm{op}}_D\otimes A\otimes A} D_{\mathrm{op}1'2''} \otimes A_{\mathrm{op}1''2'} \otimes D_{\mathrm{op}2'''3''} \otimes A_{\mathrm{op}2^4 3'} \\
&\xrightarrow{\psi_{2'2''}} D_{2''1'} \otimes A_{2'1''} \otimes D_{3''2'''} \otimes A_{3'2^4} = \big(A\bar{\otimes}D\big)_{\mathrm{op}12'} \otimes \big(A\bar{\otimes}D\big)_{\mathrm{op}2''3}\big).
\end{aligned}
$$

Thus (4.5.3) follows from:

$$
\begin{aligned}
\big((A\bar{\otimes}D)_{\mathrm{op}1''1'} &= D_{\mathrm{op}1'''1''} \otimes A_{\mathrm{op}1^4 1'} \boxtimes \mathbb{1}_2 \\
&\xrightarrow{D\otimes\Delta^{\mathrm{op}}_{1^4 2 1'}} D_{\mathrm{op}1'''1''} \otimes A_{\mathrm{op}1^4 2'} \otimes \mathbb{1}_{2''} \otimes A_{\mathrm{op}2'''1'} \\
&\xrightarrow{\Delta_{1'''2''1''}\otimes A\otimes A} D_{\mathrm{op}1'''2''} \otimes A_{\mathrm{op}1^4 2'} \otimes D_{\mathrm{op}2'''1''} \otimes A_{\mathrm{op}2^4 1'} \\
&\xrightarrow{\psi_{2'2''}} D_{\mathrm{op}1'''2''} \otimes A_{\mathrm{op}1^4 2'} \otimes D_{1''2'''} \otimes A_{1'2^4} \\
&\xrightarrow{\psi^{-1}_{2'2''}} D_{\mathrm{op}1'''2''} \otimes D_{1''2'''} \otimes A_{\mathrm{op}1^4 2'} \otimes A_{1'2^4}
\end{aligned}
$$

$$\xrightarrow{\gamma_D\otimes\varsigma_D\otimes\gamma_A\otimes\varsigma_A} B_{1'''2''}\otimes B_{1''2'''}\otimes B_{1^4 2'}\otimes B_{1'2^4}\xrightarrow{m} B_{12}\big)$$

$$= \big(D_{1''1'''}\otimes A_{\mathrm{op}1^41'}\boxtimes \mathbb{I}_2 \xrightarrow{\varepsilon\otimes\Delta^{\mathrm{op}}_{1^421'}} \mathbb{I}_{1''}\otimes A_{\mathrm{op}1'''2'}\otimes A_{\mathrm{op}2''1'}$$

$$\xrightarrow{\sim} A_{\mathrm{op}1''2'}\otimes A_{1'2''}\xrightarrow{\gamma_A\otimes\varsigma_A} B_{1''2'}\otimes B_{1'2''}\xrightarrow{m} B_{12}\big)$$

$$= \big(D_{1''1'''}\otimes A_{1'1^4}\boxtimes \mathbb{I}_2\xrightarrow{\varepsilon_D\otimes A}\mathbb{I}_{1''}\otimes A_{1'1'''}\boxtimes\mathbb{I}_2$$

$$\xrightarrow{\sim} A_{1'1''}\boxtimes\mathbb{I}_2\xrightarrow{\varepsilon_A\boxtimes\mathbb{I}}\mathbb{I}_1\boxtimes\mathbb{I}_2\xrightarrow{\eta} B_{12}\big)$$

$$= \big((A\bar{\otimes}D)_{1'1''}\boxtimes\mathbb{I}_2\xrightarrow{\varepsilon\boxtimes\mathbb{I}}\mathbb{I}_1\boxtimes\mathbb{I}_2\xrightarrow{\eta}B_{12}\big).$$

The lemma is proved. □

Now we shall describe a way to construct Hopf coalgebras.

4.5.9. Theorem. *Let C be a squared coalgebra in $\widehat{\mathcal{V}}$, and let $\gamma : C_{\mathrm{op}} \to C$ be a coalgebra homomorphism. Let $j : C_{1'1''}\boxtimes\mathbb{I}_2 \to TC_{12}$ be the difference*

$$\big(C_{\mathrm{op}1''1'}\boxtimes\mathbb{I}_2\xrightarrow{\Delta^{\mathrm{op}}_{1''21'}} C_{\mathrm{op}1''2'}\otimes C_{\mathrm{op}2''1'}\xrightarrow{\gamma\otimes C} C_{1''2'}\otimes C_{1'2''} = T^2(C)_{12}\big)$$
$$-\big(C_{1'1''}\boxtimes\mathbb{I}_2\xrightarrow{\varepsilon\boxtimes\mathbb{I}}\mathbb{I}_1\boxtimes\mathbb{I}_2 = T^0(C)_{12}\big),$$

and let $k : \mathbb{I}_1\boxtimes C_{2''2'} \to TC_{12}$ be the difference

$$\big(\mathbb{I}_1\boxtimes C_{2''2'}\xrightarrow{\Delta_{2''12'}} C_{2''1'}\otimes C_{1''2'}\xrightarrow{\gamma_{1'2''}\otimes C} C_{1'2''}\otimes C_{1''2'} = T^2(C)_{12}\big)$$
$$-\big(\mathbb{I}_1\boxtimes C_{\mathrm{op}2'2''}\xrightarrow{\mathbb{I}\boxtimes\varepsilon^{\mathrm{op}}}\mathbb{I}_1\boxtimes\mathbb{I}_2 = T^0(C)_{12}\big).$$

Then

$$H_\gamma = T(C)/(\operatorname{Im} j, \operatorname{Im} k)$$

is a bicoalgebra with a right antipode γ_ς, whose restriction to $T^1(C)_{\mathrm{op}}$ is γ. If γ is invertible, then H_γ is a Hopf coalgebra.

Proof will consist of a sequence of lemmas.

4.5.10. Lemma. *H_γ is a bicoalgebra.*

Proof. Let us prove that $\operatorname{Im} j$ and $\operatorname{Im} k$ are coideals. Then the ideal $(\operatorname{Im} j, \operatorname{Im} k)$ is a coideal, hence, H_γ is a bicoalgebra.

Let us start with j. Denote $J = C_{1'1''}\boxtimes\mathbb{I}_2$,

$$j' = \big(C_{\mathrm{op}1''1'}\boxtimes\mathbb{I}_2\xrightarrow{\Delta^{\mathrm{op}}_{1''21'}} C_{\mathrm{op}1''2'}\otimes C_{\mathrm{op}2''1'}\xrightarrow{\gamma\otimes C} C_{1''2'}\otimes C_{1'2''} = T^2(C)_{12}\big),$$

$$j'' = \big(C_{1'1''}\boxtimes\mathbb{I}_2\xrightarrow{\varepsilon\boxtimes\mathbb{I}}\mathbb{I}_1\boxtimes\mathbb{I}_2 = T^0(C)_{12}\big).$$

Then $j = j' - j''$. First we show that $\varepsilon\circ(\circledast j) = 0$. Indeed,

$$\varepsilon\circ\circledast j' = \big(C^{21}_{\mathrm{op}}\otimes\mathbb{I}^3\xrightarrow{\Delta^{231}_{\mathrm{op}}} C^{23}_{\mathrm{op}}\otimes C^{41}_{\mathrm{op}}\xrightarrow{\gamma^{23}\otimes C} C^{23}\otimes C^{14}$$
$$\xrightarrow{\varepsilon\otimes C}\mathbb{I}^2\otimes C^{13}\xrightarrow{\sim} C^{12}\xrightarrow{\varepsilon}\mathbb{I}^1\big)$$
$$= \big(C^{21}_{\mathrm{op}}\otimes\mathbb{I}^3\xrightarrow{\Delta^{231}_{\mathrm{op}}} C^{23}_{\mathrm{op}}\otimes C^{41}_{\mathrm{op}}\xrightarrow{\varepsilon^{\mathrm{op}}\otimes C}\mathbb{I}^2\otimes C^{13}\xrightarrow{\sim} C^{12}\xrightarrow{\varepsilon}\mathbb{I}^1\big)$$
$$= \big(C^{12}\otimes\mathbb{I}^3\xrightarrow{\sim} C^{12}\xrightarrow{\varepsilon}\mathbb{I}^1\big)$$
$$= \varepsilon\circ\circledast j''.$$

Let us prove commutativity of diagram (2.1.5) for TC, $W = J_{13}\boxtimes\mathbb{I}_2$, $s = \mathrm{id}_W$,

$$r = \chi\oplus\kappa : J_{13}\boxtimes\mathbb{I}_2 \to TC_{12'}\otimes J_{2''3}\oplus J_{12'}\otimes TC_{2''3},$$

where

$$
\begin{aligned}
\chi = \bigl(C_{\mathrm{op}1''1'} \boxtimes \mathbb{1}_2 \boxtimes \mathbb{1}_3 &\xrightarrow{\Delta^{\mathrm{op}}_{1''21'} \boxtimes \mathbb{1}_3} C_{\mathrm{op}1''2'} \otimes C_{\mathrm{op}2''1'} \boxtimes \mathbb{1}_3 \\
&\xrightarrow{\sim} C_{\mathrm{op}1''2'} \otimes \mathbb{1}_{2''} \otimes C_{1'2'''} \boxtimes \mathbb{1}_3 \xrightarrow{\gamma \otimes \Delta_{1'2''2'''} \boxtimes \mathbb{1}_3} \\
&C_{1''2'} \otimes C_{1'2''} \otimes C_{2'''2^4} \boxtimes \mathbb{1}_3 = T^2 C_{12'} \otimes C_{2''2'''} \boxtimes \mathbb{1}_3\bigr), \\
\kappa = \bigl(C_{1'1''} \boxtimes \mathbb{1}_2 \boxtimes \mathbb{1}_3 &\xrightarrow{\sim} C_{1'1''} \boxtimes \mathbb{1}_{2'} \otimes \mathbb{1}_{2''} \boxtimes \mathbb{1}_3 = C_{1'1''} \boxtimes \mathbb{1}_{2'} \otimes T^0 C_{2''3}\bigr).
\end{aligned}
$$

This would imply that $\operatorname{Im} j$ is a coideal. The commutativity of (2.1.5) is a consequence of the commutativity of the following three diagrams

$$
\begin{array}{ccc}
J_{13} \boxtimes \mathbb{1}_2 & \xrightarrow{\chi} & TC_{12'} \otimes J_{2''3} \\
{\scriptstyle j' \boxtimes \mathbb{1}} \downarrow & & \downarrow {\scriptstyle TC \otimes j'} \\
TC_{13} \boxtimes \mathbb{1}_2 & \xrightarrow{\Delta_{123}} & TC_{12'} \otimes TC_{2''3}
\end{array}
\qquad (4.5.8)
$$

$$
\begin{array}{ccc}
J_{13} \boxtimes \mathbb{1}_2 & \xrightarrow{\kappa} & J_{12'} \otimes TC_{2''3} \\
{\scriptstyle j'' \boxtimes \mathbb{1}} \downarrow & & \downarrow {\scriptstyle j'' \otimes TC} \\
TC_{13} \boxtimes \mathbb{1}_2 & \xrightarrow{\Delta_{123}} & TC_{12'} \otimes TC_{2''3}
\end{array}
\qquad (4.5.9)
$$

$$
\begin{array}{ccc}
J_{13} \boxtimes \mathbb{1}_2 & \xrightarrow{\chi} & TC_{12'} \otimes J_{2''3} \\
{\scriptstyle \kappa} \downarrow & & \downarrow {\scriptstyle TC \otimes j''} \\
J_{12'} \otimes TC_{2''3} & \xrightarrow{j' \otimes TC} & TC_{12'} \otimes TC_{2''3}
\end{array}
\qquad (4.5.10)
$$

The first of them (4.5.8) is obtained by the following computation.

$$
\begin{aligned}
&\bigl(J_{13} \boxtimes \mathbb{1}_2 \xrightarrow{j' \boxtimes \mathbb{1}} TC_{13} \boxtimes \mathbb{1}_2 \xrightarrow{\Delta_{123}} TC_{12'} \otimes TC_{2''3}\bigr) \\
&= \bigl(C_{\mathrm{op}1''1'} \boxtimes \mathbb{1}_2 \boxtimes \mathbb{1}_3 \xrightarrow{\Delta^{\mathrm{op}}_{1''31'} \boxtimes \mathbb{1}_2} C_{\mathrm{op}1''3'} \otimes C_{\mathrm{op}3''1'} \boxtimes \mathbb{1}_2 \\
&\qquad \xrightarrow{\gamma \otimes C \boxtimes \mathbb{1}} C_{1''3'} \otimes C_{1'3''} \boxtimes \mathbb{1}_2 \xrightarrow{m \boxtimes \mathbb{1}} TC_{13} \boxtimes \mathbb{1}_2 \xrightarrow{\Delta_{123}} TC_{12'} \otimes TC_{2''3}\bigr) \\
&= \bigl(C_{\mathrm{op}1''1'} \boxtimes \mathbb{1}_2 \boxtimes \mathbb{1}_3 \xrightarrow{\Delta^{\mathrm{op}}_{1''31'} \boxtimes \mathbb{1}_2} C_{\mathrm{op}1''3'} \otimes C_{\mathrm{op}3''1'} \boxtimes \mathbb{1}_2 \\
&\qquad \xrightarrow{\gamma \otimes C \boxtimes \mathbb{1}} C_{1''3'} \otimes C_{1'3''} \boxtimes \mathbb{1}_2 \xrightarrow{\Delta_{1''23'} \otimes C} C_{1''2'} \otimes C_{1'3''} \otimes \mathbb{1}_{2''} \otimes C_{2'''3'} \\
&\qquad \xrightarrow{C \otimes \Delta_{1'2''3''} \otimes C} C_{1''2'} \otimes C_{1'2''} \otimes C_{2'''3''} \otimes C_{2^43'} \xrightarrow{m \otimes m} TC_{12'} \otimes TC_{2''3}\bigr) \\
&= \bigl(C_{\mathrm{op}1''1'} \boxtimes \mathbb{1}_2 \boxtimes \mathbb{1}_3 \xrightarrow{\Delta^{\mathrm{op}}_{1''31'} \boxtimes \mathbb{1}_2} C_{\mathrm{op}1''3'} \otimes C_{\mathrm{op}3''1'} \boxtimes \mathbb{1}_2 \\
&\qquad \xrightarrow{\Delta^{\mathrm{op}}_{1''23'} \otimes C} C_{\mathrm{op}1''2'} \otimes \mathbb{1}_{2''} \otimes C_{\mathrm{op}2'''3'} \otimes C_{1'3''} \\
&\qquad \xrightarrow{\gamma \otimes \gamma \otimes \Delta_{1'2''3''}} C_{1''2'} \otimes C_{1'2''} \otimes C_{2'''3'} \otimes C_{2^43''} \xrightarrow{m \otimes m} TC_{12'} \otimes TC_{2''3}\bigr) \\
&= \bigl(C_{\mathrm{op}1''1'} \boxtimes \mathbb{1}_2 \boxtimes \mathbb{1}_3 \xrightarrow{\Delta^{\mathrm{op}}_{1''21'} \boxtimes \mathbb{1}_3} C_{\mathrm{op}1''2'} \otimes C_{\mathrm{op}2''1'} \boxtimes \mathbb{1}_3 \\
&\qquad \xrightarrow{C \otimes \Delta^{\mathrm{op}}_{2''31'}} C_{\mathrm{op}1''2'} \otimes \mathbb{1}_{2''} \otimes C_{\mathrm{op}2'''3'} \otimes C_{1'3''} \\
&\qquad \xrightarrow{\gamma \otimes \gamma \otimes \Delta_{1'2''3''}} C_{1''2'} \otimes C_{1'2''} \otimes C_{2'''3'} \otimes C_{2^43''} \xrightarrow{m \otimes m} TC_{12'} \otimes TC_{2''3}\bigr) \\
&\overset{(!)}{=} \bigl(C_{\mathrm{op}1''1'} \boxtimes \mathbb{1}_2 \boxtimes \mathbb{1}_3 \xrightarrow{\Delta^{\mathrm{op}}_{1''21'} \boxtimes \mathbb{1}_3} C_{\mathrm{op}1''2'} \otimes \mathbb{1}_{2''} \otimes C_{1'2'''} \boxtimes \mathbb{1}_3 \xrightarrow{\gamma \otimes \Delta_{1'2''2'''} \boxtimes \mathbb{1}}
\end{aligned}
$$

$$
\begin{aligned}
&C_{1''2'} \otimes C_{1'2''} \otimes C_{2'''2^4} \boxtimes \mathbb{I}_3 \xrightarrow{m \otimes \Delta^{\mathrm{op}}_{2^4 3 2'''}} TC_{12'} \otimes C_{\mathrm{op}2'''3'} \otimes C_{\mathrm{op}3''2''} \\
&\xrightarrow{TC \otimes \gamma \otimes C} TC_{12'} \otimes C_{2'''3'} \otimes C_{2''3''} \xrightarrow{TC \otimes m} TC_{12'} \otimes TC_{2''3}\big) \\
&= \big(C_{\mathrm{op}1''1'} \boxtimes \mathbb{I}_2 \boxtimes \mathbb{I}_3 \xrightarrow{\chi} TC_{12'} \otimes C_{\mathrm{op}2'''2''} \boxtimes \mathbb{I}_3 \xrightarrow{TC \otimes j'} TC_{12'} \otimes TC_{2''3}\big).
\end{aligned}
$$

The above equation (!) holds due to a sort of coassociativity

$$
\begin{array}{ccc}
\mathbb{I}_2 \boxtimes C_{\mathrm{op}41} \boxtimes \mathbb{I}_3 & \xrightarrow{\Delta_{124} \boxtimes \mathbb{I}_3} & C_{12'} \otimes C_{2''4} \boxtimes \mathbb{I}_3 \\
{\scriptstyle \mathbb{I}_2 \boxtimes \Delta^{\mathrm{op}}_{431}} \downarrow & & \downarrow {\scriptstyle C \otimes \Delta^{\mathrm{op}}_{432''}} \\
\mathbb{I}_2 \boxtimes C_{\mathrm{op}43'} \otimes C_{\mathrm{op}3''1} & \xrightarrow{C \otimes \Delta_{123''}} & C_{12'} \otimes C_{\mathrm{op}43'} \otimes C_{\mathrm{op}3''2''}
\end{array}
$$

(prove it!).

The check of (4.5.9) is left to the reader. To prove (4.5.10) notice that

$$
\begin{aligned}
&\big(J_{13} \boxtimes \mathbb{I}_2 \xrightarrow{\chi} TC_{12'} \otimes J_{2''3} \xrightarrow{TC \otimes j''} TC_{12'} \otimes TC_{2''3}\big) \\
&= \big(C_{\mathrm{op}1''1'} \boxtimes \mathbb{I}_2 \boxtimes \mathbb{I}_3 \xrightarrow{\Delta^{\mathrm{op}}_{1''21'} \boxtimes \mathbb{I}_3} C_{\mathrm{op}1''2'} \otimes C_{\mathrm{op}2''1'} \boxtimes \mathbb{I}_3 \\
&\qquad \xrightarrow{\gamma \otimes C \otimes \mathbb{I}} C_{1''2'} \otimes C_{1'2''} \otimes \mathbb{I}_{2'''} \boxtimes \mathbb{I}_3 \xrightarrow{m \otimes \eta} TC_{12'} \otimes TC_{2''3}\big) \\
&= \big(C_{\mathrm{op}1''1'} \boxtimes \mathbb{I}_2 \boxtimes \mathbb{I}_3 \xrightarrow{\sim} C_{\mathrm{op}1''1'} \boxtimes \mathbb{I}_{2'} \otimes \mathbb{I}_{2''} \boxtimes \mathbb{I}_3 \xrightarrow{\Delta^{\mathrm{op}}_{1''2'1'} \otimes \eta} \\
&\qquad C_{\mathrm{op}1''2'} \otimes C_{\mathrm{op}2''1'} \otimes TC_{2'''3} \xrightarrow{\gamma \otimes C \otimes TC} \\
&\qquad C_{1''2'} \otimes C_{1'2''} \otimes TC_{2'''3} \xrightarrow{m \otimes TC} TC_{12'} \otimes TC_{2''3}\big) \\
&= \big(J_{13} \boxtimes \mathbb{I}_2 \xrightarrow{\kappa} J_{12'} \otimes TC_{2''3} \xrightarrow{j' \otimes TC} TC_{12'} \otimes TC_{2''3}\big).
\end{aligned}
$$

Similarly one can prove that $\operatorname{Im} k$ is a coideal. Indeed, $-k = j' - j''$, where

$$
\begin{aligned}
j' &= \big(\mathbb{I}_1 \boxtimes C_{\mathrm{op}2'2''} \xrightarrow{\mathbb{I} \boxtimes \varepsilon^{\mathrm{op}}} \mathbb{I}_1 \boxtimes \mathbb{I}_2 = T^0(C)_{12}\big), \\
j'' &= \big(\mathbb{I}_1 \boxtimes C_{2''2'} \xrightarrow{\Delta_{2''12'}} C_{2''1'} \otimes C_{1''2'} \xrightarrow{\gamma \otimes C} C_{1'2''} \otimes C_{1''2'} = T^2(C)_{12}\big).
\end{aligned}
$$

The same commutation relations (4.5.8)–(4.5.10) hold if $J = \mathbb{I}_1 \boxtimes C_{2''2'}$ and χ, κ are set to

$$
\begin{aligned}
\chi &= \big(\mathbb{I}_1 \boxtimes \mathbb{I}_2 \boxtimes C_{3''3'} \simeq \mathbb{I}_1 \boxtimes \mathbb{I}_{2'} \otimes \mathbb{I}_{2''} \boxtimes C_{3''3'} = T^0 C_{12'} \otimes \mathbb{I}_{2''} \otimes C_{3''3'}\big), \\
\kappa &= \big(\mathbb{I}_1 \boxtimes \mathbb{I}_2 \boxtimes C_{3''3'} \xrightarrow{\mathbb{I}_1 \boxtimes \Delta_{3''23'}} \mathbb{I}_1 \boxtimes C_{3''2'} \otimes \mathbb{I}_{2''} \otimes C_{2'''3'} \\
&\qquad \xrightarrow{\mathbb{I} \boxtimes \Delta^{\mathrm{op}}_{2'2''3''} \otimes C} \mathbb{I}_1 \boxtimes C_{\mathrm{op}2'2''} \otimes C_{\mathrm{op}2'''3''} \otimes C_{2^4 3'} \xrightarrow{\mathbb{I} \boxtimes C \otimes \gamma \otimes C} \\
&\qquad \mathbb{I}_1 \boxtimes C_{2''2'} \otimes C_{2'''3''} \otimes C_{2^4 3'} = \mathbb{I}_1 \boxtimes C_{2''2'} \otimes T^2 C_{2'''3}\big).
\end{aligned}
$$

The check is left to the reader. □

By Lemma 4.5.8 there are bialgebra homomorphisms $\gamma' : T(C)^{\mathrm{op}}_{\mathrm{op}} \to H_\gamma = B$ and $\varsigma' : T(C) \to H_\gamma$, the natural projection, such that (4.5.5), (4.5.6) hold. Recall that $T(C)^{\mathrm{op}}_{\mathrm{op}} \simeq T(C_{\mathrm{op}})$ by the proof of Lemma 4.5.7. The projection $P\varsigma' : T(C)^{\mathrm{op}}_{\mathrm{op}} \to$

$H_{\gamma_{\mathrm{op}}^{\mathrm{op}}}$ has the kernel $\operatorname{Ker} P\varsigma' = (\operatorname{Im} Pj, \operatorname{Im} Pk)$,

$$Pj = \big(C_{\mathrm{op}2''2'} \boxtimes \mathbb{I}_1 \xrightarrow{\Delta^{\mathrm{op}}_{2''12'}} C_{\mathrm{op}2''1'} \otimes C_{\mathrm{op}1''2'} \xrightarrow{\gamma_{2''1'}\otimes C} C_{2''1'} \otimes C_{2'1''} = (T^2C)_{21}\big)$$
$$- \big(C_{2'2''} \boxtimes \mathbb{I}_1 \xrightarrow{\varepsilon\boxtimes\mathbb{I}} \mathbb{I}_2 \boxtimes \mathbb{I}_1 = (T^0C)_{21}\big),$$
$$Pk = \big(\mathbb{I}_2 \boxtimes C_{1''1'} \xrightarrow{\Delta_{1''21'}} C_{1''2'} \otimes C_{2''1'} \xrightarrow{\gamma_{2'1''}\otimes C} C_{2'1''} \otimes C_{2''1'} = (T^2C)_{21}\big)$$
$$- \big(\mathbb{I}_2 \boxtimes C_{\mathrm{op}1'1''} \xrightarrow{\mathbb{I}\boxtimes\varepsilon^{\mathrm{op}}} \mathbb{I}_2 \boxtimes \mathbb{I}_1 = (T^0C)_{21}\big).$$

Indeed, if A, D are subobjects of B and $C = \operatorname{Im}(m : A\bar{\otimes}D \to B)$ is their product, then

$$\begin{aligned}&\operatorname{Im}(m^{\mathrm{op}} : PD\bar{\otimes}PA \to PB)\\ &= \operatorname{Im}\big(PD_{1'2''} \otimes PA_{1''2'} \xrightarrow{\psi_{2'2''}} D_{2''1'} \otimes A_{2'1''} = (A\bar{\otimes}D)_{21} \xrightarrow{m^{21}} B_{21}\big)\\ &= PC.\end{aligned}$$

Therefore, P applied to the left (resp. right, two-sided) ideal of B generated by A is the right (resp. left, two-sided) ideal of $B^{\mathrm{op}}_{\mathrm{op}}$ generated by PA.

4.5.11. Lemma. $\gamma' \circ Pj = 0$, $\gamma' \circ Pk = 0$.

PROOF. Let us prove that $\gamma' \circ Pj = 0$ using (4.5.7).

$$\begin{aligned}\gamma' \circ Pj' =& \big(C_{\mathrm{op}2''2'} \boxtimes \mathbb{I}_1 \xrightarrow{\Delta^{\mathrm{op}}_{2''12'}} C_{\mathrm{op}2''1'} \otimes C_{\mathrm{op}1''2'} \xrightarrow{\gamma_{2''1'}\otimes C} C_{2''1'} \otimes C_{\mathrm{op}1''2'}\\ &\xrightarrow{\psi^{-1}_{2'2''}} C_{\mathrm{op}1'2''} \otimes C_{\mathrm{op}1''2'} \xrightarrow{\gamma\otimes\gamma} C_{1'2''} \otimes C_{1''2'} \xrightarrow{m} B_{12}\big)\\ =& \big(C_{\mathrm{op}2''2'} \boxtimes \mathbb{I}_1 \xrightarrow{\gamma_{2''2'}\boxtimes\mathbb{I}} C_{2''2'} \boxtimes \mathbb{I}_1 \xrightarrow{\Delta_{2''12'}} C_{2''1'} \otimes C_{1''2'}\\ &\xrightarrow{\psi^{-1}_{2'2''}} C_{\mathrm{op}1'2''} \otimes C_{1''2'} \xrightarrow{\gamma\otimes C} C_{1'2''} \otimes C_{1''2'} \xrightarrow{m} B_{12}\big)\\ =& \big(C_{\mathrm{op}2''2'} \boxtimes \mathbb{I}_1 \xrightarrow{\gamma_{2''2'}\boxtimes\mathbb{I}} C_{2''2'} \boxtimes \mathbb{I}_1 \xrightarrow{\psi^{-1}_{2'2''}\boxtimes\mathbb{I}} C_{2''2'} \boxtimes \mathbb{I}_1\\ &\xrightarrow{\Delta_{2''12'}} C_{2''1'} \otimes C_{1''2'} \xrightarrow{\gamma_{1'2''}\otimes C} C_{1'2''} \otimes C_{1''2'} \xrightarrow{m} B_{12}\big)\\ =& \big(C_{\mathrm{op}2''2'} \boxtimes \mathbb{I}_1 \xrightarrow{\gamma_{2''2'}\boxtimes\mathbb{I}} C_{2''2'} \boxtimes \mathbb{I}_1 \xrightarrow{\psi^{-1}_{2'2''}\boxtimes\mathbb{I}} C_{\mathrm{op}2'2''} \boxtimes \mathbb{I}_1\\ &\xrightarrow{\varepsilon^{\mathrm{op}}\boxtimes\mathbb{I}} \mathbb{I}_2 \boxtimes \mathbb{I}_1 \xrightarrow{\eta} B_{12}\big)\\ =& \big(C_{2'2''} \boxtimes \mathbb{I}_1 \xrightarrow{\varepsilon\boxtimes\mathbb{I}} \mathbb{I}_1 \boxtimes \mathbb{I}_2 \xrightarrow{\eta} B_{12}\big)\\ =& \gamma' \circ Pj'',\end{aligned}$$

if we prove that

$$\big(C^{21}_{\mathrm{op}} \xrightarrow{\gamma^{21}} C^{21} \xrightarrow{\psi^{-1(12)}} C^{12}_{\mathrm{op}} \xrightarrow{\varepsilon^{\mathrm{op}}} \mathbb{I}\big) = \big(C^{12} \xrightarrow{\varepsilon} \mathbb{I}\big). \tag{4.5.11}$$

By Proposition 4.1.6 $P\gamma : C \to P_{{}^t\varsigma^{-1}}C$ is a homomorphism of coalgebras. Thus (4.5.11) follows from

$$\varepsilon^{{}^t\varsigma^{-1}} = \big(C^{12}_{\mathrm{op}} \xrightarrow{\psi^{-1(12)}_{\varsigma}} C^{12}_{\mathrm{op}} \xrightarrow{\varepsilon^{\varsigma}} \mathbb{I}\big),$$

which we shall prove now.

4.5.12. Lemma. $\varepsilon^{\varsigma} = \big(C^{12}_{\mathrm{op}} \xrightarrow{\psi^{12}_{\varsigma}} C^{12}_{\mathrm{op}} \xrightarrow{\varepsilon^{{}^t\varsigma^{-1}}} \mathbb{I}\big).$

PROOF. It suffices to consider $C = X \boxtimes X^\vee$, $X \in \mathcal{V}$. We have to prove that

$$(X^\vee \otimes X \xrightarrow{\psi_{X^\vee,X}} X^\vee \otimes X \xrightarrow{X^\vee \otimes {}^t\zeta^{-1}_{X^\vee}} X^\vee \otimes X^{\vee\vee} \xrightarrow{\mathrm{ev}} \mathbb{I})$$
$$= (X^\vee \otimes X \xrightarrow{X^\vee \otimes \zeta} X^\vee \otimes X^{\vee\vee} \xrightarrow{\mathrm{ev}} \mathbb{I}). \quad (4.5.12)$$

Using the definition of ψ we expand the left hand side to the upper path connecting the left top corner with the right bottom corner in the following commutative diagram

$$\begin{array}{ccccccc}
X^\vee \otimes X & & & & & & \\
\downarrow{\scriptstyle \zeta_{X^\vee}\otimes\zeta} & & & & & & \\
X^{\vee\vee\vee} \otimes X^{\vee\vee} & \xrightarrow{j_+} & (X^\vee \otimes X^{\vee\vee})^\vee & \xrightarrow{j_+^{-1\,t}} & (X^\vee \otimes X)^{\vee\vee} & \xrightarrow{\zeta^{-1}} & X^\vee \otimes X \\
\downarrow{\scriptstyle \zeta^{-1}_{X^\vee}\otimes X^{\vee\vee}} & & \downarrow{\scriptstyle (X^\vee\otimes {}^t\zeta^{-1}_{X^\vee})^t} & & \downarrow{\scriptstyle ({}^{tt}\zeta^{-1}_{X^\vee}\otimes X)^{tt}} & & \downarrow{\scriptstyle {}^{tt}\zeta^{-1}_{X^\vee}\otimes X} \\
X^\vee \otimes X^{\vee\vee} & \xrightarrow{j_+} & (X^\vee \otimes X)^\vee & \xrightarrow{j_+^{-1\,t}} & ({}^\vee X \otimes X)^{\vee\vee} & & {}^\vee X \otimes X \\
\downarrow{\scriptstyle \mathrm{ev}} & & \downarrow{\scriptstyle \mathrm{coev}^t} & & \downarrow{\scriptstyle \mathrm{ev}^{tt}} & & \downarrow{\scriptstyle \mathrm{ev}} \\
\mathbb{I} & \xleftarrow{d^{-1}} & \mathbb{I}^\vee & \xleftarrow{d^t} & \mathbb{I}^{\vee\vee} & \xrightarrow{\zeta_{\mathbb{I}}^{-1}} & \mathbb{I}
\end{array}$$

The left column equals to the right hand side of (4.5.12). Notice now that the lower row equals to id : $\mathbb{I} \to \mathbb{I}$ by normalisation condition on $\zeta_{\mathbb{I}}$. □

Similarly one can show that $\gamma' \circ Pk = 0$, proving Lemma 4.5.11. □

Therefore, the homomorphism $\gamma' : T(C)^{\mathrm{op}}_{\mathrm{op}} \to H_\gamma$ factorises as

$$\gamma' = (T(C)^{\mathrm{op}}_{\mathrm{op}} \xrightarrow{P\zeta'} H_{\gamma}{}^{\mathrm{op}}_{\mathrm{op}} \xrightarrow{\gamma_\zeta} H_\gamma)$$

for some bicoalgebra homomorphism γ_ζ. Formulae (4.5.5), (4.5.6) imply that γ_ζ is a right antipode of H_γ. If γ is invertible, then so is γ_ζ. Thus, Theorem 4.5.9 is proved.

4.6. Reconstruction of rigid monoidal categories

Now we formulate conditions on monoidal construction on monoidal construction data sufficient for generating rigid monoidal categories. This allows to construct examples of Hopf coalgebras of practical interest.

4.6.1. Definition. *Rigid monoidal construction data* are monoidal construction data $p : \mathcal{P} \to \bar{\mathcal{V}}$ such that for each $X \in \operatorname{Ob}\mathcal{P}$ there exist composite tensor monomials of morphisms from $p \operatorname{Mor}\mathcal{P}$ of the form

$$f = 1 \otimes f_1 \otimes 1 \circ \dots \circ 1 \otimes f_a \otimes 1 : X \otimes (Y_1 \otimes \dots \otimes Y_k) \to \mathbb{I}, \quad (4.6.1)$$
$$g = 1 \otimes g_1 \otimes 1 \circ \dots \circ 1 \otimes g_b \otimes 1 : (W_1 \otimes \dots \otimes W_l) \otimes X \to \mathbb{I}, \quad (4.6.2)$$
$$h = 1 \otimes h_1 \otimes 1 \circ \dots \circ 1 \otimes h_c \otimes 1 : \mathbb{I} \to (U_1 \otimes \dots \otimes U_m) \otimes X, \quad (4.6.3)$$
$$j = 1 \otimes j_1 \otimes 1 \circ \dots \circ 1 \otimes j_d \otimes 1 : \mathbb{I} \to X \otimes (V_1 \otimes \dots \otimes V_n) \quad (4.6.4)$$

(where p is dropped out from notation), which are non-degenerate in X. That is, the morphisms $X \to {}^\vee(Y_1 \otimes \dots \otimes Y_k)$ and $X \to (W_1 \otimes \dots \otimes W_l)^\vee$ induced by f and

g are monic, and the morphisms ${}^\vee(U_1 \otimes \cdots \otimes U_m) \to X$ and $(V_1 \otimes \cdots \otimes V_n)^\vee \to X$ induced by h and j are epic.

4.6.2. Theorem. *Let $p : \mathcal{P} \to \bar{\mathcal{V}}$ be rigid construction data. Then the monoidal category $\mathcal{C}$ constructed from p is rigid.*

PROOF. We can modify the monoidal construction data as in Section 2.8 and Section 3.4 approaching $\mathcal{C}$. Using Step 7 on page 78 we can extend the initial $\mathcal{O} = \operatorname{Ob}\mathcal{P}$ so that it becomes closed under the operation $\otimes : \mathcal{O} \times \mathcal{O} \to \mathcal{O}$. The morphisms (4.6.1)–(4.6.4) with the required properties still exist for so extended $\mathcal{O}$. By Step 10 on page 79 we may assume that tensor monomials $1 \otimes pf_i \otimes 1$ come from some $1 \cdot f_i \cdot 1 \in \operatorname{Mor}\mathcal{P}$. At Step 4 (page 62) we add composite monomials to $\operatorname{Mor}\mathcal{P}$. Therefore, it suffices to prove the theorem in the following assumptions. For each $X \in \operatorname{Ob}\mathcal{P}$ there are morphisms $f : XY \to \varnothing$, $g : WX \to \varnothing$, $h : \varnothing \to UX$, $j : \varnothing \to XV$ such that the corresponding morphisms in $\mathcal{V}$

$$pf : pX \otimes pY \to \mathbb{1}, \qquad pg : pW \otimes pX \to \mathbb{1},$$
$$ph : \mathbb{1} \to pU \otimes pX, \qquad pj : \mathbb{1} \to pX \otimes pV$$

are non-degenerate in X, that is, $pX \to {}^\vee pY$ and $pX \to pW^\vee$ are monic, and ${}^\vee pU \to pX$ and $pV^\vee \to pX$ are epic. We assume also that $\mathcal{O} = \operatorname{Ob}\mathcal{P}$ is closed under $\otimes$, and $\operatorname{Mor}\mathcal{P}$ is closed under composition. To simplify notation we identify XY and $X \otimes Y$ via h as at Step 7 and drop out p from notation.

For any $X \in \mathcal{O}$ there are morphisms in $\operatorname{Mor}\mathcal{P}$ built out of chosen f, g, h, j:

$$d_X = \big(Y \xrightarrow{l_Y^{-1}} \mathbb{1} \otimes Y \xrightarrow{h \otimes Y} (U \otimes X) \otimes Y \xrightarrow{a} U \otimes (X \otimes Y) \xrightarrow{U \otimes f} U \otimes \mathbb{1} \xrightarrow{r_U} U\big),$$

$$s_X = \big(W \xrightarrow{r_W^{-1}} W \otimes \mathbb{1} \xrightarrow{W \otimes j} W \otimes (X \otimes V) \xrightarrow{a^{-1}} (W \otimes X) \otimes V \xrightarrow{g \otimes V} \mathbb{1} \otimes V \xrightarrow{l_V} V\big).$$

In $\mathcal{V}$ they can be decomposed as

$$d_X = \big(Y \xrightarrow{y} X^\vee \xhookrightarrow{u} U\big), \qquad s_X = \big(W \xrightarrow{w} {}^\vee X \xhookrightarrow{v} V\big).$$

Using the full sequence of steps in the proof of Theorem 3.4.1 we produce from $p : \mathcal{P} \to \bar{\mathcal{V}}$ an abelian monoidal category $\mathcal{C}$, a monoidal functor $F : \mathcal{C} \to \mathcal{V}$ and an embedding $\mathcal{P} \hookrightarrow \bar{\mathcal{C}}$ which gives a morphism $p \to \bar{F}$. The induced $H_p \to H_F = H$ is an isomorphism and $\mathcal{C}$ is equivalent to ${}^H\mathcal{V}$. In particular, by Steps 5 and 6 we add objects $X^\vee$, ${}^\vee X$ and morphisms y, u, w, v, so they belong to $\mathcal{C}$.

Applying Step 6 to the morphism $X \otimes \ker y : X \otimes \operatorname{Ker} y \to X \otimes Y$, whose cokernel is $X \otimes y : X \otimes Y \to X \otimes X^\vee$, we added $\mathrm{ev} : X \otimes X^\vee \to \mathbb{1}$ to $\operatorname{Mor}\mathcal{C}$, for $f = \big(X \otimes Y \xrightarrow{X \otimes y} X \otimes X^\vee \xrightarrow{\mathrm{ev}} \mathbb{1}\big)$.

Applying Step 5 to $\operatorname{coker} u \otimes X : U \otimes X \to \operatorname{Coker} u \otimes X$, whose kernel is $u \otimes X : X^\vee \otimes X \to U \otimes X$, we added $\mathrm{coev} : \mathbb{1} \to X^\vee \otimes X$ to $\operatorname{Mor}\mathcal{C}$, for $h = \big(\mathbb{1} \xrightarrow{\mathrm{coev}} X^\vee \otimes X \xrightarrow{u \otimes X} U \otimes X\big)$.

Applying Step 6 to $\ker w \otimes X : \operatorname{Ker} w \otimes X \to W \otimes X$, whose cokernel is $w \otimes X : W \otimes X \to {}^\vee X \otimes X$, we added $\mathrm{ev} : {}^\vee X \otimes X \to \mathbb{1}$ to $\operatorname{Mor}\mathcal{C}$, for $g = \big(W \otimes X \xrightarrow{w \otimes X} {}^\vee X \otimes X \xrightarrow{\mathrm{ev}} \mathbb{1}\big)$.

Applying Step 5 to $X \otimes \operatorname{coker} v : X \otimes V \to X \otimes \operatorname{Coker} v$, whose kernel is $X \otimes v : X \otimes {}^\vee X \to X \otimes V$, we added $\mathrm{coev} : \mathbb{1} \to X \otimes {}^\vee X$ to $\operatorname{Mor}\mathcal{C}$, for $j =$

$(\mathbb{I} \xrightarrow{\text{coev}} X \otimes {}^\vee X \xrightarrow{X \otimes v} X \otimes V)$. The standard relations between those evaluations and coevaluations hold, because they hold in $\mathcal{V}$.

Therefore, any object $X \in \mathcal{O}$ has duals $X^\vee$, ${}^\vee X$ as well as evaluation and coevaluation morphisms in the category $\mathcal{C}$ obtained from $\mathcal{P}$ by Steps 0–10. Let us prove that the monoidal category $\mathcal{C}$ is rigid. We shall achieve it by finding in $\mathcal{C}$ dual objects to any object added at these steps.

Step 0. Set $\mathbb{I}^\vee = {}^\vee\mathbb{I} = \mathbb{I}$ and use $r_\mathbb{I} : \mathbb{I} \otimes \mathbb{I} \to \mathbb{I}$ as evaluations and $r_\mathbb{I}^{-1} : \mathbb{I} \to \mathbb{I} \otimes \mathbb{I}$ as coevaluations.

Step 1. Take $\oplus_i X_i^\vee$ and $\oplus_i {}^\vee X_i$ for dual objects to $\oplus_i X_i$ with obvious evaluations and coevaluations.

Step 5. Assume that we added to $\mathcal{P}$ the kernel $i : K \to M$ of a morphism $f : M \to N \in \mathcal{C}$, such that the dual objects to M and N are already found in $\mathcal{C}$. We have to find $K^\vee$, ${}^\vee K$ in $\mathcal{C}$. The transposed morphisms f^t and ${}^t f$ are in $\mathcal{C}$ by definition:

$$f^t = (N^\vee \xrightarrow{l^{-1}} \mathbb{I} \otimes N^\vee \xrightarrow{\text{coev} \otimes N^\vee} (M^\vee \otimes M) \otimes N^\vee \xrightarrow{M^\vee \otimes f \otimes N^\vee} (M^\vee \otimes N) \otimes N^\vee$$
$$\xrightarrow{a^{-1}} M^\vee \otimes (N \otimes N^\vee) \xrightarrow{M^\vee \otimes \text{ev}} M^\vee \otimes \mathbb{I} \xrightarrow{r} M^\vee),$$
$${}^t f = ({}^\vee N \xrightarrow{r^{-1}} {}^\vee N \otimes \mathbb{I} \xrightarrow{{}^\vee N \otimes \text{coev}} {}^\vee N \otimes (M \otimes {}^\vee M) \xrightarrow{{}^\vee N \otimes f \otimes {}^\vee M} {}^\vee N \otimes (N \otimes {}^\vee M)$$
$$\xrightarrow{a} ({}^\vee N \otimes N) \otimes {}^\vee M \xrightarrow{\text{ev} \otimes {}^\vee M} \mathbb{I} \otimes {}^\vee M \xrightarrow{l} {}^\vee M).$$

Hence, $K^\vee = \operatorname{Coker} f^t$, ${}^\vee K = \operatorname{Coker} {}^t f$ are added at Step 6 as well as $(i^t : M^\vee \to K^\vee) = \operatorname{coker} f^t$, $({}^t i : {}^\vee M \to {}^\vee K) = \operatorname{coker} {}^t f$. Now we construct (co)evaluations for K. Applying Step 6 to $K \otimes f^t : K \otimes N^\vee \to K \otimes M^\vee$, whose cokernel is $K \otimes i^t : K \otimes M^\vee \to K \otimes K^\vee$, we added $\text{ev} : K \otimes K^\vee \to \mathbb{I}$ to morphisms, for

$$(K \otimes M^\vee \xrightarrow{K \otimes i^t} K \otimes K^\vee \xrightarrow{\text{ev}} \mathbb{I}) = (K \otimes M^\vee \xrightarrow{i \otimes M^\vee} M \otimes M^\vee \xrightarrow{\text{ev}} \mathbb{I})$$

is already there. Applying Step 6 to ${}^t f \otimes K : {}^\vee N \otimes K \to {}^\vee M \otimes K$, whose cokernel is ${}^t i \otimes K : {}^\vee M \otimes K \to {}^\vee K \otimes K$, we added $\text{ev} : {}^\vee K \otimes K \to \mathbb{I}$ to morphisms, for

$$({}^\vee M \otimes K \xrightarrow{{}^t i \otimes K} {}^\vee K \otimes K \xrightarrow{\text{ev}} \mathbb{I}) = ({}^\vee M \otimes K \xrightarrow{{}^\vee M \otimes i} {}^\vee M \otimes M \xrightarrow{\text{ev}} \mathbb{I})$$

is already there. Applying Step 5 to $K^\vee \otimes f : K^\vee \otimes M \to K^\vee \otimes N$, which has the kernel $K^\vee \otimes i : K^\vee \otimes K \to K^\vee \otimes M$, we added $\text{coev} : \mathbb{I} \to K^\vee \otimes K$ to morphisms of $\mathcal{C}$, for

$$(\mathbb{I} \xrightarrow{\text{coev}} K^\vee \otimes K \xrightarrow{K^\vee \otimes i} K^\vee \otimes M) = (\mathbb{I} \xrightarrow{\text{coev}} M^\vee \otimes M \xrightarrow{i^t \otimes M} K^\vee \otimes M)$$

is already there. Applying Step 5 to $f \otimes {}^\vee K : M \otimes {}^\vee K \to N \otimes {}^\vee K$, which has the kernel $i \otimes {}^\vee K : K \otimes {}^\vee K \to M \otimes {}^\vee K$, we added $\text{coev} : \mathbb{I} \to K \otimes {}^\vee K$ to morphisms of $\mathcal{C}$, for

$$(\mathbb{I} \xrightarrow{\text{coev}} K \otimes {}^\vee K \xrightarrow{i \otimes {}^\vee K} M \otimes {}^\vee K) = (\mathbb{I} \xrightarrow{\text{coev}} M \otimes {}^\vee M \xrightarrow{M \otimes {}^t i} M \otimes {}^\vee K)$$

is already there. Hence, K has dual objects in $\mathcal{C}$.

Step 6. Assume that we added to $\mathcal{P}$ the cokernel $j : N \to L$ of a morphism $f : M \to N$ such that M and N have dual objects in $\mathcal{C}$. Then $L^\vee$, ${}^\vee L$ are added at Step 5 as the kernels $(j^t : L^\vee \to N^\vee) = \ker f^t$, $({}^t j : {}^\vee L \to {}^\vee N) = \ker {}^t f$. Applying Step 6 to $f \otimes L^\vee : M \otimes L^\vee \to N \otimes L^\vee$, whose cokernel is $j \otimes L^\vee : N \otimes L^\vee \to L \otimes L^\vee$, we added $\text{ev} : L \otimes L^\vee \to \mathbb{I}$ to $\operatorname{Mor} \mathcal{C}$. Applying Step 6 to ${}^\vee L \otimes f : {}^\vee L \otimes M \to {}^\vee L \otimes N$, whose cokernel is ${}^{\vee}L \otimes j : {}^{\vee}L \otimes N \to {}^{\vee}L \otimes L$, we added $\mathrm{ev} : {}^{\vee}L \otimes L \to \mathbb{1}$ to $\operatorname{Mor} \mathcal{C}$. Applying Step 5 to $f^t \otimes L : N^{\vee} \otimes L \to M^{\vee} \otimes L$, whose kernel is $j^t \otimes L : L^{\vee} \otimes L \to N^{\vee} \otimes L$, we added $\mathrm{coev} : \mathbb{1} \to L^{\vee} \otimes L$ to $\operatorname{Mor} \mathcal{C}$. Applying Step 5 to $L \otimes {}^t f : L \otimes {}^{\vee}N \to L \otimes {}^{\vee}M$, whose kernel is $L \otimes {}^t j : L \otimes {}^{\vee}L \to L \otimes {}^{\vee}N$, we added $\mathrm{coev} : \mathbb{1} \to L \otimes {}^{\vee}L$ to $\operatorname{Mor} \mathcal{C}$.

Step 7. If X, Y have dual objects in $\mathcal{C}$, then $X \otimes Y$ has dual objects $Y^{\vee} \otimes X^{\vee}$ and ${}^{\vee}Y \otimes {}^{\vee}X$ with standard evaluations and coevaluations.

Therefore, $\mathcal{C}$ is rigid. □

4.6.3. Proposition. *Let H be a bicoalgebra in $\widehat{\mathcal{V}}$. If the monoidal category ${}^H\mathcal{V}$ of H-comodules is rigid, then H is a Hopf coalgebra.*

PROOF. We may assume that H is reconstructed from a small monoidal category $\mathcal{C}$ and a fibre functor $\omega : \mathcal{C} \to \mathcal{V}$. Since $\mathcal{C}$ is small, we can define morphisms $\gamma' : C^{\mathcal{C}}_{\mathrm{op}} \to H$, ${}'\gamma : C^{\mathcal{C}}_{\mathrm{op}} \to H$ by commutative diagrams holding for any $Z \in \operatorname{Ob} \mathcal{C}$:

$$\begin{array}{ccc} Z^{\vee} \boxtimes Z & \xrightarrow{Z^{\vee} \boxtimes \zeta} & Z^{\vee} \boxtimes Z^{\vee\vee} \\ {\scriptstyle Pi_Z}\big\downarrow & & \big\downarrow{\scriptstyle i_{Z^{\vee}}} \\ C^{\mathcal{C}}_{\mathrm{op}} & \xrightarrow{\gamma'} & H \end{array} \tag{4.6.5}$$

$$\begin{array}{ccc} Z^{\vee} \boxtimes Z & \xrightarrow{{}^t\zeta \boxtimes Z} & {}^{\vee}Z \boxtimes Z \\ {\scriptstyle Pi_Z}\big\downarrow & & \big\downarrow{\scriptstyle i_{{}^{\vee}Z}} \\ C^{\mathcal{C}}_{\mathrm{op}} & \xrightarrow{{}'\gamma} & H \end{array} \tag{4.6.6}$$

The upper rows are coalgebra homomorphisms $X^{\vee} \boxtimes \zeta : P_{\zeta}(X \boxtimes X^{\vee}) \to X^{\vee} \boxtimes X^{\vee\vee}$, ${}^t\zeta \boxtimes X : P_{\zeta}(X \boxtimes X^{\vee}) \to {}^{\vee}X \boxtimes X$. Hence, γ' and ${}'\gamma$ are coalgebra homomorphisms as well. The coalgebra H is the quotient of $C^{\mathcal{C}}$. Let us prove that γ', ${}'\gamma$ factor through $C^{\mathcal{C}}_{\mathrm{op}} \twoheadrightarrow H_{\mathrm{op}}$.

Let $f : X \to Y \in \operatorname{Mor} \mathcal{C}$. The associated relation in H is

$$\operatorname{rel} f = \big(X \boxtimes Y^{\vee} \xrightarrow{-f \boxtimes Y^{\vee} + X \boxtimes f^t} Y \boxtimes Y^{\vee} \oplus X \boxtimes X^{\vee} \xrightarrow{i_Y + i_X} C^{\mathcal{C}}\big).$$

We have to check vanishing of

$$\begin{aligned} &\big(Y^{\vee} \boxtimes X \xrightarrow{-P \operatorname{rel} f} C^{\mathcal{C}}_{\mathrm{op}} \xrightarrow{\gamma'} H\big) \\ &= \big(Y^{\vee} \boxtimes X \xrightarrow{Y^{\vee} \boxtimes f - f^t \boxtimes X} Y^{\vee} \boxtimes Y \oplus X^{\vee} \boxtimes X \xrightarrow{Pi_Y + Pi_X} C^{\mathcal{C}}_{\mathrm{op}} \xrightarrow{\gamma'} H\big) \\ &= \big(Y^{\vee} \boxtimes X \xrightarrow{Y^{\vee} \boxtimes f - f^t \boxtimes X} Y^{\vee} \boxtimes Y \oplus X^{\vee} \boxtimes X \xrightarrow{Y^{\vee} \boxtimes \zeta_Y \oplus X^{\vee} \boxtimes \zeta_X} \\ &\qquad\qquad Y^{\vee} \boxtimes Y^{\vee\vee} \oplus X^{\vee} \boxtimes X^{\vee\vee} \xrightarrow{i_{Y^{\vee}} + i_{X^{\vee}}} H\big) \\ &= \big(Y^{\vee} \boxtimes X \xrightarrow{Y^{\vee} \boxtimes \zeta_X} Y^{\vee} \boxtimes X^{\vee\vee} \xrightarrow{Y^{\vee} \boxtimes f^{tt} - f^t \boxtimes X^{\vee\vee}} \\ &\qquad\qquad Y^{\vee} \boxtimes Y^{\vee\vee} \oplus X^{\vee} \boxtimes X^{\vee\vee} \xrightarrow{i_{Y^{\vee}} + i_{X^{\vee}}} H\big) \\ &= \big(Y^{\vee} \boxtimes X \xrightarrow{Y^{\vee} \boxtimes \zeta_X} Y^{\vee} \boxtimes X^{\vee\vee} \xrightarrow{\operatorname{rel} f^t} H\big) = 0 \end{aligned}$$

since $f^t : Y^\vee \to X^\vee$ is in $\mathcal{C}$. Hence, $\gamma' = \big(C^{\mathcal{C}}_{\mathrm{op}} \to H_{\mathrm{op}} \xrightarrow{\gamma_\varsigma} H\big)$ for some coalgebra morphism γ_ς. Similarly, $'\gamma = \big(C^{\mathcal{C}}_{\mathrm{op}} \to H_{\mathrm{op}} \xrightarrow{{}_\varsigma\gamma} H\big)$ for some coalgebra morphism ${}_\varsigma\gamma$.

By diagram (4.6.5) (resp. (4.6.6)) the natural H-comodule structure on $Z^\vee \in \mathcal{C}$ (resp. ${}^\vee Z \in \mathcal{C}$) coincides with the one obtained from the H_{op}-comodule structure through γ_ς (resp. ${}_\varsigma\gamma$). Hence, we can apply the computations made in the proof of Theorem 4.3.4. Namely, since evaluations and coevaluations for arbitrary Z are morphisms of $\mathcal{C}$, equations (4.3.1a)–(4.3.2b) hold. This means that H is a Hopf coalgebra. □

In particular, we proved the converse to Theorem 4.3.4.

4.6.4. Remark. Even if there is no isomorphism $\zeta : X \to X^{\vee\vee}$ in $\mathcal{V}$, the category obtained from rigid monoidal construction data is rigid. However, we can not talk about the antipode in this case.

CHAPTER 5

Quasitriangular Hopf coalgebras

We want to discuss braided monoidal categories. Naturally, we assume that $\mathcal{V}$ is braided. It was shown in Propositions 3.5.1, 3.5.2 that a monoidal functor $(\circledast, \phi, r_{\mathbb{1}}^{-1}) : (\mathcal{V} \boxtimes \mathcal{V}, \bar{\otimes}, \mathbb{1} \boxtimes \mathbb{1}) \to (\mathcal{V}, \otimes, \mathbb{1})$ maps bicoalgebras H to braided bialgebras $\bar{H} = (\circledast H, \bar{\Delta}, \varepsilon, \bar{m}, \bar{\eta})$. The structure responsible for the braiding in $^B\mathcal{V}$ is the R-matrix. Bicoalgebras which admit an R-matrix will be called quasitriangular. We are going to establish correspondence between $\Bbbk$-linear exact monoidal functors $\omega : \mathcal{C} \to \mathcal{V}$ which do not preserve braiding and quasitriangular bicoalgebras. Quasitriangular Hopf coalgebra is a natural implementation of the idea of a quantum braided group [15]. Although there are two R-matrices – R_+ and R_-, they are linearly expressed one through another with the help of braiding in $\mathcal{V}$. We also discuss ribbon structures in the categories $^H\mathcal{V}$.

5.1. R-matrices in Hopf coalgebras

5.1.1. Definition. A *quasitriangular Hopf coalgebra* in $\widehat{\mathcal{V}}$ is a Hopf coalgebra H in $\widehat{\mathcal{V}}$ equipped with bilinear forms $R_+ : \bar{H} \otimes \bar{H} \to \mathbb{1}$, $R_- : \bar{H} \otimes \bar{H} \to \mathbb{1} \in \operatorname{Mor} \widehat{\mathcal{V}}$ called the R-matrices such that

$$R_+ = \bigl(\bar{H} \otimes \bar{H} \xrightarrow{\Omega} \bar{H} \otimes \bar{H} \xrightarrow{R_-} \mathbb{1}\bigr), \tag{5.1.1}$$

where Ω denotes the morphism $1^1 \otimes (c^{32} \cdot c^{23}) \otimes 1^4 : H^{12} \otimes H^{34} \to H^{12} \otimes H^{34}$,

$$\varepsilon = \bigl(\bar{H} \simeq \mathbb{1} \otimes \bar{H} \xrightarrow{\bar{\eta} \otimes \bar{H}} \bar{H} \otimes \bar{H} \xrightarrow{R_+} \mathbb{1}\bigr), \tag{5.1.2a}$$

$$\varepsilon = \bigl(\bar{H} \simeq \bar{H} \otimes \mathbb{1} \xrightarrow{\bar{H} \otimes \bar{\eta}} \bar{H} \otimes \bar{H} \xrightarrow{R_+} \mathbb{1}\bigr), \tag{5.1.2b}$$

$$\bigl(\bar{H} \otimes \bar{H} \otimes \bar{H} \xrightarrow{\bar{m} \otimes \bar{H}} \bar{H} \otimes \bar{H} \xrightarrow{R_+} \mathbb{1}\bigr)$$
$$= \bigl(\bar{H} \otimes \bar{H} \otimes \bar{H} \xrightarrow{\bar{H} \otimes \bar{H} \otimes \bar{\Delta}} \bar{H} \otimes \bar{H} \otimes \bar{H} \otimes \bar{H} \tag{5.1.3a}$$
$$\xrightarrow{\bar{H} \otimes R_+ \otimes \bar{H}} \bar{H} \otimes \mathbb{1} \otimes \bar{H} \simeq \bar{H} \otimes \bar{H} \xrightarrow{R_+} \mathbb{1}\bigr),$$

$$\bigl(\bar{H} \otimes \bar{H} \otimes \bar{H} \xrightarrow{\bar{H} \otimes \bar{m}} \bar{H} \otimes \bar{H} \xrightarrow{R_-} \mathbb{1}\bigr)$$
$$= \bigl(\bar{H} \otimes \bar{H} \otimes \bar{H} \xrightarrow{\bar{\Delta} \otimes \bar{H} \otimes \bar{H}} \bar{H} \otimes \bar{H} \otimes \bar{H} \otimes \bar{H} \tag{5.1.3b}$$
$$\xrightarrow{\bar{H} \otimes c^{-1} \otimes \bar{H}} \bar{H} \otimes \bar{H} \otimes \bar{H} \otimes \bar{H} \xrightarrow{R_- \otimes R_-} \mathbb{1} \otimes \mathbb{1} \simeq \mathbb{1}\bigr),$$

$$
\begin{aligned}
&\big(H_{1'2'} \otimes H_{1''2''} \simeq H_{1'2'} \otimes H_{1''2''} \otimes \mathbb{1}_{1'''} \xrightarrow{\Delta_{1'1'''2'} \otimes H} H_{1'1'''} \otimes H_{1''2''} \otimes H_{1^4 2'} \\
&\quad \xrightarrow{c^{-1}_{1''1'''} \otimes l^{-1}_H} H_{1'1''} \otimes H_{1'''2''} \otimes \mathbb{1}_{1^4} \otimes H_{1^5 2'} \xrightarrow{H \otimes \Delta_{1'''1^4 2''} \otimes H} \\
&\quad H_{1'1''} \otimes H_{1'''1^4} \otimes H_{1^5 2''} \otimes H_{1^6 2'} \xrightarrow{R_+ \otimes m} \mathbb{1}_{1'} \otimes H_{1''2} \simeq H_{12}\big) \\
&= \big(H_{1'2'} \otimes H_{1''2''} \simeq H_{1'2''} \otimes \mathbb{1}_{2'} \otimes H_{1''2'''} \xrightarrow{H \otimes \Delta_{1''2'2'''}} H_{1'2'''} \otimes H_{1''2'} \otimes H_{2''2^4} \\
&\quad \xrightarrow{r_H^{-1} \otimes c^{-1}_{2''2'''}} H_{1'2'''} \otimes H_{1''2'} \otimes \mathbb{1}_{2''} \otimes H_{2^4 2^5} \xrightarrow{\Delta_{1'2''2'''} \otimes H \otimes H} \\
&\quad H_{1'2''} \otimes H_{1''2'} \otimes H_{2'''2^4} \otimes H_{2^5 2^6} \xrightarrow{m \otimes R_+} H_{12'} \otimes \mathbb{1}_{2''} \simeq H_{12}\big).
\end{aligned}
\tag{5.1.4}
$$

It is easier to look at these conditions in graphical form. Equation (5.1.3a) becomes

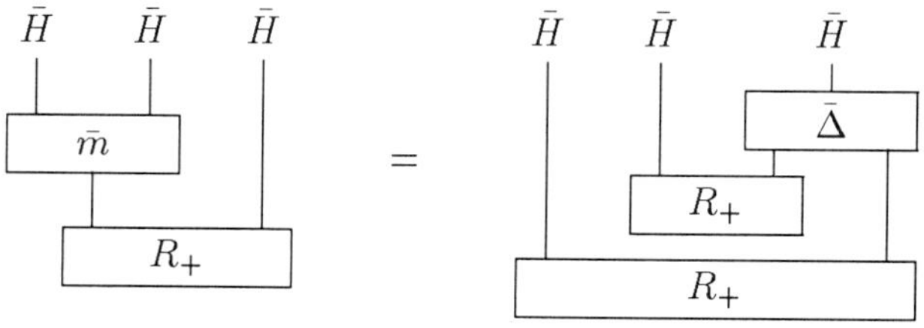

(5.1.5)

(5.1.3b) becomes

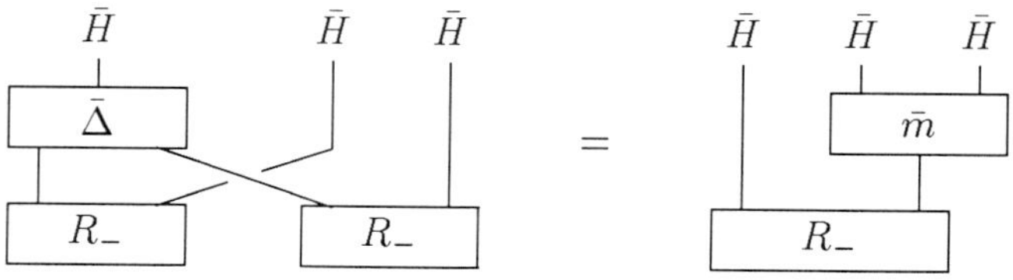

(5.1.6)

and (5.1.4) becomes

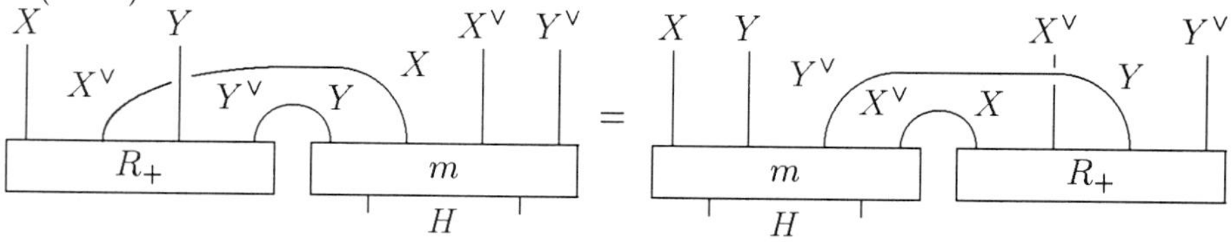

(5.1.7)

With a certain effort one can see that these properties are the dual ones to the equations for R-matrix written by Drinfeld [**10**].

5.1.2. Theorem. *Let (H, R_+, R_-) be a quasitriangular Hopf coalgebra. Then the categories of comodules ${}^H\mathcal{V}$ and ${}^H\widehat{\mathcal{V}}$ are braided and the braiding $R_{X,Y} : X \otimes Y \to Y \otimes X$ is given by*

$$
\begin{aligned}
R_{X,Y} &= \big(X \otimes Y \xrightarrow{\bar\delta_X \otimes \bar\delta_Y} \bar H \otimes X \otimes \bar H \otimes Y \xrightarrow{(432)^\sim_+} \bar H \otimes \bar H \otimes Y \otimes X \\
&\qquad\qquad \xrightarrow{R_+ \otimes Y \otimes X} \mathbb{1} \otimes Y \otimes X \simeq Y \otimes X\big) \\
&= \big(X \otimes Y \xrightarrow{\bar\delta_X \otimes \bar\delta_Y} \bar H \otimes X \otimes \bar H \otimes Y \xrightarrow{(432)^\sim_-} \bar H \otimes \bar H \otimes Y \otimes X \\
&\qquad\qquad \xrightarrow{R_- \otimes Y \otimes X} \mathbb{1} \otimes Y \otimes X \simeq Y \otimes X\big),
\end{aligned}
$$

or

$R_{X,Y} =$ 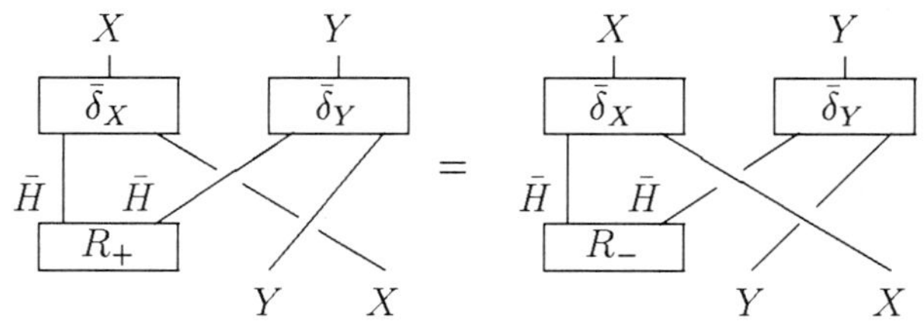

. (5.1.8)

PROOF. Equation (5.1.8) can be rewritten in the form

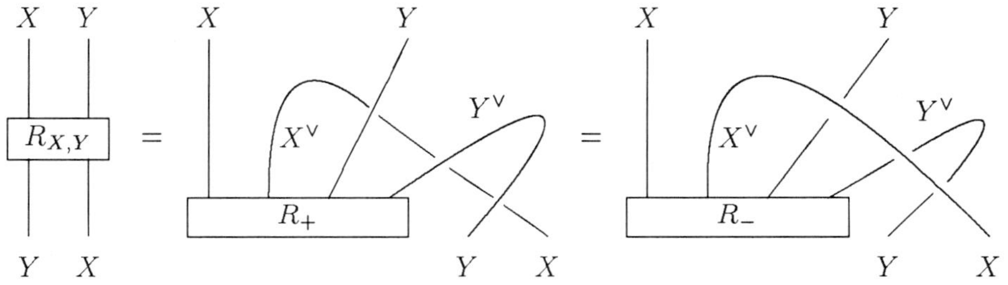

for $X, Y \in {}^H\mathcal{V}$, or

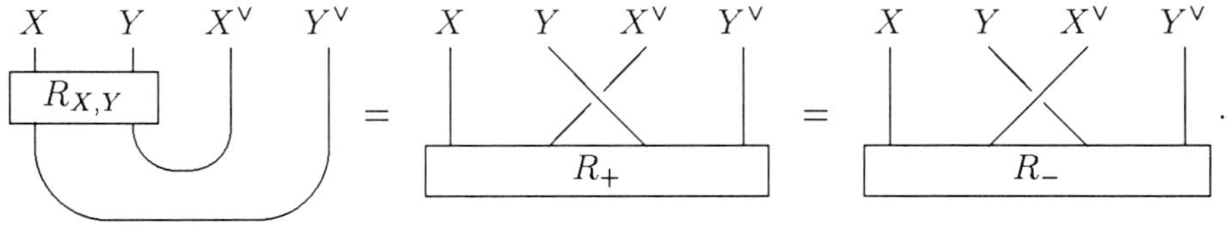

(5.1.9)

Hence, (5.1.1) implies that both expressions for $R_{X,Y}$ coincide.

It is clear from (5.1.8) that $R \circ f \otimes g = g \otimes f \circ R$ for any pair of morphisms $f, g \in {}^H\widehat{\mathcal{V}}$.

From (5.1.2a), (5.1.2b) one easily gets

$$R_{\mathbb{1},Y} = \big(\mathbb{1} \otimes Y \xrightarrow[\sim]{l_Y} Y \xrightarrow[\sim]{r_Y^{-1}} Y \otimes \mathbb{1}\big),$$

$$R_{X,\mathbb{1}} = \big(X \otimes \mathbb{1} \xrightarrow[\sim]{r_X} X \xrightarrow[\sim]{l_X^{-1}} \mathbb{1} \otimes X\big).$$

The hexagon equation

$$R_{X\otimes Y,Z} = \big(X \otimes Y \otimes Z \xrightarrow{X \otimes R_{Y,Z}} X \otimes Z \otimes Y \xrightarrow{R_{X,Z} \otimes Y} Z \otimes X \otimes Y\big) \qquad (5.1.10)$$

follows from (5.1.3a). Indeed,

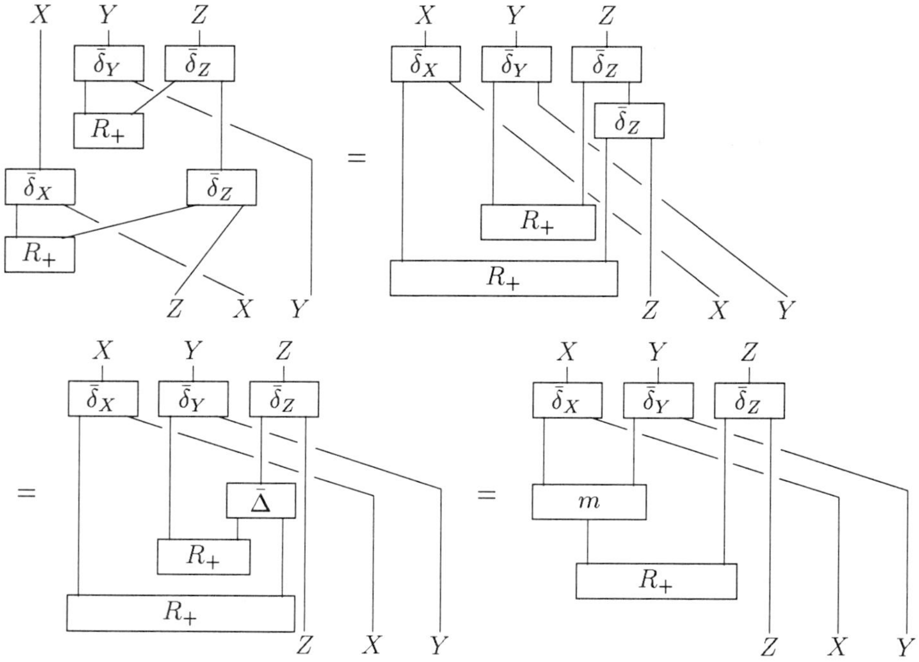

by (5.1.5).

The other hexagon condition

$$R_{Z,X\otimes Y} = \left(Z\otimes X\otimes Y \xrightarrow{R_{Z,X}\otimes Y} X\otimes Z\otimes Y \xrightarrow{X\otimes R_{Z,Y}} X\otimes Y\otimes Z\right) \tag{5.1.11}$$

follows from (5.1.3b). Indeed,

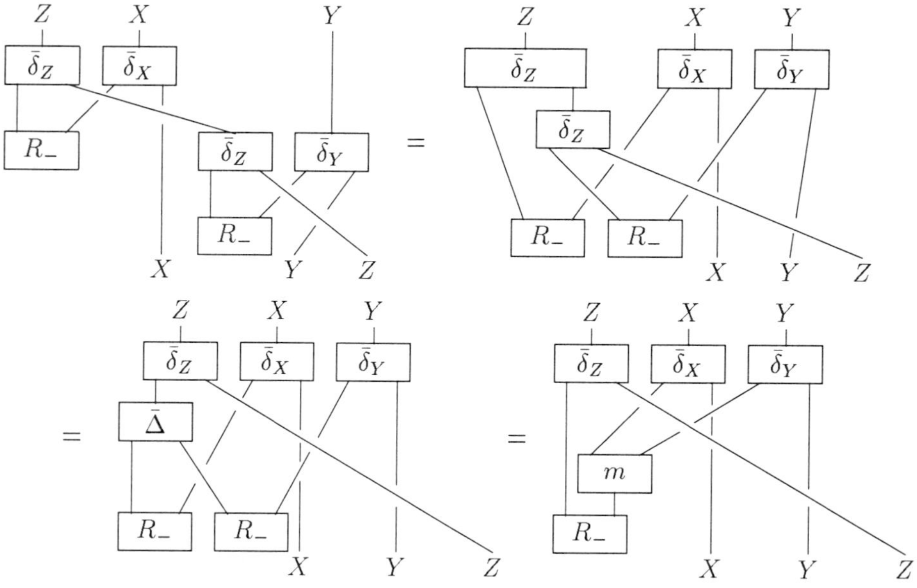

by (5.1.6).

Now let us prove that $R_{X,Y} : X \otimes Y \to Y \otimes X$ is a morphism of H-comodules. This is expressed by the equation

$$\begin{aligned}&\big(X_{1'} \otimes Y_{1''} \boxtimes \mathbb{1}_2 \xrightarrow{R_{X,Y}\boxtimes \mathbb{1}} Y_{1'} \otimes X_{1''} \boxtimes \mathbb{1}_2 \xrightarrow{Y\otimes \delta_X} Y_{1'} \otimes H_{1''2'} \otimes \mathbb{1}_{2''} \otimes X_{2'''}\\ &\quad \xrightarrow{\delta_Y \otimes H \otimes X} H_{1'2''} \otimes H_{1''2'} \otimes Y_{2'''} \otimes X_{2^4} \xrightarrow{m\otimes Y\otimes X} H_{12'} \otimes Y_{2''} \otimes X_{2'''}\big)\\ &= \big(X_{1'} \otimes Y_{1''} \boxtimes \mathbb{1}_2 \xrightarrow{X\otimes \delta_Y} X_{1'} \otimes H_{1''2'} \otimes \mathbb{1}_{2''} \otimes Y_{2'''} \xrightarrow{\delta_X \otimes H \otimes Y}\\ &\quad H_{1'2''} \otimes H_{1''2'} \otimes X_{2'''} \otimes Y_{2^4} \xrightarrow{m\otimes X\otimes Y} H_{12'} \otimes X_{2''} \otimes Y_{2'''}\\ &\quad \xrightarrow{H\otimes R_{X,Y}} H_{12'} \otimes Y_{2''} \otimes X_{2'''}\big),\end{aligned}$$

or

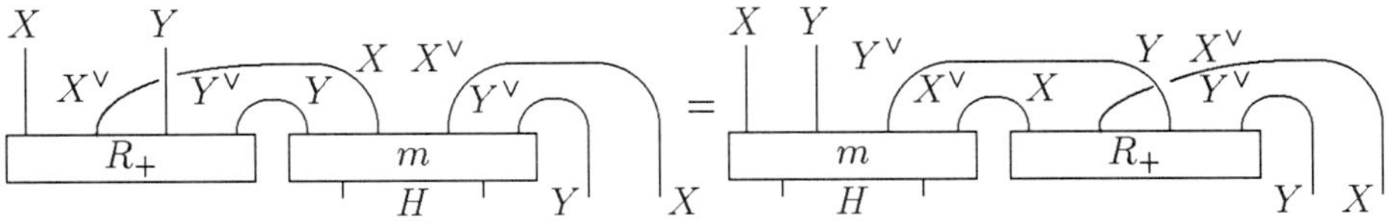

which is equivalent to (5.1.7).

Therefore, $R_{X,Y}$ is a functorial isomorphism. Rigidity of ${}^H\mathcal{V}$ implies that $R_{X,Y}$ is invertible. Indeed, its inverse is given by

$$\begin{aligned}R^{-1}_{X,Y} = \big(X \otimes Y &\xrightarrow{r^{-1}} X \otimes Y \otimes \mathbb{1} \xrightarrow{X\otimes Y\otimes \mathrm{coev}} X \otimes Y \otimes X^\vee \otimes X\\ &\xrightarrow{X\otimes R_{Y,X^\vee}\otimes X} X \otimes X^\vee \otimes Y \otimes X \xrightarrow{\mathrm{ev}\otimes Y\otimes X} \mathbb{1} \otimes Y \otimes X \xrightarrow{l} Y \otimes X\big)\end{aligned}$$

as one can easily see from the picture:

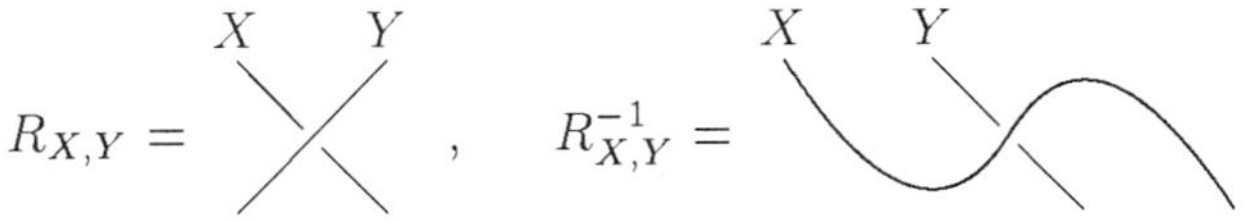

Thus, ${}^H\mathcal{V}$ is braided. □

5.1.3. Remark. One can define quasitriangular bicoalgebras in the same way as in Definition 5.1.1, adding one condition – invertibility of (5.1.8).

5.1.4. Remark. Equation (5.1.4) can be written in 4 equivalent forms. By (5.1.1) one can replace c^{-1}, R_+ by c, R_- in the left and/or, independently, in the right hand side of (5.1.4).

5.1.5. Proposition. *Let H be a bicoalgebra (e.g. a Hopf coalgebra). If ${}^H\mathcal{V}$ has a braiding $R_{X,Y} : X \otimes Y \to Y \otimes X$, then (H, R_+, R_-) is quasitriangular, where R_+, R_- are determined by*

$$\begin{aligned}&\big(X \otimes X^\vee \otimes Y \otimes Y^\vee \xrightarrow{\ddot{\imath}_X\otimes \ddot{\imath}_Y} \bar{H} \otimes \bar{H} \xrightarrow{R_\pm} \mathbb{1}\big)\\ &= \big(X \otimes X^\vee \otimes Y \otimes Y^\vee \xrightarrow{X\otimes c^{\pm 1}\otimes Y^\vee} X \otimes Y \otimes X^\vee \otimes Y^\vee \xrightarrow{R_{X,Y}\otimes X^\vee\otimes Y^\vee} \qquad (5.1.12)\\ &\quad Y \otimes X \otimes X^\vee \otimes Y^\vee \xrightarrow{Y\otimes \mathrm{ev}\otimes Y^\vee} Y \otimes \mathbb{1} \otimes Y^\vee \simeq Y \otimes Y^\vee \xrightarrow{\mathrm{ev}} \mathbb{1}\big).\end{aligned}$$

PROOF. Choose any small full subcategory $\mathcal{O} \subset {}^H\mathcal{V}$ equivalent to ${}^H\mathcal{V}$. Define $R'_\pm : (\circledast C^{\mathcal{O}}) \otimes (\circledast C^{\mathcal{O}}) \to \mathbb{1}$ by (5.1.12). Since c and $R_{X,Y}$ are functorial isomorphisms, the subspace $\circledast \operatorname{Rel} f \subset \circledast C^{\mathcal{O}}$ is orthogonal to $\circledast C^{\mathcal{O}}$ in any sense for any $f : X \to Y \in \mathcal{O}$. Hence, $R'_\pm$ determine bilinear forms $R_\pm$ on the quotient $\bar{H} \simeq \circledast C^{\mathcal{O}} / (\sum_f \circledast \operatorname{Rel} f)$.

Since the definition (5.1.12) of $R_\pm$ is equivalent to (5.1.9), we can recover $R_{X,Y}$ from $R_\pm$ through (5.1.8). The proof of Theorem 5.1.2 shows that properties (5.1.1)–(5.1.4) of $R_\pm$ are equivalent to R being a braiding. Hence, the converse to that theorem holds as well. □

5.2. Braiding for comodules over a braided Hopf algebra

It seems that there is no gadget which would make the *whole* category of comodules over a braided Hopf algebra into a braided category. That is why the framework of braided Hopf algebras seems insufficient for the ideas like quantum braided groups. However, braided Hopf algebras, which come from quasitriangular squared Hopf coalgebras, make an exception.

Let H be a quasitriangular Hopf coalgebra, and let $\bar{H} = \circledast H$. The category ${}^{\bar{H}}\mathcal{V}$ of comodules over the braided Hopf algebra $\bar{H}$ is equivalent to the tensor product ${}^{H}\mathcal{V} \boxtimes \mathcal{V}$ of two braided rigid monoidal categories by Theorem 2.7.1. This allows to introduce a braiding in ${}^{\bar{H}}\mathcal{V}$.

Indeed, the tensor product

$$\underline{\otimes} : ({}^{H}\mathcal{V} \boxtimes \mathcal{V}) \times ({}^{H}\mathcal{V} \boxtimes \mathcal{V}) \xrightarrow{\boxtimes} {}^{H}\mathcal{V} \boxtimes \mathcal{V} \boxtimes {}^{H}\mathcal{V} \boxtimes \mathcal{V}$$
$$\xrightarrow{1 \boxtimes P \boxtimes 1} {}^{H}\mathcal{V} \boxtimes {}^{H}\mathcal{V} \boxtimes \mathcal{V} \boxtimes \mathcal{V} \xrightarrow{\circledast \boxtimes \circledast} {}^{H}\mathcal{V} \boxtimes \mathcal{V},$$

$$(A \underline{\otimes} B)_{12} = A_{1'2'} \otimes B_{1''2''} \qquad \text{for } A, B \in {}^{H}\mathcal{V} \otimes \mathcal{V},$$
$$(M \boxtimes X) \underline{\otimes} (N \boxtimes Y) = (M \otimes N) \boxtimes (X \otimes Y) \qquad \text{for } M, N \in {}^{H}\mathcal{V},\ X, Y \in \mathcal{V},$$

with the corresponding associativity and unit constraints makes $({}^{H}\mathcal{V} \boxtimes \mathcal{V}, \underline{\otimes}, \mathbb{1} \boxtimes \mathbb{1})$ into a monoidal category similarly to Section 1.6.

5.2.1. Proposition. *The functorial isomorphism* $\psi : A^{12} \otimes B^{34} \to A^{13} \otimes B^{24}$, $A, B \in {}^{H}\mathcal{V} \boxtimes \mathcal{V}$, *determined by*

$$\psi = 1 \otimes c \otimes 1 : M \otimes X \otimes N \otimes Y \to M \otimes N \otimes X \otimes Y \tag{5.2.1}$$

for $M, N \in {}^{H}\mathcal{V}$, $X, Y \in \mathcal{V}$, *gives a monoidal equivalence*

$$(\circledast, \psi, r_{\mathbb{1}}^{-1}) : ({}^{H}\mathcal{V} \boxtimes \mathcal{V}, \underline{\otimes}, \mathbb{1} \boxtimes \mathbb{1}) \to ({}^{\bar{H}}\mathcal{V}, \otimes, \mathbb{1}). \tag{5.2.2}$$

PROOF. The functor $\circledast : {}^{H}\mathcal{V} \boxtimes \mathcal{V} \to {}^{\bar{H}}\mathcal{V}$ is an equivalence by Theorem 2.7.1. If we forget about the $\bar{H}$-comodule structure, we have a monoidal functor

$$(\circledast, \psi, r_{\mathbb{1}}^{-1}) : ({}^{H}\mathcal{V} \boxtimes \mathcal{V}, \underline{\otimes}, \mathbb{1} \boxtimes \mathbb{1}) \to (\mathcal{V}, \otimes, \mathbb{1})$$

similarly to Proposition 3.5.1. To lift it to the monoidal functor (5.2.2) we only have to prove that (5.2.1) is a morphism of $\bar{H}$-comodules.

This follows from the commutative diagram

$$\begin{array}{ccccc}
M\otimes X\otimes N\otimes Y & \xrightarrow{\delta_M\otimes X\otimes\delta_N\otimes Y} & \bar{H}\otimes M\otimes X\otimes\bar{H}\otimes N\otimes Y & \xrightarrow{(234)^{\sim}_{+}} & \bar{H}\otimes\bar{H}\otimes M\otimes X\otimes N\otimes Y \\
\downarrow{\scriptstyle M\otimes c\otimes Y} & & & & \downarrow{\scriptstyle \bar{H}\otimes\bar{H}\otimes M\otimes c\otimes Y} \\
M\otimes N\otimes X\otimes Y & \xrightarrow{\delta_M\otimes\delta_N\otimes X\otimes Y} & \bar{H}\otimes M\otimes\bar{H}\otimes N\otimes X\otimes Y & \xrightarrow{\bar{H}\otimes c\otimes N\otimes X\otimes Y} & \bar{H}\otimes\bar{H}\otimes M\otimes N\otimes X\otimes Y
\end{array}$$

The commutativity of the above follows from a more general diagram, valid for arbitrary comodules $M \in {}^{\bar{C}}\mathcal{V}$, $N \in {}^{\bar{D}}\mathcal{V}$ over ordinary coalgebras $\bar{C}, \bar{D} \in \widehat{\mathcal{V}}$ and for objects $X, Y \in \mathcal{V}$, which, in turn, follows from the particular case $\bar{C} = M \otimes M^\vee$, $\bar{D} = N \otimes N^\vee$. The last case is proved graphically in the following picture

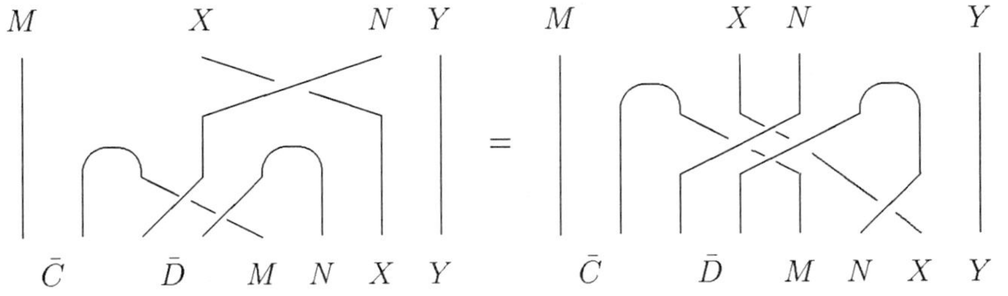

□

Each of the four isomorphisms

$$c_{\pm\pm} : (M \boxtimes X) \underline{\otimes} (N \boxtimes Y) = (M \otimes N) \boxtimes (X \otimes Y) \xrightarrow{R^{\pm 1} \boxtimes c^{\pm 1}} (N \otimes M) \boxtimes (Y \otimes X) = (N \boxtimes Y) \underline{\otimes} (M \boxtimes X)$$

gives a braiding in $({}^H\mathcal{V} \boxtimes \mathcal{V}, \underline{\otimes}, \mathbb{1} \boxtimes \mathbb{1})$. Thus, the category of $\bar{H}$-comodules is braided as well.

5.2.2. Corollary. *The category* $({}^{\bar{H}}\mathcal{V}, \otimes, \mathbb{1})$ *has braidings* $c'_{\pm\pm}$ *such that*

$$c'_{\pm\pm} = \big((M \otimes X) \otimes (N \otimes Y) \xrightarrow{1 \otimes c \otimes 1} M \otimes N \otimes X \otimes Y \xrightarrow{R^{\pm 1} \otimes c^{\pm 1}} N \otimes M \otimes Y \otimes X \xrightarrow{1 \otimes c^{-1} \otimes 1} (N \otimes Y) \otimes (M \otimes X)\big)$$

for $M, N \in {}^H\mathcal{V}$, $X, Y \in \mathcal{V}$.

5.3. Constructing braided categories

5.3.1. Definition. *Braided construction data* are monoidal construction data $p : \mathcal{P} \to \overline{\mathcal{V}}$ containing morphisms $R_{X,Y} : XY \to YX \in \operatorname{Mor} \mathcal{P}$ for any pair $X, Y \in \mathcal{O} = \operatorname{Ob} \mathcal{P}$, such that $pR_{X,Y} : pX \otimes pY \to pY \otimes pX$ is invertible and the following requirements hold (we omit p from the notation): for any $f : X_1 \dots X_k \to Y_1 \dots Y_n \in \operatorname{Mor} \mathcal{P}$ and any $Z \in \operatorname{Ob} \mathcal{P}$ the equations hold in $\mathcal{V}$

$$\begin{aligned}
&\big(Z \otimes X_1 \otimes \dots \otimes X_k \xrightarrow{R_{Z,X_1} \otimes 1} X_1 \otimes Z \otimes X_2 \otimes \dots \otimes X_k \xrightarrow{1 \otimes R_{Z,X_2} \otimes 1} \dots \\
&\xrightarrow{1 \otimes R_{Z,X_{k-1}} \otimes 1} X_1 \otimes \dots \otimes X_{k-1} \otimes Z \otimes X_k \xrightarrow{1 \otimes R_{Z,X_k}} \\
&X_1 \otimes \dots \otimes X_k \otimes Z \xrightarrow{f \otimes Z} Y_1 \otimes \dots \otimes Y_n \otimes Z\big) \\
&= \big(Z \otimes X_1 \otimes \dots \otimes X_k \xrightarrow{Z \otimes f} Z \otimes Y_1 \otimes \dots \otimes Y_n \xrightarrow{R_{Z,Y_1} \otimes 1} \\
&Y_1 \otimes Z \otimes Y_2 \otimes \dots \otimes Y_n \xrightarrow{1 \otimes R_{Z,Y_2} \otimes 1} \dots \xrightarrow{1 \otimes R_{Z,Y_{n-1}} \otimes 1} \\
&Y_1 \otimes \dots \otimes Y_{n-1} \otimes Z \otimes Y_n \xrightarrow{1 \otimes R_{Z,Y_n}} Y_1 \otimes \dots \otimes Y_n \otimes Z\big),
\end{aligned} \tag{5.3.1}$$

$$
\begin{aligned}
&\big(X_1\otimes\cdots\otimes X_k\otimes Z \xrightarrow{1\otimes R_{X_k,Z}} X_1\otimes\cdots\otimes X_{k-1}\otimes Z\otimes X_k \xrightarrow{1\otimes R_{X_{k-1},Z}\otimes 1} \dots \\
&\qquad\xrightarrow{1\otimes R_{X_2,Z}\otimes 1} X_1\otimes Z\otimes X_2\otimes\cdots\otimes X_k \xrightarrow{R_{X_1,Z}\otimes 1}\\
&\qquad Z\otimes X_1\otimes\cdots\otimes X_k \xrightarrow{Z\otimes f} Z\otimes Y_1\otimes\cdots\otimes Y_n\big)\\
&= \big(X_1\otimes\cdots\otimes X_k\otimes Z \xrightarrow{f\otimes Z} Y_1\otimes\cdots\otimes Y_n\otimes Z \xrightarrow{1\otimes R_{Y_n,Z}}\\
&\qquad Y_1\otimes\cdots\otimes Y_{n-1}\otimes Z\otimes Y_n \xrightarrow{1\otimes R_{Y_{n-1},Z}\otimes 1}\dots \xrightarrow{1\otimes R_{Y_2,Z}\otimes 1}\\
&\qquad Y_1\otimes Z\otimes Y_2\otimes\cdots\otimes Y_n \xrightarrow{R_{Y_1,Z}\otimes 1} Z\otimes Y_1\otimes\cdots\otimes Y_n\big).
\end{aligned} \tag{5.3.2}
$$

If $k=0$ or $n=0$ the composite $Z\otimes\mathbb{I}\xrightarrow{r_Z} Z\xrightarrow{l_Z^{-1}}\mathbb{I}\otimes Z$ and $\mathbb{I}\otimes Z\xrightarrow{l_Z} Z\xrightarrow{r_Z^{-1}} Z\otimes\mathbb{I}$ are used above.

In particular, the above equations for $f=R_{X,Y}$ imply that R satisfies the Yang–Baxter (braid) equation $R^{12}R^{23}R^{12}=R^{23}R^{12}R^{23}$.

5.3.2. Theorem. *The monoidal category $\mathcal{C}$ constructed from braided construction data p is braided. The bicoalgebra H_p is quasitriangular.*

PROOF. Repeating Steps 0–10 we get in the limit the monoidal construction data $\bar{\omega}:\overline{\mathcal{C}}\to\overline{\mathcal{V}}$. Let us find such morphisms $R_{X,Y}:XY\to YX$ in $\overline{\mathcal{C}}$ for all objects $X,Y\in\mathcal{C}$ that $\overline{\mathcal{C}}$ is braided construction data. Then $\mathcal{C}\subset\overline{\mathcal{C}}$ will be a braided category with the braiding $\bar{\omega}R_{X,Y}:\omega X\otimes\omega Y\to\omega Y\otimes\omega X$. We shall not distinguish between XY and $X\otimes Y$ due to identification morphisms $h_{X,Y}$.

Step 0. Set $R_{Z,\mathbb{I}}=\big(Z\otimes\mathbb{I}\xrightarrow{r_Z} Z\xrightarrow{l_Z^{-1}}\mathbb{I}\otimes Z\big)$, $R_{\mathbb{I},Z}=\big(\mathbb{I}\otimes Z\xrightarrow{l_Z} Z\xrightarrow{r_Z^{-1}} Z\otimes\mathbb{I}\big)$. Then (5.3.1), (5.3.2) hold for $f=i:\mathbb{I}\to\varnothing$.

Step 1. Extend $R_{X,Y}$ linearly.

Steps 2–4. The added morphisms obey (5.3.1), (5.3.2).

Step 5. Suppose we have a morphism $u:M\to N\in\mathcal{P}$ and we included its kernel $i:K\to M$ into $\mathcal{C}$. Then $i\otimes Z$ is the kernel of $u\otimes Z$ and by Step 5 we know that there exists $R_{Z,K}$ in $\mathcal{C}$ such that the right part of the diagram

$$
\begin{array}{ccccccc}
Z\otimes X & \xrightarrow{Z\otimes g} & Z\otimes K & \xrightarrow{Z\otimes i} & Z\otimes M & \xrightarrow{Z\otimes u} & Z\otimes N\\
{\scriptstyle R_{Z,X}}\big\downarrow & & {\scriptstyle R_{Z,K}}\big\downarrow & & \big\downarrow{\scriptstyle R_{Z,M}} & & \big\downarrow{\scriptstyle R_{Z,N}}\\
X\otimes Z & \xrightarrow{g\otimes Z} & K\otimes Z & \xrightarrow{i\otimes Z} & M\otimes Z & \xrightarrow{u\otimes Z} & N\otimes Z
\end{array}
$$

commutes. For any $g:X\to K\in\mathcal{V}$ such that $\big(X\xrightarrow{g}K\xrightarrow{i}M\big)\in\mathcal{C}$ one easily deduce commutativity of the left square. Therefore, (5.3.1) holds for morphisms $f=i$ and $f=g$, added at this step. One deals with (5.3.2) symmetrically.

Step 6 is dual to Step 5 (revert all arrows).

Step 7. Suppose we know $R_{X,Z}$, $R_{Y,Z}$, $R_{Z,X}$, $R_{Z,Y}$. Then $R_{X\otimes Y,Z}$ and $R_{Z,X\otimes Y}$ are defined by (5.1.10), (5.1.11). This amounts to say that (5.3.1), (5.3.2) hold for $f=h_{X,Y}:XY\to X\otimes Y$.

Step 8. It is easy to check (5.3.1), (5.3.2) for $f = r : X\mathbb{1} \to X$ or $f = l : \mathbb{1}X \to X$ with the above definition of $R_{Z,\mathbb{1}}$, $R_{\mathbb{1},Z}$.

Step 9 gives nothing to check.

Step 10. If (5.3.1), (5.3.2) hold for f and g, they hold for $f \otimes g$ as well.

Therefore, $\overline{\mathcal{C}}$ is braided, hence, $\mathcal{C}$ is braided. By Proposition 5.1.5 H_p is quasitriangular. □

5.4. Ribbon Hopf coalgebras

Let us assume that $\mathcal{V}$ is braided. Fix the isomorphism $\zeta = u_1^2 : X \to X^{\vee\vee}$. We want to discuss ribbon structures in the categories ${}^H\mathcal{V}$. A ribbon structure in $\mathcal{V}$ is not required.

5.4.1. Definition. A *ribbon Hopf coalgebra* is a quasitriangular Hopf coalgebra H equipped with a linear form $\Theta : \bar{H} \to \mathbb{1} \in \widehat{\mathcal{V}}$ such that

$$\big(\mathbb{1} \xrightarrow{\bar{\eta}} \bar{H} \xrightarrow{\Theta} \mathbb{1}\big) = \mathrm{id}_{\mathbb{1}}, \tag{5.4.1}$$

$$\big(H^{12} \xrightarrow{c^{12}} H^{21} \xrightarrow{\gamma_\zeta^{12}} H^{12} \xrightarrow{\Theta} \mathbb{1}\big) = \Theta, \tag{5.4.2}$$

$$\begin{aligned}\big(\bar{H} \otimes \bar{H} \xrightarrow{\bar{\Delta}\otimes\bar{\Delta}} \bar{H} \otimes \bar{H} \otimes \bar{H} \otimes \bar{H} &\xrightarrow{\bar{\Delta}\otimes\Theta\otimes\bar{H}\otimes\Theta} \bar{H} \otimes \bar{H} \otimes \mathbb{1} \otimes \bar{H} \otimes \mathbb{1} \\ &\xrightarrow{\sim} \bar{H} \otimes \bar{H} \otimes \bar{H} \xrightarrow{\bar{H}\otimes c} \bar{H} \otimes \bar{H} \otimes \bar{H} \xrightarrow{\bar{H}\otimes\bar{\Delta}\otimes\bar{H}} \\ &\bar{H} \otimes \bar{H} \otimes \bar{H} \otimes \bar{H} \xrightarrow{R_+\otimes R_-} \mathbb{1} \otimes \mathbb{1} \xrightarrow{\sim} \mathbb{1}\big) \\ = \big(\bar{H} \otimes \bar{H} \xrightarrow{\bar{m}} \bar{H} \xrightarrow{\Theta} \mathbb{1}\big),&\end{aligned} \tag{5.4.3}$$

$$\begin{aligned}&\big(H_{12} \xrightarrow{\sim} H_{1'2} \otimes \mathbb{1}_{1''} \xrightarrow{\Delta_{1'1''2}} H_{1'1''} \otimes H_{1'''2} \xrightarrow{\Theta\otimes H} \mathbb{1}_{1'} \otimes H_{1''2} \xrightarrow{\sim} H_{12}\big) \\ &= \big(H_{12} \xrightarrow{\sim} H_{12''} \otimes \mathbb{1}_{2'} \xrightarrow{\Delta_{12'2''}} H_{12'} \otimes H_{2''2'''} \xrightarrow{H\otimes\Theta} H_{12'} \otimes \mathbb{1}_{2''} \xrightarrow{\sim} H_{12}\big).\end{aligned} \tag{5.4.4}$$

5.4.2. Theorem. *The category ${}^H\mathcal{V}$ of comodules over a quasitriangular Hopf coalgebra H is ribbon if and only if H is ribbon.*

PROOF. Assume that H is ribbon. For any comodule $X \in {}^H\mathcal{V}$ define

$$\theta_X = \big(X \xrightarrow{\bar{\delta}} \bar{H} \otimes X \xrightarrow{\Theta\otimes X} \mathbb{1} \otimes X \simeq X\big) \in \mathcal{V}. \tag{5.4.5}$$

This is a functorial endomorphism $\theta : \mathcal{U} \to \mathcal{U}$. By (5.4.1) $\theta_{\mathbb{1}} = \mathrm{id}_{\mathbb{1}}$.

The property $\theta_X^t = \theta_{X^\vee} : X^\vee \to X^\vee$ can be written as

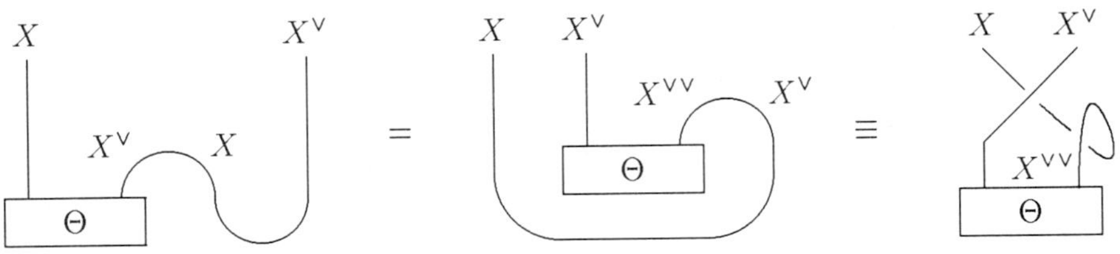

or

$$
\begin{aligned}
&\left(X \otimes X^\vee \xrightarrow{i_X^{12}} \bar{H} \xrightarrow{\Theta} \mathbb{1}\right) \\
&= \left(X \otimes X^\vee \xrightarrow{c} X^\vee \otimes X \xrightarrow{X^\vee \otimes \zeta} X^\vee \otimes X^{\vee\vee} \xrightarrow{i_{X^\vee}} \bar{H} \xrightarrow{\Theta} \mathbb{1}\right) \\
&= \left(X \otimes X^\vee \xrightarrow{c} X^\vee \otimes X \xrightarrow{i_X^{21}} H^{21} \xrightarrow{\gamma_\zeta^{12}} H^{12} \xrightarrow{\Theta} \mathbb{1}\right) \\
&= \left(X \otimes X^\vee \xrightarrow{i_X^{12}} H^{12} \xrightarrow{c} H^{21} \xrightarrow{\gamma_\zeta^{12}} H^{12} \xrightarrow{\Theta} \mathbb{1}\right).
\end{aligned}
$$

Thus this property follows from (5.4.2).

The balancing property

$$\theta_{X\otimes Y} = \left(X \otimes Y \xrightarrow{R^2_{X,Y}} X \otimes Y \xrightarrow{\theta_X \otimes \theta_Y} X \otimes Y\right) \tag{5.4.6}$$

can be written as

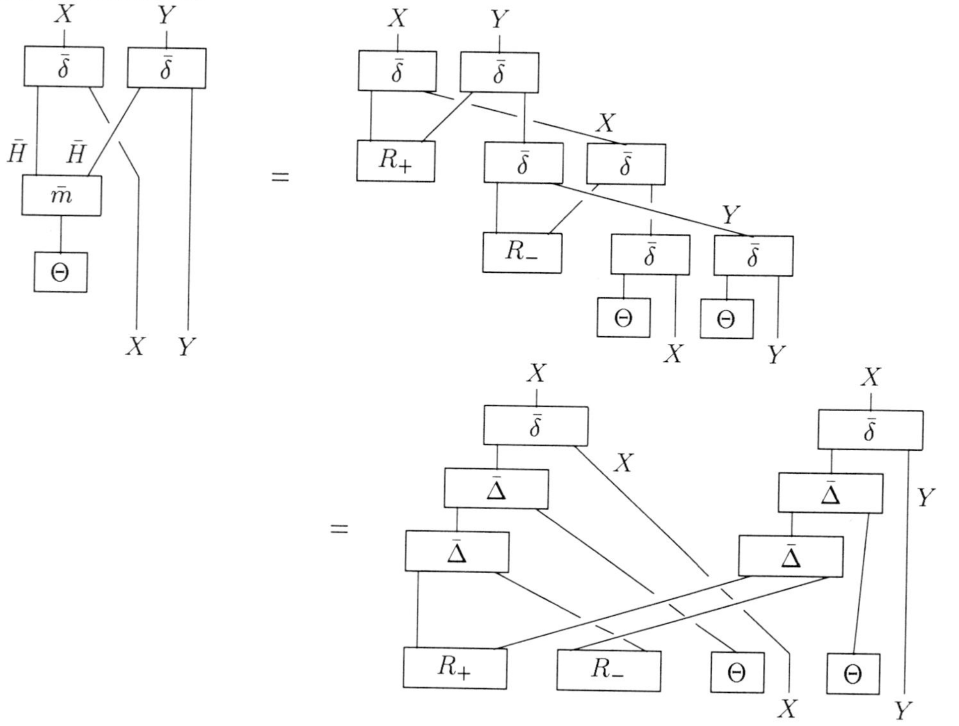

This is equivalent to

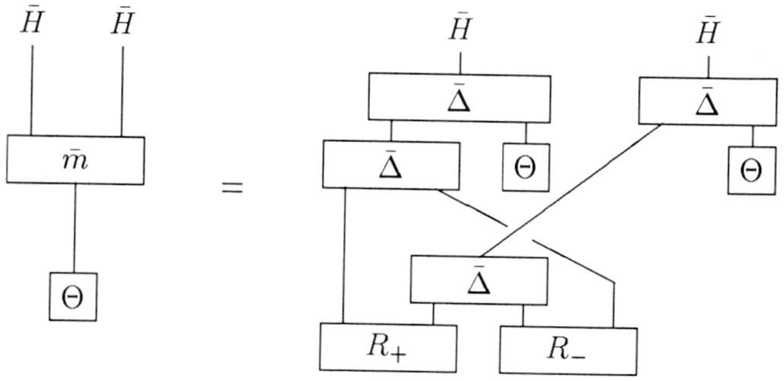

which is precisely (5.4.3).

To be H-comodule morphism, θ must satisfy the equation

$$\bigl(X \boxtimes \mathbb{1} \xrightarrow{\delta} H_{12'} \otimes X_{2''} \xrightarrow{H\otimes\theta} H_{12'} \otimes X_{2''}\bigr) = \bigl(X \boxtimes \mathbb{1} \xrightarrow{\theta\boxtimes\mathbb{1}} X \boxtimes \mathbb{1} \xrightarrow{\delta} H_{12'} \otimes X_{2''}\bigr),$$

or,

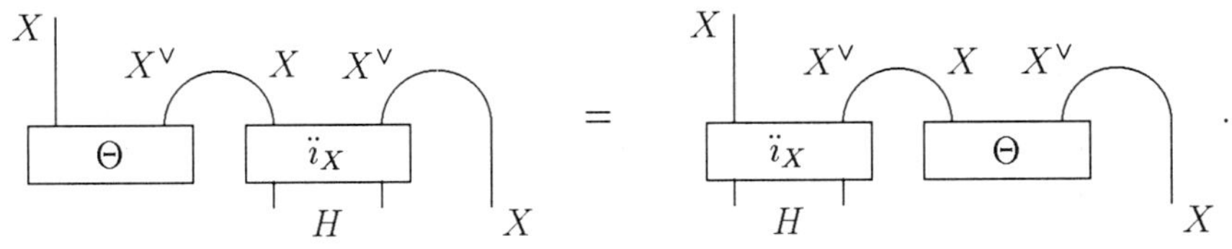

This is equivalent to

$$\begin{aligned}
&\bigl(X_1 \boxtimes X_2^\vee \simeq X_{1'} \otimes \mathbb{1}_{1''} \boxtimes X_2^\vee \xrightarrow[\Delta_{1'1''2}]{X\otimes\mathrm{coev}\boxtimes X^\vee} X_{1'} \otimes X^\vee_{1''} \otimes X_{1'''} \boxtimes X_2^\vee \\
&\qquad \xrightarrow{\ddot{i}_X \otimes \ddot{i}_X} H_{1'1''} \otimes H_{1'''2} \xrightarrow{\Theta\otimes H} \mathbb{1}_{1'} \otimes H_{1''2} \simeq H_{12}\bigr) \\
&= \bigl(X_1 \boxtimes X_2^\vee \simeq X_1 \boxtimes \mathbb{1}_{2'} \otimes X^\vee_{2''} \xrightarrow[\Delta_{12'2''}]{X\boxtimes\mathrm{coev}\otimes X^\vee} X_1 \boxtimes X^\vee_{2'} \otimes X_{2''} \otimes X^\vee_{2'''} \\
&\qquad \xrightarrow{\ddot{i}_X \otimes \ddot{i}_X} H_{12'} \otimes H_{2''2'''} \xrightarrow{H\otimes\Theta} H_{12'} \otimes \mathbb{1}_{2''} \simeq H_{12}\bigr),
\end{aligned}$$

which is nothing else but (5.4.4).

Therefore, ${}^H\mathcal{V}$ is ribbon, and it is shown in Proposition 1.2.11 and Corollary 1.2.12 that θ is invertible.

Vice versa, assume that ${}^H\mathcal{V}$ is ribbon with the twist θ. Then the diagram

$$\begin{array}{ccc}
X \otimes X^\vee & \xrightarrow{\theta\otimes X^\vee} & X \otimes X^\vee \\
{\scriptstyle \circledast\ddot{i}_X}\big\downarrow & & \big\downarrow{\scriptstyle \mathrm{ev}} \\
\bar{H} & \xrightarrow{\exists\Theta} & \mathbb{1}
\end{array}$$

determines a unique functional Θ. The automorphism θ can be restored from Θ by (5.4.5). We already have shown that the ribbon properties of θ are equivalent to (5.4.1)–(5.4.4). Theorem is proved. □

5.5. Ribbon reconstruction

Recall that $w_k \in \mathfrak{S}_k$ denotes the permutation of highest length.

5.5.1. Definition. *Ribbon construction data* are braided rigid monoidal construction data $p : \mathcal{P} \to \overline{\mathcal{V}}$ such that for every object $X \in \operatorname{Ob}\mathcal{P}$ an endomorphism $\theta_X : X \to X$ is chosen so that
(a) for any $f : X_1 \dots X_k \to Y_1 \dots Y_n \in \operatorname{Mor}\mathcal{P}$

$$\begin{aligned}
&f \circ ((w_k)^{\sim}_{R})^2 \circ (\theta_{X_1} \otimes \dots \otimes \theta_{X_k}) \\
&\quad = ((w_n)^{\sim}_{R})^2 \circ (\theta_{Y_1} \otimes \dots \otimes \theta_{Y_n}) \circ f : X_1 \otimes \dots \otimes X_k \to Y_1 \otimes \dots \otimes Y_n, \qquad (5.5.1)
\end{aligned}$$

here p is omitted;
(b) for any object $X \in \operatorname{Ob}\mathcal{P}$ there exists a composite tensor monomial $f : X \otimes Y_1 \otimes \dots \otimes Y_n \to \mathbb{1}$, made of morphisms from $\mathcal{P}$ (see (4.6.1)), non-degenerate in X and

such that

$$\begin{array}{ccc}
X \otimes Y_1 \otimes \cdots \otimes Y_n & \xrightarrow{X\otimes[((w_n)^{\sim}_{R})^2\circ(\theta_{Y_1}\otimes\cdots\otimes\theta_{Y_n})]} & X \otimes Y_1 \otimes \cdots \otimes Y_n \\
{\scriptstyle \theta_{X\otimes Y_1\otimes\cdots\otimes Y_n}}\big\downarrow & & \big\downarrow{\scriptstyle f} \\
X \otimes Y_1 \otimes \cdots \otimes Y_n & \xrightarrow{\quad f \quad} & \mathbb{I}
\end{array}$$

5.5.2. Theorem. *Let $p : \mathcal{P} \to \overline{\mathcal{V}}$ be ribbon construction data, then the constructed monoidal category $\mathcal{C}$ has a unique ribbon structure such that the ribbon twist coincides with θ_X for $X \in \operatorname{Ob}\mathcal{P}$.*

PROOF. Let us deduce from (a) the existence of a functorial endomorphism $\theta : \mathrm{Id}_{\mathcal{C}} \to \mathrm{Id}_{\mathcal{C}}$ such that $\theta_{\mathbb{I}} = \mathrm{id}_{\mathbb{I}}$ and the balancing property (5.4.6) holds. We shall do it step by step.

Step 0. Set $\theta_{\mathbb{I}} = \mathrm{id}_{\mathbb{I}}$, then (5.5.1) holds for $f = i : \mathbb{I} \to \varnothing$.

Step 1. Extend θ linearly.

Steps 2–4. Equation (5.5.1) holds for added morphisms.

Step 5. Let $u : M \to N \in \mathcal{P}$ and $(i : K \to M) = \ker u$. The unique morphism $\theta_K : K \to K \in \mathcal{V}$ which makes the central square commute in the diagram

$$\begin{array}{ccccccc}
X & \xrightarrow{g} & K & \xrightarrow{i} & M & \xrightarrow{u} & N \\
{\scriptstyle \theta_X}\big\downarrow & & \exists\big\downarrow{\scriptstyle \theta_K} & & \big\downarrow{\scriptstyle \theta_M} & & \big\downarrow{\scriptstyle \theta_N} \\
X & \xrightarrow{g} & K & \xrightarrow{i} & M & \xrightarrow{u} & N
\end{array} \tag{5.5.2}$$

is included in $\mathcal{C}$ by Step 5. For any g such that $\big(X \xrightarrow{g} K \xrightarrow{i} M\big)$ is in $\mathcal{C}$ both paths from the upper X to the lower M are equal, hence, the left square commutes.

Step 6 is dual to Step 5.

Step 7. Define θ for $X \otimes Y$ by (5.4.6). Then $f = h_{X,Y} : XY \to X \otimes Y$ satisfies (5.5.1).

Step 8. Easy to see that $r : X\mathbb{I} \to X$, $l : \mathbb{I}X \to X$ satisfy (5.5.1).

Step 9. Equation (5.5.1) for $f = a : X(YZ) \to (XY)Z$ follows from the diagram

$$\begin{array}{ccccccc}
X \otimes (Y \otimes Z) & \xrightarrow{\theta_X\otimes\theta_Y\otimes\theta_Z} & X \otimes Y \otimes Z & \xrightarrow{X\otimes R^2} & X \otimes Y \otimes Z & \xrightarrow{(321)^{\sim}_{R}} & Y \otimes Z \otimes X \\
{\scriptstyle a}\big\downarrow & & \Big\| & & & & \big\downarrow{\scriptstyle (123)^{\sim}_{R}} \\
(X \otimes Y) \otimes Z & \xrightarrow{\theta_X\otimes\theta_Y\otimes\theta_Z} & X \otimes Y \otimes Z & & & & X \otimes (Y \otimes Z) \\
 & & {\scriptstyle R^2\otimes Z}\big\downarrow & & & & \big\downarrow{\scriptstyle a} \\
 & & X \otimes Y \otimes Z & \xrightarrow{(123)^{\sim}_{R}} & Z \otimes X \otimes Y & \xrightarrow{(321)^{\sim}_{R}} & (X \otimes Y) \otimes Z
\end{array}$$

whose commutativity is straightforward (draw the corresponding braided diagram).

Step 10. Given $g : X \to Y$, $h : V \to W \in \mathcal{P}$, we have to check that $g \otimes h$ commutes with θ. This follows from the diagram

$$\begin{array}{ccccc} X \otimes V & \xrightarrow{\theta_X \otimes \theta_V} & X \otimes V & \xrightarrow{R^2} & X \otimes V \\ {\scriptstyle g\otimes h}\downarrow & & \downarrow{\scriptstyle g\otimes h} & & \downarrow{\scriptstyle g\otimes h} \\ Y \otimes W & \xrightarrow{\theta_Y \otimes \theta_W} & Y \otimes W & \xrightarrow{R^2} & Y \otimes W \end{array}$$

Now we have to check that θ is self-adjoint, that is $\theta_{X^\vee} = \theta^t_X$. First let us do it for $X \in \operatorname{Ob}\mathcal{P}$. By the property (b) from Definition 5.5.1 there is an object $Y = Y_1 \otimes \cdots \otimes Y_n \in \mathcal{C}$ and a pairing $f : X \otimes Y \to \mathbb{I}$, non-degenerate in X such that

$$\big(X \otimes Y \xrightarrow{\theta_X \otimes Y} X \otimes Y \xrightarrow{f} \mathbb{I}\big) = \big(X \otimes Y \xrightarrow{X \otimes \theta_Y} X \otimes Y \xrightarrow{f} \mathbb{I}\big). \tag{5.5.3}$$

The pairing f can be presented as

$$f = \big(X \otimes Y \xrightarrow{X \otimes y} X \otimes X^\vee \xrightarrow{\mathrm{ev}} \mathbb{I}\big)$$

for some epimorphism y. Hence, (5.5.3) can be presented as

$$\begin{aligned} &\big(X \otimes Y \xrightarrow{X \otimes y} X \otimes X^\vee \xrightarrow{X \otimes \theta^t_X} X \otimes X^\vee \xrightarrow{\mathrm{ev}} \mathbb{I}\big) \\ &\qquad = \big(X \otimes Y \xrightarrow{X \otimes y} X \otimes X^\vee \xrightarrow{X \otimes \theta_{X^\vee}} X \otimes X^\vee \xrightarrow{\mathrm{ev}} \mathbb{I}\big). \end{aligned}$$

Thus, $\theta^t_X = \theta_{X^\vee}$.

Now let us deduce the same for other objects included in $\mathcal{C}$.

Step 0. Obviously, $\theta_{\mathbb{I}^\vee} = \mathrm{id}_{\mathbb{I}^\vee} = \mathrm{id}^t_{\mathbb{I}} = \theta^t_{\mathbb{I}}$.

Step 1 follows by additivity.

Step 5. Assume that $\theta_{X^\vee} = \theta^t_X$ for $X = M, N$. From (5.5.2) we get the diagram with exact rows

$$\begin{array}{ccccccc} N^\vee & \xrightarrow{u^t} & M^\vee & \xrightarrow{i^t} & K^\vee & \longrightarrow & 0 \\ {\scriptstyle \theta^t_N}\downarrow{\scriptstyle =\theta_{N^\vee}} & & {\scriptstyle \theta^t_M}\downarrow{\scriptstyle =\theta_{M^\vee}} & & {\scriptstyle \theta^t_K}\downarrow & & \\ N^\vee & \xrightarrow[u^t]{} & M^\vee & \xrightarrow[i^t]{} & K^\vee & \longrightarrow & 0 \end{array}$$

and the right column equals to $\theta_{K^\vee}$ as well.

Step 6 is dual to Step 5.

Step 7. We defined $(X \otimes Y)^\vee$ as $Y^\vee \otimes X^\vee$. Hence, the equation $\theta_{(X\otimes Y)^\vee} = \theta^t_{X \otimes Y}$ follows from

$$\begin{array}{ccccc} Y^\vee \otimes X^\vee & \xrightarrow{\theta_{Y^\vee} \otimes \theta_{X^\vee}} & Y^\vee \otimes X^\vee & \xrightarrow{R^2} & Y^\vee \otimes X^\vee \\ \| & & \| & & \| \\ Y^\vee \otimes X^\vee & \xrightarrow{\theta^t_Y \otimes \theta^t_X} & Y^\vee \otimes X^\vee & \xrightarrow{R^{2\,t}} & Y^\vee \otimes X^\vee \end{array}$$

Therefore, $\mathcal{C}$ is ribbon. □

APPENDIX A

Symmetric monoidal 2-categories

A.1. A review of 2-categories

We shall not use weak 2-categories, so we shall not define them. But the reader can recover this definition, which differs slightly from the notion of bicategory [**4**], from the definition of a weak 3-category below. We recall the definitions of strict 2-categories and associated notions.

A.1.1. Definition. A (strict) *2-category* $\mathcal{C}$ consists of

1. a class of objects $\operatorname{Ob}\mathcal{C}$;
2. for any pair of objects $X, Y \in \operatorname{Ob}\mathcal{C}$ a category $\mathcal{C}(X,Y)$;
3. (a) for any object $X \in \operatorname{Ob}\mathcal{C}$ a functor $\bullet^0 : \mathbf{1} \to \mathcal{C}(X,X)$, object $\mapsto 1_X$;
   (b) for any triple of objects $X, Y, Z \in \operatorname{Ob}\mathcal{C}$ a functor

$$\bullet^2 : \mathcal{C}(X,Y) \times \mathcal{C}(Y,Z) \to \mathcal{C}(X,Z), \qquad (F,G) \mapsto F\bullet G = G \circ F;$$

   such that the following functors are equal
4. $F\bullet 1 = F = 1\bullet F$, $F\bullet(G\bullet H) = (F\bullet G)\bullet H$.

A.1.2. Definition. A *weak 2-functor* (a homomorphism in [**4**]) between 2-categories $\mathcal{C}$ and $\mathcal{C}'$ consists of

1. a function $F : \operatorname{Ob}\mathcal{C} \to \operatorname{Ob}\mathcal{C}'$;
2. a functor $F = F_{X,Y} : \mathcal{C}(X,Y) \to \mathcal{C}'(FX,FY)$ for each pair of objects $X, Y \in \operatorname{Ob}\mathcal{C}$;
3. (a) an isomorphism $\phi_0 : 1_{FX} \to F1_X$;
   (b) an invertible (natural) transformation

$$\begin{array}{ccc}
\mathcal{C}(X,Y) \times \mathcal{C}(Y,Z) & \xrightarrow{\ \bullet^2\ } & \mathcal{C}(X,Z) \\
{\scriptstyle F_{X,Y}\times F_{Y,Z}}\big\downarrow & \overset{\phi_2}{\nearrow\!\!\!\nearrow} & \big\downarrow{\scriptstyle F_{X,Z}} \\
\mathcal{C}'(FX,FY) \times \mathcal{C}'(FY,FZ) & \xrightarrow{\ \bullet^2\ } & \mathcal{C}'(FX,FZ)
\end{array}$$

   for each triple $X, Y, Z \in \operatorname{Ob}\mathcal{C}$;
   such that
4. (a) for any object $M \in \mathcal{C}(X,Y)$ the composites

$$FM = FM\bullet 1_{FY} \xrightarrow{FM\bullet\phi_0} FM\bullet F1_Y \xrightarrow{\phi_2} F(M\bullet 1_Y) = FM \tag{A.1.1}$$

$$FM = 1_{FX}\bullet FM \xrightarrow{\phi_0\bullet FM} F1_X\bullet FM \xrightarrow{\phi_2} F(1_X\bullet M) = FM \tag{A.1.2}$$

   are identity morphisms in $\mathcal{C}'(FM,FM)$;
   (b) For any objects $W, X, Y, Z \in \operatorname{Ob}\mathcal{C}$ and any object

$$(K,L,M) \in \mathcal{C}(W,X) \times \mathcal{C}(X,Y) \times \mathcal{C}(Y,Z)$$

there is an equation

$$
\begin{aligned}
(FK{\bullet}(FL{\bullet}FM) &\xrightarrow{FK\bullet\phi_2} FK{\bullet}F(L{\bullet}M) \xrightarrow{\phi_2} F(K{\bullet}(L{\bullet}M))) \\
&= ((FK{\bullet}FL){\bullet}FM \xrightarrow{\phi_2\bullet FM} F(K{\bullet}L){\bullet}FM \xrightarrow{\phi_2} F((K{\bullet}M){\bullet}M)).
\end{aligned}
$$

A.1.3. Definition. A *weak 2-transformation* (pseudo-natural transformation [**12**]) $\lambda : (F, \phi_2, \phi_0) \to (G, \psi_2, \psi_0) : \mathcal{C} \to \mathcal{C}'$ is

1. a family of 1-morphisms $\lambda_X : FX \to GX$, $X \in \operatorname{Ob}\mathcal{C}$;
2. for any 1-morphism $f : X \to Y$ in $\mathcal{C}$ a 2-isomorphism in $\mathcal{C}'$

$$\lambda_f : Ff{\bullet}\lambda_Y \xrightarrow{\sim} \lambda_X{\bullet}Gf : FX \to GY$$

which is an isomorphism of functors

$$\lambda_- : F - {\bullet}\lambda_Y \to \lambda_X{\bullet}G- : \mathcal{C}(X,Y) \to \mathcal{C}'(FX, GY),$$

such that

3. (a) for any object $X \in \operatorname{Ob}\mathcal{C}$

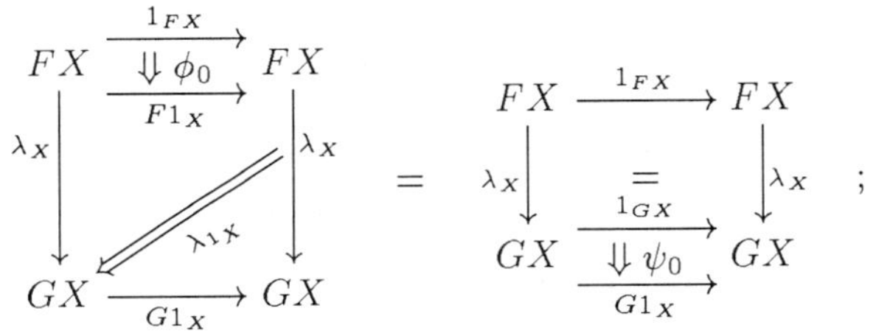

(b) for any pair of composable 1-morphisms $f, g \in \mathcal{C}_1$

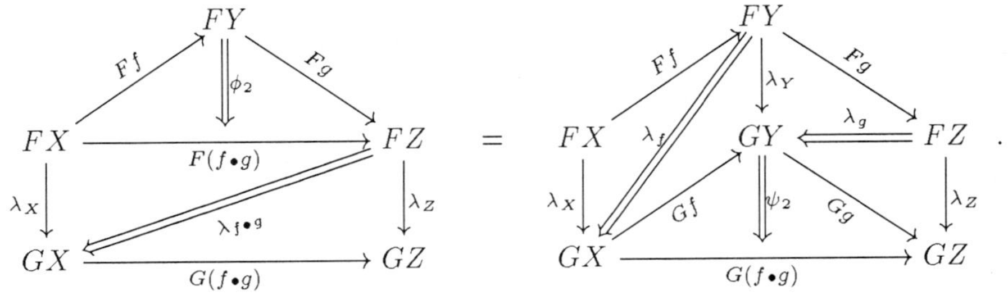

A weak 2-transformation $\lambda = (\lambda_X)$ for which λ_X are equivalences is called a *2-natural equivalence*.

A.1.4. Definition. A *modification* $m : \lambda \to \mu : F \to G : \mathcal{C} \to \mathcal{C}'$ is

1. a family of 2-morphisms $m_X : \lambda_X \to \mu_X$, $X \in \operatorname{Ob}\mathcal{C}$ such that
2. for any 1-morphism $f : X \to Y$ in $\mathcal{C}$

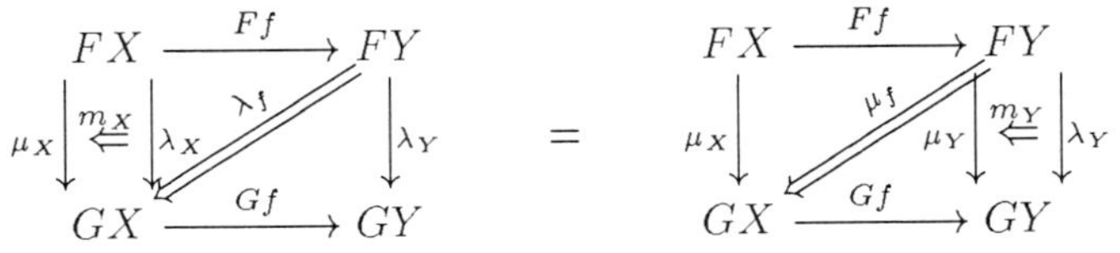

A.1.5. Compositions. For any 2-categories $\mathcal{C}$, $\mathcal{C}'$ the collection of weak 2-functors from $\mathcal{C}$ to $\mathcal{C}'$, weak 2-transformations between them and modifications between them with obvious compositions form a 2-category (e.g. [**12**]). Composition of two weak 2-functors $(F, \phi_2, \phi_0) : \mathcal{C} \to \mathcal{C}'$ and $(G, \psi_2, \psi_0) : \mathcal{C}' \to \mathcal{C}''$ is given by the map $GF : \operatorname{Ob} \mathcal{C} \to \operatorname{Ob} \mathcal{C}''$, the functor

$$\mathcal{C}(X,Y) \xrightarrow{F_{X,Y}} \mathcal{C}'(FX,FY) \xrightarrow{G_{FX,FY}} \mathcal{C}''(GFX,GFY),$$

the isomorphism

$$1_{GFX} \xrightarrow{\psi_0} G1_{FX} \xrightarrow{G\phi_0} GF1_X,$$

and the invertible transformation – the pasting of

$$\begin{array}{ccc}
\mathcal{C}(X,Y) \times \mathcal{C}(Y,Z) & \xrightarrow{\bullet^2} & \mathcal{C}(X,Z) \\
\Big\downarrow{\scriptstyle F_{X,Y} \times F_{Y,Z}} & \overset{\phi_2}{\nearrow\!\!\!\nearrow} & \Big\downarrow{\scriptstyle F_{X,Z}} \\
\mathcal{C}'(FX,FY) \times \mathcal{C}'(FY,FZ) & \xrightarrow{\bullet^2} & \mathcal{C}'(FX,FZ) \\
\Big\downarrow{\scriptstyle G_{FX,FY} \times G_{FY,FZ}} & \overset{\psi_2}{\nearrow\!\!\!\nearrow} & \Big\downarrow{\scriptstyle G_{FX,FZ}} \\
\mathcal{C}''(GFX,GFY) \times \mathcal{C}''(GFY,GFZ) & \xrightarrow{\bullet^2} & \mathcal{C}''(GFX,GFZ)
\end{array}$$

It is well defined and strictly associative [**4**].

If $\lambda : F \to F' : \mathcal{C} \to \mathcal{C}'$ is a weak 2-transformation and $G : \mathcal{C}' \to \mathcal{C}''$ is a weak 2-functor, denote by $\lambda {\bullet} G = G \circ \lambda : GF \to GF' : \mathcal{C} \to \mathcal{C}''$ the 2-transformation

$$(G \circ \lambda)_X = G(\lambda_X) : GFX \to GF'X, \qquad X \in \operatorname{Ob} \mathcal{C},$$

$$(G \circ \lambda)_f = (GFf {\bullet} G\lambda_Y \xrightarrow{\psi_2} G(Ff {\bullet} \lambda_Y) \xrightarrow{G\lambda_f} G(\lambda_X {\bullet} F'f) \xrightarrow{\psi_2^{-1}} G\lambda_X {\bullet} GF'f)$$

for $f : X \to Y \in \mathcal{C}$.

If $F : \mathcal{C} \to \mathcal{C}'$ is a weak 2-functor and $\mu : G \to G' : \mathcal{C}' \to \mathcal{C}''$ is a weak 2-transformation, denote by $\mu \circ F = F {\bullet} \mu : GF \to G'F : \mathcal{C} \to \mathcal{C}''$ the weak 2-transformation

$$(\mu \circ F)_X = \mu_{FX} : GFX \to G'FX,$$

$$(\mu \circ F)_f = \mu_{Ff} : GFf {\bullet} \mu_{FY} \to \mu_{FX} {\bullet} G'Ff : GFX \to G'FY.$$

If, furthermore, $H : \mathcal{C}'' \to \mathcal{C}'''$ is another weak 2-functor, we have $(F {\bullet} \mu) {\bullet} H = F {\bullet} (\mu {\bullet} H)$. Whenever equations $(F_1 {\bullet} F_2) {\bullet} \mu = F_1 {\bullet} (F_2 {\bullet} \mu)$, $(\mu {\bullet} H_1) {\bullet} H_2 = \mu {\bullet} (H_1 {\bullet} H_2)$ make sense, they hold true.

A.2. 2-pasting schemes

Following Power [**24**], we define a 2-pasting scheme as a non-empty, connected, finite, oriented graph G together with a (homotopy class of) embedding(s) of $|G|$ – 1-dimensional cell complex, realization of G – into the plane, which satisfies the following conditions:

(a) there are two distinct vertices $s(G)$, $t(G)$ in the boundary of the exterior face E (a face is a connected component of $\mathbb{R}^2 - |G|$), such that for each vertex v there are directed paths from $s(G)$ to v and from v to $t(G)$;
(b) G contains no directed cycle.

Then the following holds [**24**]:

1. For each interior face f there exist distinct vertices $s(f)$ and $t(f)$ and direct paths $\partial_s(f)$, $\partial_t(f) : s(f) \to t(f)$ such that the boundary of f is $\partial f = \overline{\partial_s(f) \cup \partial_t(f)}$. $\overline{p}$ is the path p with the opposite orientation. The paths $\partial_s(f)$ and $\partial_t(f)$ touch only at $s(f)$ and $t(f)$.
2. There are two directed paths $\partial_s(G)$, $\partial_t(G) : s(G) \to t(G)$ such that $\partial E = \partial_s(G) \cup \overline{\partial_t(G)}$.

Let P be the set of oriented paths in G from $s(G)$ to $t(G)$. Denote by F the set of interior faces of G. Assume that the number of interior faces $n = \#F$ is positive.

A.2.1. Definition. An *admissible order* $<_A$ on F is such a linear order $a_1 <_A \cdots <_A a_n$ on F, that there exist a sequence $A_0, A_1, \ldots, A_n$ of paths from P with the following properties:

1. $A_0 = \partial_s(G)$, $A_n = \partial_t(G)$;
2. for any $0 < i \leqslant n$ $A_{i-1} \cap \partial a_i = \partial_s a_i$, $A_i = (A_{i-1} - \partial_s a_i) \cup \partial_t a_i$.

Clearly, the paths A_i are uniquely determined by an admissible order $<_A$. The sequence of paths should be viewed as a combinatorial homotopy. The admissible order is the order in which 2-cells are swept over by the homotopy. The set of all admissible orders is denoted $\mathcal{O}$.

A.2.2. Example. For a 2-pasting scheme G presented on the picture

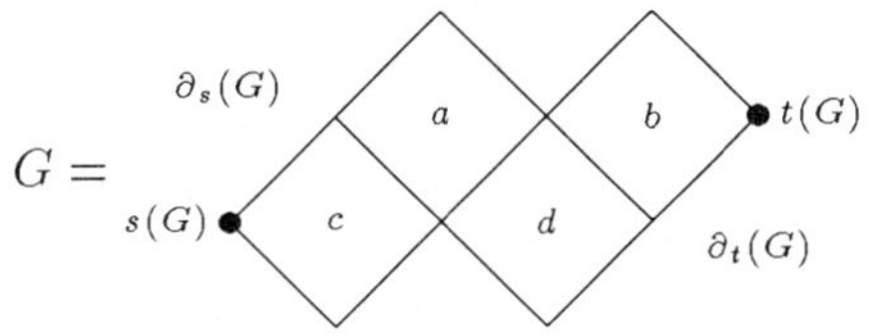

the following orders are admissible

$$D_1 : a < c < b < d,$$
$$D_2 : a < b < c < d,$$
$$D_3 : a < b < d < c,$$
$$D_4 : b < a < c < d,$$
$$D_5 : b < a < d < c.$$

A.2.3. Proposition. *$\mathcal{O}$ is not empty.*

PROOF. If $n > 0$, there exists a face $f \in F$ such that $\partial_s(G) \cap \partial f = \partial_s f$ (see Power [**24**, Proposition 2.10]). Hence, there is a pasting scheme $G' = G - \partial_s f$. It has a smaller number of faces and we proceed by induction. □

There is a linear order on the set $\mathcal{O}$ – the *lexicographic order*. For any two distinct admissible orders A and B there is an integer $0 < m < n$ such that

$$A_0 = B_0, a_1 = b_1, \ldots, a_{m-1} = b_{m-1}, A_{m-1} = B_{m-1}, a_m \neq b_m.$$

Then by definition A precedes B, $A \prec B$, iff $\partial_s a_m \subset A_{m-1}$ precedes $\partial_s b_m \subset B_{m-1}$ along the path $A_{m-1} = B_{m-1}$. In Example A.2.2 the admissible orders are listed in the lexicographic order: $D_1 \prec D_2 \prec D_3 \prec D_4 \prec D_5$.

Let us make $\mathcal{O}$ into a 2-dimensional cell complex. The set of 0-cells $\mathcal{O}_0$ is $\mathcal{O}$. If both orders

$$a_1 <_A \cdots <_A a_m <_A a_{m+1} <_A \cdots <_A a_n,$$
$$a_1 <_B \cdots <_B a_{m+1} <_B a_m <_B \cdots <_B a_n$$

are admissible (that is, if $A_{m-1} \cap \partial a_{m+1} = \partial_s a_{m+1}$) they are ends of a 1-cell from $\mathcal{O}_1$, and all 1-cells are as above. This 1-cell is directed from $<_A$ to $<_B$ iff $A \prec B$ lexicographically. See the picture.

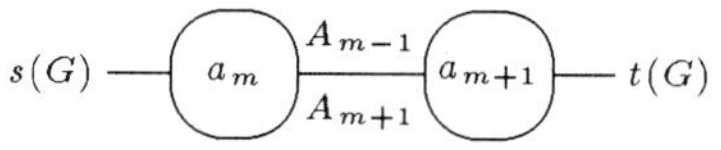

(A.2.1)

2-cells are of two kinds: squares and hexagons. Whenever for $i+1 < j$ the following 4 orders are admissible

$$a_1 <_A \cdots <_A a_i <_A a_{i+1} <_A \cdots <_A a_j <_A a_{j+1} <_A \cdots <_A a_n$$
$$a_1 <_B \cdots <_B a_{i+1} <_B a_i <_B \cdots <_B a_j <_B a_{j+1} <_B \cdots <_B a_n$$
$$a_1 <_C \cdots <_C a_i <_C a_{i+1} <_C \cdots <_C a_{j+1} <_C a_j <_C \cdots <_C a_n$$
$$a_1 <_D \cdots <_D a_{i+1} <_D a_i <_D \cdots <_D a_{j+1} <_D a_j <_D \cdots <_D a_n$$

(that is, when $A_{i-1} \cap \partial a_{i+1} = \partial_s a_{i+1}$, $A_{j-1} \cap \partial a_{j+1} = \partial_s a_{j+1}$ and situation (A.2.1) occurs twice) there is a square in $\mathcal{O}_2$ with edges AB, AC, BD, CD. With respect to the orientation of edges one of the four vertices A, B, C, D is a source, say A. Then the edges are oriented as AB, AC, BD, CD.

Whenever 6 orders obtained by permutation of a_i, a_{i+1}, a_{i+2} are admissible (that is, when $A_{i-1} \cap \partial a_{i+1} = \partial_s a_{i+1}$, $A_{i-1} \cap \partial a_{i+2} = \partial_s a_{i+2}$,) there is a hexagon in $\mathcal{O}_2$ whose edges correspond to transpositions. Taking into account the orientation of edges we see that one of the 6 vertices is a source, say A. This means that $\partial_s a_i \subset A_{i-1}$ precedes $\partial_s a_{i+1} \subset A_{i-1}$ which, in turn, precedes $\partial_s a_{i+2} \subset A_{i-1}$ on the path A_{i-1}. See the picture.

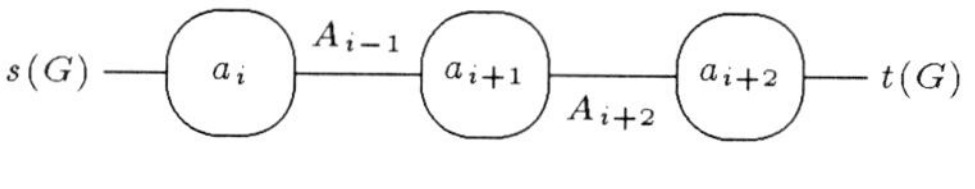

(A.2.2)

The orientation of edges is fixed by the inequalities

$$\begin{array}{ccccccc} A & \prec s_i.A \prec & s_{i+1}s_i.A & \prec & s_is_{i+1}s_i.A \\ \| & & & & \| \\ A & \prec s_{i+1}.A \prec & s_is_{i+1}.A & \prec & s_{i+1}s_is_{i+1}.A \end{array}$$

In Example A.2.2 we have the 2-complex $\mathcal{O}$

$$D_1 \prec D_2 \begin{array}{c} \prec D_3 \prec \\ \\ \prec D_4 \prec \end{array} D_5 \quad =$$

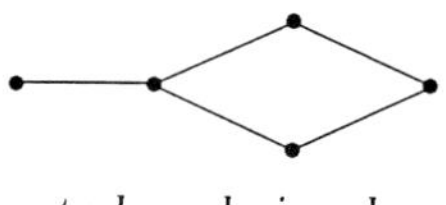

A.2.4. Proposition. *The 2-complex $\mathcal{O}$ is connected and simply connected.*

PROOF. Introduce a distance between linear orders on F by

$$l(A, B) = \#\{(f, g) \in F \times F \mid f <_A g, f >_B g\}.$$

The symmetric group $\mathfrak{S}_n = \operatorname{Aut} F$ acts simply transitively on the set of linear orders. If $\sigma.A = B$, then $l(A, B) = l(\sigma)$ is the length of σ.

A.2.5. Lemma. *If admissible orders A and B are distinct, there exists an integer $0 < m < n$ such that $B_{m-1} \cap \partial b_{m+1} = \partial_s b_{m+1}$ and $l(A, s_m.B) = l(A, B) - 1$, where*

$s_m(b_m) = b_{m+1}$, $s_m(b_{m+1}) = b_m$, $s_m(b_i) = b_i$ *for* $i \neq m, m+1$. *The order* $s_m.B$ *is admissible.*

PROOF. We have two orders $a_1 <_A \cdots <_A a_n$ and $b_1 <_B \cdots <_B b_n$ on F. Let $k < n-1$ be such that

$$a_n = b_n, a_{n-1} = b_{n-1}, \ldots, a_{k+2} = b_{k+2}, a_{k+1} \neq b_{k+1}.$$

For some $m \leqslant k$ it holds $b_m = a_{k+1}$. Now consider the order $C = s_m.B$

$$b_1 <_C \cdots <_C b_{m+1} <_C b_m <_C \cdots <_C b_n.$$

We have

$$B_{k+1} \cap \partial b_m = B_{k+1} \partial a_{k+1} = \partial_t a_{k+1} = \partial_t b_m.$$

This implies for $m \leqslant l \leqslant k+1$

$$B_l \cap \partial b_m = \partial_t b_m.$$

For $l = m+1$ this equation means that $<_C$ is admissible. Since

$$b_{m+1} <_A b_m = a_{k+1}, \qquad b_{m+1} >_B b_m, \qquad b_{m+1} <_C b_m$$

the equation $l(A, C) = l(A, B) - 1$ follows. □

A.2.6. Corollary. *For any pair* $A, B \in \mathcal{O}$ *there is a sequence of* $C_i \in \mathcal{O}$, $0 \leqslant i \leqslant p$, $A = C^0$, $B = C^p$, *such that* C^i *and* C^{i+1} *are connected by a 1-cell and* $l(A, C^{i+1}) = l(A, C^i) + 1$.

We have to prove that any loop in the 1-skeleton $\mathcal{O}_1$ is contractible in the 2-skeleton $\mathcal{O}_2$. Let L be a loop based at $A \in \mathcal{O}$, $L = (L^i)_{0 \leqslant i \leqslant p}$, $L^0 = A = L^p$. The *height* of L is defined as

$$h(L) = \max_{0 \leqslant i \leqslant p} l(A, L^i).$$

A.2.7. Lemma. *Any loop* L *in* $\mathcal{O}$ *based at* A *of height* h *is contractible in* $\mathcal{O}_2$ *within the class of loops based at* A *of height* $\leqslant h$.

PROOF. We prove this statement by induction on h. Consider a loop L based at A, $h(L) = h$. For each $0 < i < p$ connect A with L^i by a sequence γ^i as in Corollary A.2.6. It is enough to prove that all loops of height $\leqslant h$

$$(A, \gamma^i, L^i, L^{i+1}, \overline{\gamma^{i+1}}, A)$$

are contractible. Set $B = L^{i+1}$ if $l(A, L^i) < l(A, L^{i+1})$, otherwise, if $l(A, L^i) > l(A, L^{i+1})$, set $B = L^i$. So we can assume that our loop is made of two paths in $\mathcal{O}$

$$\gamma = (A, \ldots, C, B), \qquad \delta = (A, \ldots, D, B)$$

with strictly increasing $l(A, \gamma^i)$ and $l(A, \delta^i)$. If $C = D$, we can contract this loop to the loop $(A, \ldots, C = D, \ldots, A)$ of height $< h$, and we apply the induction assumption.

If $C \neq D$, there exist $i \neq j$ such that

$$B_{i-1} \cap \partial b_{i+1} = \partial_s b_{i+1}, \qquad B_{j-1} \cap \partial b_{j+1} = \partial_s b_{j+1},$$

and $C = s_i.B$, $D = s_j.B$. We can assume that $i < j$ and study the following two cases separately.

1) $i+1<j$. Then we can interchange b_i with b_{i+1} and b_j with b_{j+1} independently. Set $E = s_i s_j.B = s_j s_i.B$. Since

$$b_i >_A b_{i+1}, \quad b_i <_B b_{i+1}, \quad b_j >_A b_{j+1}, \quad b_j <_B b_{j+1}$$

we have $l(A,E) = l(A,B) - 2$.

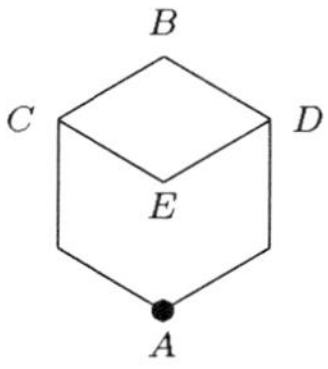

The loop contracts along the square $BCED$ to the loop

$$(A, \ldots, \gamma^r, \ldots, C, E, D, \ldots, \delta^r, \ldots, A)$$

which has height $< h$ and, thus, is contractible by induction hypothesis.

2) $i+1=j$. Since

$$B_{i-1} \cap \partial b_{i+1} = \partial_s b_{i+1}, \qquad B_{i-1} \cap \partial b_{i+2} = \partial_s b_{i+2},$$

a part of G looks as shown below.

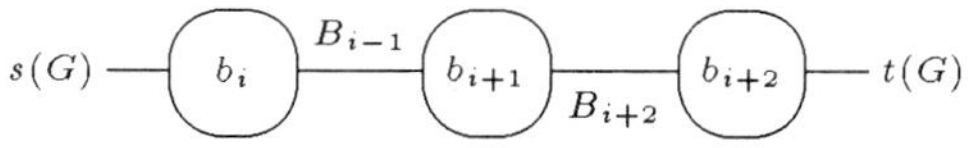

The combinatorial homotopy can sweep these cells in any order. Applying products of s_i and s_{i+1} to B we get the following admissible orders:

$$C = s_i.B, \quad E = s_{i+1}s_i.B, \quad F = s_i s_{i+1} s_i.B, \quad G = s_i s_{i+1}.B, \quad D = s_{i+1}.B.$$

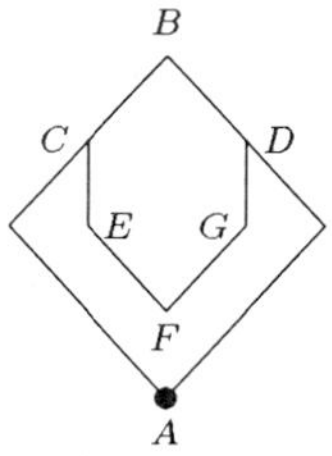

Contracting our loop along the hexagon $BCEFGD$ we shall get a loop

$$(A, \ldots, \gamma^r, \ldots, C, E, F, G, D, \ldots, \delta^r, \ldots, A)$$

of height $< h$ since

$$b_i <_B b_{i+1} <_B b_{i+2}, \qquad b_i >_A b_{i+1} >_A b_{i+2}.$$

By the induction hypothesis it is contractible. □

This finishes the proof of the proposition. □

A.3. Weak 2-pasting

The weak 2-pasting refers to the following setup. Assume that we have a 2-pasting scheme G and a labelling: 0-cells are labelled with 2-categories, 1-cells are labelled with weak 2-functors and 2-cells with weak 2-transformations. To any directed path p from $s(G)$ to $t(G)$ corresponds a uniquely determined weak 2-functor $|p| : |s(G)| \to |t(G)|$ – the composite. With the scheme

$$\xrightarrow{G_1} \xrightarrow{\dots} \xrightarrow{G_k} \begin{array}{c} \xrightarrow{F} \\ \Downarrow \lambda \\ \xrightarrow[F']{} \end{array} \xrightarrow{H_1} \xrightarrow{\dots} \xrightarrow{H_l}$$

is associated a unique weak 2-transformation

$$G_1 \bullet \dots \bullet G_k \bullet \lambda \bullet H_1 \bullet \dots \bullet H_l : G_1 \bullet \dots \bullet G_k \bullet F \bullet H_1 \bullet \dots \bullet H_l \to G_1 \bullet \dots \bullet G_k \bullet F' \bullet H_1 \bullet \dots \bullet H_l .$$

Thus, to any admissible order $<_A$ corresponds a unique weak 2-transformation $| <_A | : |\partial_s(G)| \to |\partial_t(G)|$ – the composition of the above weak 2-transformations taken in the order $<_A$. Weak 2-functors between $|s(G)|$ and $|t(G)|$, their weak 2-transformations and modifications form a strict 2-category $\mathrm{Bicat}(|s(G)|, |t(G)|)$ (e.g. [**12**]).

A.3.1. Lemma. *To an edge $AB \in \mathcal{O}_1$, $A \prec B$, corresponds an invertible modification $m_{A,B} : | <_A | \to | <_B |$.*

PROOF. We have to construct a modification for the pasting scheme

$$\xrightarrow{E_1} \dots \xrightarrow{E_p} \begin{array}{c} \longrightarrow \\ \Downarrow \nu \\ \longrightarrow \end{array} \xrightarrow{H_1} \dots \xrightarrow{H_l} \begin{array}{c} \longrightarrow \\ \Downarrow \kappa \\ \longrightarrow \end{array} \xrightarrow{K_1} \dots \xrightarrow{K_q}$$

between $| <_A |$ and $| <_B |$ for two orders: $\nu <_A \kappa$ and $\kappa <_B \nu$. Choose some k, $0 \leqslant k \leqslant l$, and simplify the pasting scheme to

$$\mathcal{C} \begin{array}{c} \xrightarrow{F} \\ \Downarrow \lambda \\ \xrightarrow[F']{} \end{array} \mathcal{C}' \begin{array}{c} \xrightarrow{G} \\ \Downarrow \mu \\ \xrightarrow[G']{} \end{array} \mathcal{C}'',$$

where

$$\lambda = E_1 \bullet \dots \bullet E_p \bullet \nu \bullet H_1 \bullet \dots \bullet H_k, \qquad \mu = H_{k+1} \bullet \dots \bullet H_l \bullet \kappa \bullet K_1 \bullet \dots \bullet K_q .$$

We shall construct a modification

$$m : (\lambda \bullet G) \bullet (F' \bullet \mu) \to (F \bullet \mu) \bullet (\lambda \bullet G') : F \bullet G \to F' \bullet G' : \mathcal{C} \to \mathcal{C}''$$

and later we shall show that it does not depend on the choice of k.

Using the full form of weak 2-functors (G, ψ_2, ψ_0), (G', ψ_2', ψ_0'), we find the first 2-transformation as

$$[(\lambda \bullet G) \bullet (F' \bullet \mu)]_X = \left(GFX \xrightarrow{G\lambda_X} GF'X \xrightarrow{\mu_{F'X}} G'F'X \right);$$

for a 1-morphism $f : X \to Y \in \mathcal{C}$ the modification $[(\lambda \bullet G) \bullet (F' \bullet \mu)]_f$ is the pasting of

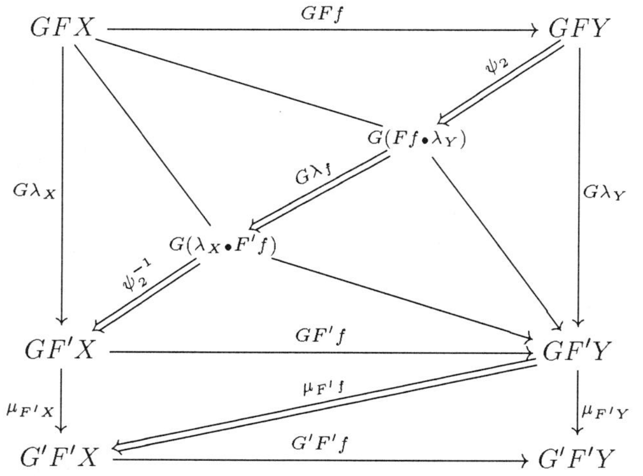

(A.3.1)

the second 2-transformation is

$$[(F \bullet \mu) \bullet (\lambda \bullet G')]_X = (GFX \xrightarrow{\mu_{FX}} G'FX \xrightarrow{G'\lambda_X} G'F'X);$$

the corresponding modification $[(F \bullet \mu) \bullet (\lambda \bullet G')]_f$ is the pasting of

(A.3.2)

The family

$$m_X = \begin{array}{c} GFX \xrightarrow{G\lambda_X} GF'X \\ \mu_{FX} \downarrow \quad \overset{\mu_{\lambda_X}}{\Rightarrow} \quad \downarrow \mu_{F'X} \\ G'FX \xrightarrow{G'\lambda_X} G'F'X \end{array}$$

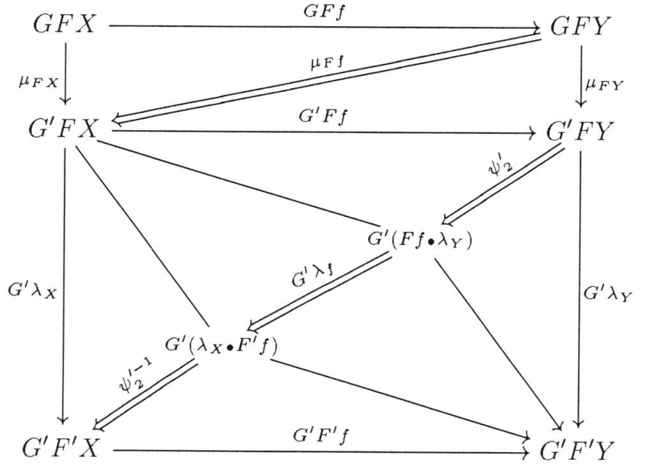

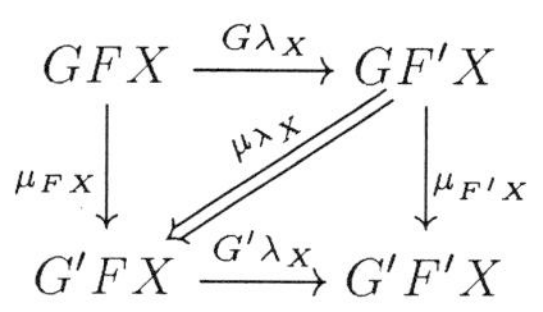

is the modification we have been looking for. Indeed, we have to check the equation for $f : X \to Y \in \mathcal{C}$

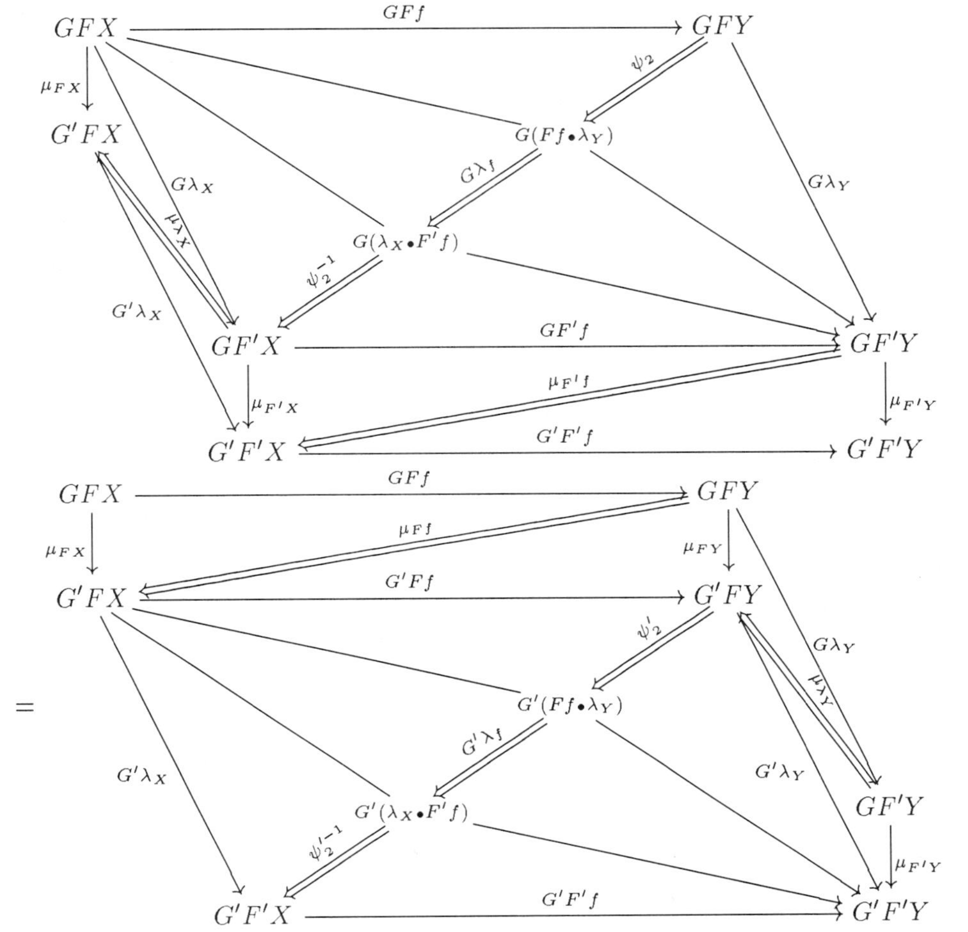

Double use of axiom A.1.3.3(b) of weak 2-transformations reduces this equation to the following:

$$\begin{array}{c}
GFX \xrightarrow[G(\lambda_X \bullet F'f)]{\overset{G(Ff \bullet \lambda_Y)}{G\lambda_f \Downarrow}} GF'Y \\
\mu_{FX} \downarrow \quad \overset{\mu(\lambda_X \bullet F'f)}{} \quad \downarrow \mu_{F'Y} \\
G'FX \xrightarrow[G'(\lambda_X \bullet F'f)]{} G'F'Y
\end{array}
\quad = \quad
\begin{array}{c}
GFX \xrightarrow{G(Ff \bullet \lambda_Y)} GF'Y \\
\mu_{FX} \downarrow \quad \overset{\mu(Ff \bullet \lambda_Y)}{} \quad \downarrow \mu_{F'Y} \\
G'FX \xrightarrow[G'(\lambda_X \bullet F'f)]{\overset{G'(Ff \bullet \lambda_Y)}{G'\lambda_f \Downarrow}} G'F'Y
\end{array}$$

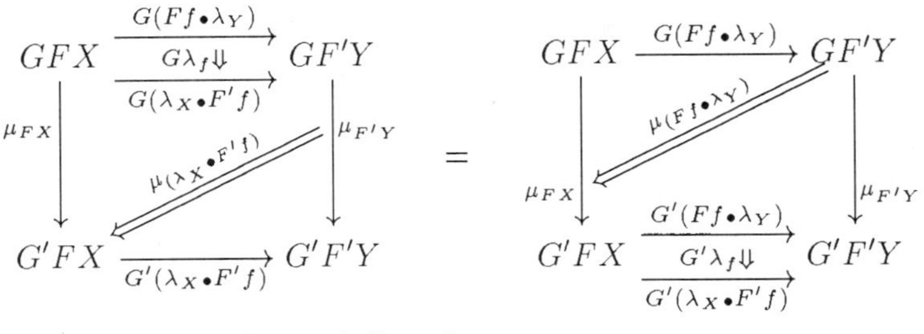

And this follows from functoriality of μ_-.

To check independence of the choice of k above we have to consider the 2-pasting scheme

$$\mathcal{C} \underset{F'}{\overset{F}{\rightrightarrows}} \Downarrow \lambda \; \mathcal{C}' \xrightarrow{H} \mathcal{C}'' \underset{G'}{\overset{G}{\rightrightarrows}} \Downarrow \mu \; \mathcal{C}''', \qquad (A.3.3)$$

and show that the two modifications obtained as above from

$$\mathcal{C} \underset{F'}{\overset{F}{\underset{\lambda\Downarrow}{\rightrightarrows}}} \mathcal{C}' \underset{H\bullet G'}{\overset{H\bullet G}{\underset{H\bullet\mu\Downarrow}{\rightrightarrows}}} \mathcal{C}'''$$

or from

$$\mathcal{C} \underset{F'\bullet H}{\overset{F\bullet H}{\underset{\lambda\bullet H\Downarrow}{\rightrightarrows}}} \mathcal{C}'' \underset{G'}{\overset{G}{\underset{\mu\Downarrow}{\rightrightarrows}}} \mathcal{C}'''$$

coincide. And it is indeed the case, both modifications are

$$m_X = \begin{array}{ccc} GHFX & \xrightarrow{GH\lambda_X} & GHF'X \\ {\scriptstyle \mu_{HFX}}\downarrow & \overset{\mu_{H\lambda_X}}{\swarrow\!\!\!\swarrow} & \downarrow{\scriptstyle \mu_{HF'X}} \\ G'HFX & \xrightarrow{G'H\lambda_X} & G'HF'X \end{array} . \qquad \text{(A.3.4)}$$

□

A.3.2. Lemma. *$m_{A,B}$ is natural in weak 2-transformations substituted for 2-cells.*

PROOF. Consider the scheme (A.3.3) and vary λ, μ. That is, consider modifications $l : \lambda \to \lambda'$, $k : \mu \to \mu'$ in

$$\mathcal{C} \underset{F'}{\overset{F}{\rightrightarrows}} {\scriptstyle \lambda \Downarrow \overset{l}{\Rrightarrow} \Downarrow \lambda'} \; \mathcal{C}' \xrightarrow{H} \mathcal{C}'' \underset{G'}{\overset{G}{\rightrightarrows}} {\scriptstyle \mu \Downarrow \overset{k}{\Rrightarrow} \Downarrow \mu'} \; \mathcal{C}''' .$$

We have to prove the equations

$$\begin{array}{ccc} |\lambda < \mu| & \xrightarrow{m} & |\lambda > \mu| \\ {\scriptstyle l\bullet\mu}\downarrow & = & \downarrow{\scriptstyle l\bullet\mu} \\ |\lambda' < \mu| & \xrightarrow{m} & |\lambda' > \mu| \end{array} , \qquad \begin{array}{ccc} |\lambda < \mu| & \xrightarrow{m} & |\lambda > \mu| \\ {\scriptstyle \lambda\bullet k}\downarrow & = & \downarrow{\scriptstyle \lambda\bullet k} \\ |\lambda < \mu'| & \xrightarrow{m} & |\lambda > \mu'| \end{array} \qquad \text{(A.3.5)}$$

Explicitly they are, respectively,

$$\begin{array}{ccc} GHFX & \xrightarrow{GH\lambda_X} & GHF'X \\ {\scriptstyle \mu_{HFX}}\downarrow & \overset{\mu_{H\lambda_X}}{\swarrow\!\!\!\swarrow} & \downarrow{\scriptstyle \mu_{HF'X}} \\ G'HFX & \underset{G'H\lambda'_X}{\overset{G'H\lambda_X}{\underset{\Downarrow G'Hl_X}{\rightrightarrows}}} & G'HF'X \end{array} = \begin{array}{ccc} GHFX & \underset{GH\lambda'_X}{\overset{GH\lambda_X}{\underset{\Downarrow GHl_X}{\rightrightarrows}}} & GHF'_X \\ {\scriptstyle \mu_{HFX}}\downarrow & \overset{\mu_{H\lambda'_X}}{\swarrow\!\!\!\swarrow} & \downarrow{\scriptstyle \mu_{HF'X}} \\ G'HFX & \xrightarrow[G'H\lambda'_X]{} & G'HF'X \end{array}$$

$$\begin{array}{ccc} GHFX & \xrightarrow{GH\lambda_X} & GHF'X \\ {\scriptstyle \mu'_{HFX}}\downarrow \overset{k_{HFX}}{\Leftarrow} \downarrow{\scriptstyle \mu_{HFX}} & \overset{\mu_{H\lambda_X}}{\swarrow\!\!\!\swarrow} & \downarrow{\scriptstyle \mu_{HF'X}} \\ G'HFX & \xrightarrow{G'H\lambda_X} & G'HF'X \end{array} = \begin{array}{ccc} GHFX & \xrightarrow{GH\lambda_X} & GHF'_X \\ {\scriptstyle \mu'_{HFX}}\downarrow & \overset{\mu'_{H\lambda_X}}{\swarrow\!\!\!\swarrow} & {\scriptstyle \mu'_{HF'X}}\downarrow \overset{k_{HF'X}}{\Leftarrow} \downarrow{\scriptstyle \mu_{HF'X}} \\ G'HFX & \xrightarrow[G'H\lambda_X]{} & G'HF'X \end{array}$$

The first equation follows from functoriality of the μ_{-} component of the weak 2-transformation μ, the second follows from the property A.1.4.2 of the modification k.

The general case follows from the above. □

A.3.3. Lemma (cf. [13]???). *Consider the pasting scheme*

$$\mathcal{C} \underset{F'}{\overset{F}{\underset{a\Downarrow}{\rightrightarrows}}} \mathcal{C}' \underset{G'}{\overset{G}{\underset{b\Downarrow}{\rightrightarrows}}} \mathcal{C}'' \underset{H'}{\overset{H}{\underset{c\Downarrow}{\rightrightarrows}}} \mathcal{C}'''$$

Then the following holds:

1) Let $d = (a{\bullet}G){\bullet}(F'{\bullet}b) = |a < b|$, $e = (F{\bullet}b){\bullet}(a{\bullet}G') = |b < a|$,

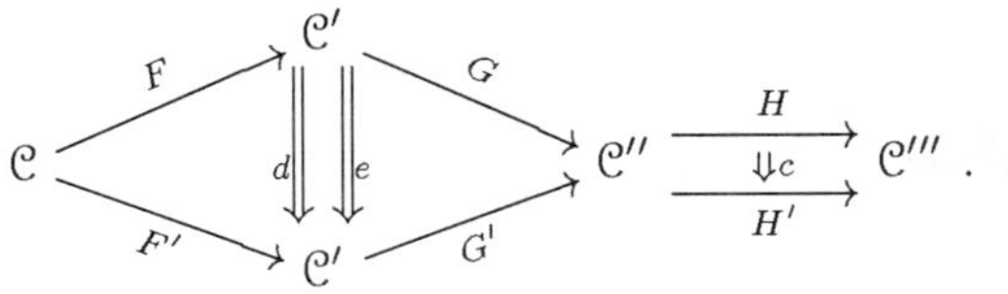

Then

$$\begin{array}{ccc} |a<b<c| & \xrightarrow[\sim]{\eta_2\bullet 1} & |d<c| \\ {\scriptstyle m_{b,c}}\downarrow & & \\ |a<c<b| & = & \downarrow{\scriptstyle m_{d,c}} \\ {\scriptstyle m_{a,c}}\downarrow & & \\ |c<a<b| & \xrightarrow[\sim]{1\bullet\eta_2'} & |c<d| \end{array}, \qquad \begin{array}{ccc} |b<a<c| & \xrightarrow[\sim]{\eta_2\bullet 1} & |e<c| \\ {\scriptstyle m_{a,c}}\downarrow & & \\ |b<c<a| & = & \downarrow{\scriptstyle m_{e,c}} \\ {\scriptstyle m_{b,c}}\downarrow & & \\ |c<b<a| & \xrightarrow[\sim]{1\bullet\eta_2'} & |c<e| \end{array},$$

where $H = (H, \eta_2, \eta_0)$, $H' = (H', \eta_2', \eta_0')$.

2) Let $f = (b{\bullet}H){\bullet}(G'{\bullet}c) = |b < c|$, $g = (G{\bullet}c){\bullet}(b{\bullet}H') = |c < b|$,

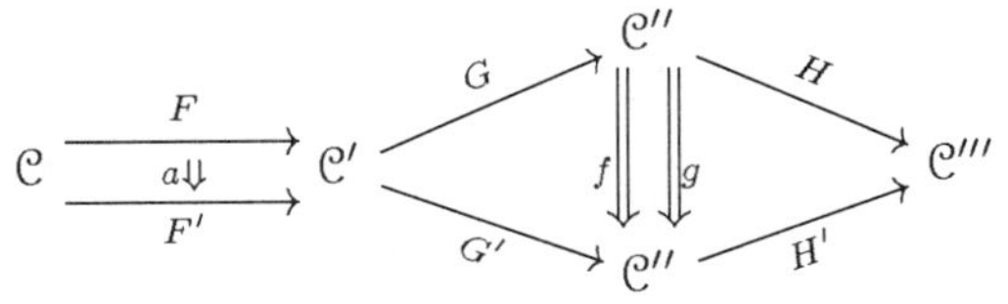

Then

$$\begin{array}{ccc} |a<b<c| & = & |a<f| \\ {\scriptstyle m_{a,b}}\downarrow & & \\ |b<a<c| & = & \downarrow{\scriptstyle m_{a,f}} \\ {\scriptstyle m_{a,c}}\downarrow & & \\ |b<c<a| & = & |f<a| \end{array}, \qquad \begin{array}{ccc} |a<c<b| & = & |a<g| \\ {\scriptstyle m_{a,c}}\downarrow & & \\ |c<a<b| & = & \downarrow{\scriptstyle m_{a,g}} \\ {\scriptstyle m_{a,b}}\downarrow & & \\ |c<b<a| & = & |g<a| \end{array}.$$

PROOF. 1) The first equation reads

$$\begin{array}{ccc} HGa{\bullet}HbF'{\bullet}cF'G' & \xrightarrow{\eta_2\bullet 1} & Hd{\bullet}cF'G' \\ {\scriptstyle a\bullet m_{b,c}}\downarrow & & \\ HGa{\bullet}cGF'{\bullet}H'bF' & = & \downarrow{\scriptstyle m_{d,c}} \\ {\scriptstyle m_{a,c}\bullet b}\downarrow & & \\ cGF{\bullet}H'Ga{\bullet}H'bF' & \xrightarrow{1\bullet\eta_2'} & cGF{\bullet}H'd \end{array}$$

Substituting the definitions of $m_{-,-}$ we get

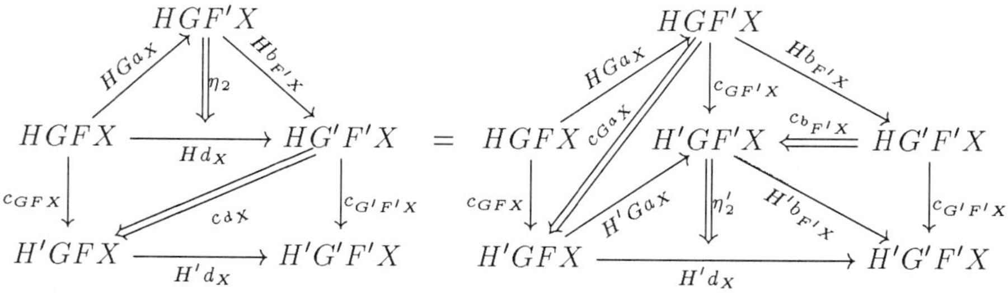

which holds by property A.1.3.3(b) of the weak 2-transformation c.

The second equation reads

$$\begin{array}{ccc}
HbF{\scriptstyle\bullet}HG'a{\scriptstyle\bullet}cG'F' & \xrightarrow{\eta_2\bullet 1} & He{\scriptstyle\bullet}cG'F' \\
\downarrow{\scriptstyle b\bullet m_{a,c}} & & \\
HbF{\scriptstyle\bullet}cG'F{\scriptstyle\bullet}H'G'a & = & \downarrow{\scriptstyle m_{e,c}} \\
\downarrow{\scriptstyle m_{b,c}\bullet a} & & \\
cGF{\scriptstyle\bullet}H'bF{\scriptstyle\bullet}H'G'a & \xrightarrow{1\bullet\eta'_2} & cGF{\scriptstyle\bullet}H'e
\end{array}$$

It also holds by property A.1.3.3(b) of the weak 2-transformation c.

2) The first equation reads

$$\begin{array}{ccc}
HGa{\scriptstyle\bullet}HbF'{\scriptstyle\bullet}cG'F' & = & HGa{\scriptstyle\bullet}fF' \\
\downarrow{\scriptstyle H(m_{a,b})\bullet c} & & \\
HbF{\scriptstyle\bullet}HG'a{\scriptstyle\bullet}cG'F' & = & \downarrow{\scriptstyle m_{a,f}} \\
\downarrow{\scriptstyle b\bullet m_{a,c}} & & \\
HbF{\scriptstyle\bullet}cG'F{\scriptstyle\bullet}H'G'a & = & fF{\scriptstyle\bullet}H'G'a
\end{array}$$

When definitions of $m_{-,-}$ are substituted, this equation becomes the definition (cf. (A.3.1)) of f_{a_X}.

The second equation reads

$$\begin{array}{ccc}
HGa{\scriptstyle\bullet}cGF'{\scriptstyle\bullet}H'bF' & = & HGa{\scriptstyle\bullet}gF' \\
\downarrow{\scriptstyle m_{a,c}\bullet b} & & \\
cGF{\scriptstyle\bullet}H'Ga{\scriptstyle\bullet}H'bF' & = & \downarrow{\scriptstyle m_{a,g}} \\
\downarrow{\scriptstyle c\bullet H'(m_{a,b})} & & \\
cGF{\scriptstyle\bullet}H'bF{\scriptstyle\bullet}H'G'a & = & gF{\scriptstyle\bullet}H'G'a
\end{array}$$

When definitions of $m_{-,-}$ are substituted, this becomes the definition (cf. (A.3.2)) of g_{a_X}. □

A.3.4. Theorem. *The groupoid whose objects are weak 2-transformations* $| <_A |$, $A \in \mathcal{O}$, *and morphisms are compositions of* $m_{A,B}$ *and* $m_{A,B}^{-1}$ *is contractible (equivalent to* **1***). In other words, any two* $| <_C |$, $| <_D |$ *are isomorphic via a unique modification* $m_{C,D}$ *– a product of various* $m_{A,B}$, $m_{A,B}^{-1}$. *Moreover,* $m_{C,D}$ *is natural in 2-cells.*

PROOF. Connectedness of $\mathcal{O}_1$ implies that the groupoid is equivalent to a category with one object. Since $\mathcal{O}$ is simply connected by Proposition A.2.4, we have to check that squares and hexagons of $\mathcal{O}_2$ give equations between products of modifications $m_{A,B}$.

1) When $ABDC$ is a square in $\mathcal{O}_2$, $A \prec B \prec D$, $A \prec C \prec D$ we have the following 2-pasting scheme in $\mathrm{Bicat}(|s(G)|, |t(G)|)$

$$|A_0| \xrightarrow{|a_1|} \cdots \xrightarrow{|a_{i-1}|} |A_{i-1}| \begin{array}{c} \xrightarrow{|a_i|} |A_i| \xrightarrow{|a_{i+1}|} \\ {}_{|a_{i+1}|}\!\!\searrow\;\Downarrow m_{i,i+1}\;\nearrow\!{}_{|a_i|} \\ |B_i| \end{array} |A_{i+1}| \xrightarrow{|a_{i+2}|} \cdots$$

$$\cdots \xrightarrow{|a_{j-1}|} |A_{j-1}| \begin{array}{c} \xrightarrow{|a_j|} |A_j| \xrightarrow{|a_{j+1}|} \\ {}_{|a_{j+1}|}\!\!\searrow\;\Downarrow m_{j,j+1}\;\nearrow\!{}_{|a_j|} \\ |C_j| \end{array} |A_{j+1}| \xrightarrow{|a_{j+2}|} \cdots \xrightarrow{|a_n|} |A_n|.$$

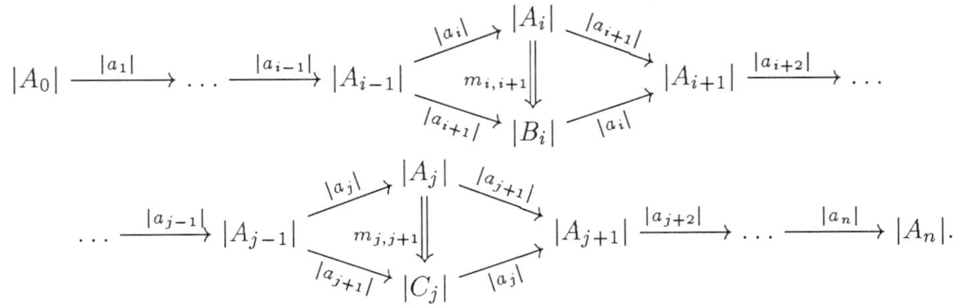

But this is a strict 2-category, hence, $m_{B,D} \circ m_{A,B} = m_{C,D} \circ m_{A,C}$.

2) When there is a hexagon in $\mathcal{O}_2$ the pasting scheme contains a part of the form (A.2.2). The construction of Lemma A.3.1 shows that it suffices to consider the pasting scheme

$$\mathcal{C} \begin{array}{c}\xrightarrow{F}\\ a \Downarrow \\ \xrightarrow[F']{}\end{array} \mathcal{C}' \begin{array}{c}\xrightarrow{G}\\ b \Downarrow \\ \xrightarrow[G']{}\end{array} \mathcal{C}'' \begin{array}{c}\xrightarrow{H}\\ c \Downarrow \\ \xrightarrow[H']{}\end{array} \mathcal{C}'''$$

and to prove the equation between compositions of modifications

$$\left(|a<b<c| \xrightarrow{m_{a,b}} |b<a<c| \xrightarrow{m_{a,c}} |b<c<a| \xrightarrow{m_{b,c}} |c<b<a|\right)$$
$$= \left(|a<b<c| \xrightarrow{m_{b,c}} |a<c<b| \xrightarrow{m_{a,c}} |c<a<b| \xrightarrow{m_{a,b}} |c<b<a|\right)$$

This equation has two proofs. The first is the diagram written in notations of Lemma A.3.3:

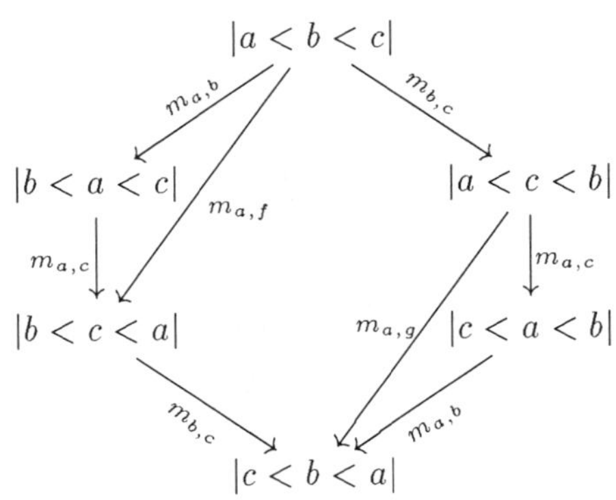

Commutativity of the triangles is shown in Lemma A.3.3 2). The square commutes by Lemma A.3.2, namely, by equation (A.3.5) with $k = m_{b,c} : f \to g$.

The second proof is the diagram

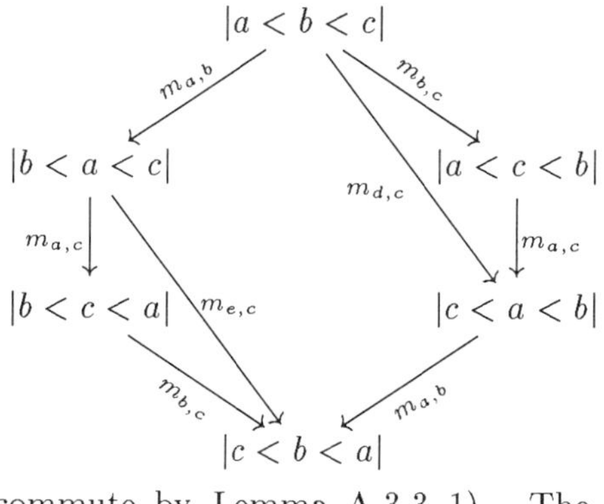

where the triangles commute by Lemma A.3.3 1). The square commutes by Lemma A.3.2, namely, by equation (A.3.5) with $l = m_{a,b} : d \to e$. □

A.4. Weak 3-categories

A.4.1. Trees. Let $\mathcal{O}$ be the category of finite linearly ordered sets and order preserving maps from Section 1.2. Denote by $[n]$ the category of paths of the quiver $\{0 \to 1 \to \cdots \to n\}$. It corresponds to the linearly ordered set $[0, n] \cap \mathbb{Z}$. Following Batanin [**3**] we define a *plane staged forest* as a functor $f : [n] \to \mathcal{O}$. We say that f has n levels. A *plane staged tree* is a plane staged forest $t : [n] \to \mathcal{O}$ with $t(n) \simeq [0]$. The set of *leaves* of a plane staged forest f is $f(0)$. The set of *roots* of $f : [n] \to \mathcal{O}$ is $f(n)$.

Plane staged forests $f_1 : [n] \to \mathcal{O}$, $f_2 : [m] \to \mathcal{O}$ are composable if $f_1(n) = f_2(0)$. The composite is $f : [n + m] \to \mathcal{O}$ such that

$$
\begin{aligned}
f(k \to k+1) &= f_1(k \to k+1), && \text{for } 0 \leqslant k < n,\\
f(k \to k+1) &= f_2(k - n \to k - n + 1), && \text{for } n \leqslant k < n + m,
\end{aligned}
$$

A *rooted tree* is an (abstract) tree with one distinguished vertex called the *root*. With a rooted tree is associated the most coarse partial order on the set of vertices, such that if $a \neq b$ are two vertices connected by an edge, and a is closer to the root, then $a < b$. A *rooted forest*, which is a finite family of rooted trees, has a similar partial order.

A vertex v of a rooted tree is an l-vertex if it has l *entries* – vertices connected by an edge with v, which are higher than v. The root is an l-vertex if it has l adjacent edges. A non-root v is an l-vertex if it has $l + 1$ adjacent edge.

A *plane rooted tree* is a tree with one vertex or a rooted tree drawn in the plane with one of the edges ending in the root distinguished, called the *leftmost root edge*. A *plane rooted forest* is a linearly ordered finite family of plane rooted trees. The minimal tree with respect to this order is called the leftmost tree, the maximal is the rightmost.

A *measured forest* is a plane rooted forest with linearly ordered (by height) vertices, so that the linear order is a refinement of the partial order associated with the rooted forest structure. A *measured tree* is a measured forest consisting of one tree.

A.4.2. Definition. A *weak 3-category* (which is locally a strict 2-category) $\mathcal{C}$ consists of the following data

0. a class of objects $\mathcal{C}_0 = \operatorname{Ob} \mathcal{C}$;

1. for any pair of objects $A, B \in \mathcal{C}_0$ a 2-category $\mathcal{C}(A, B)$;
2. for any $n \geqslant 0$ and any $(n+1)$-tuple $(X_0, X_1, \dots, X_n) \in (\mathcal{C}_0)^{n+1}$ a weak 2-functor

$$\bullet^n : \mathcal{C}(X_0, X_1) \times \mathcal{C}(X_1, X_2) \times \cdots \times \mathcal{C}(X_{n-1}, X_n) \to \mathcal{C}(X_0, X_n)$$

called the n-fold composition. For $n = 1$ the 2-functor $\bullet^1$ is the identity functor. For $n = 0$ we assume that $\bullet^0 : \mathbf{1} \to \mathcal{C}(X, X)$ is a strict 2-functor. $\mathbf{1}$ is a 2-category with one object, one 1-morphism and one 2-morphism. $\bullet^0(\text{object})$ is denoted $1_X \in \operatorname{Ob} \mathcal{C}(X, X)$;

Since the composition of weak 2-functors is strictly associative, to any plane staged tree T with n leaves is associated a weak 2-functor

$$[T]_1 : \mathcal{C}(X_0, X_1) \times \cdots \times \mathcal{C}(X_{n-1}, X_n) \to \mathcal{C}(X_0, X_n),$$

so that each vertex of the tree with k entries is replaced with $\bullet^k$. In particular, for a 1-level tree t^n with n leaves we have $[t^n]_1 = \bullet^n$;
3. for any plane staged tree T with k levels and n leaves, a weak 2-natural equivalence, denoted

$$[t^k]_2 : [T]_1 \to [t^n]_1 : \mathcal{C}(X_0, X_1) \times \cdots \times \mathcal{C}(X_{n-1}, X_n) \to \mathcal{C}(X_0, X_n).$$

Consider a measured tree T' with k leaves. To T' corresponds a unique 2-natural equivalence

$$[T']_2 : [T]_1 \to [t^n]_1 : \mathcal{C}(X_0, X_1) \times \cdots \times \mathcal{C}(X_{n-1}, X_n) \to \mathcal{C}(X_0, X_n)$$

where an l-vertex of T' is replaced with $[t^l]_2$ and 2-transformations for higher vertices precede the lower ones. More generally, to a measured forest F with k leaves corresponds a unique 2-natural equivalence

$$[F]_2 : [T]_1 \to [\tilde{T}]_1 : \mathcal{C}(X_0, X_1) \times \cdots \times \mathcal{C}(X_{n-1}, X_n) \to \mathcal{C}(X_0, X_n)$$

where the number of levels in $\tilde{T}$ is the number of trees in F. For any two measured forests F' and F'' isomorphic as plane rooted forests, which differ only by the linear order of vertices, there is a unique invertible modification

$$m_{F',F''} : [F']_2 \to [F'']_2 : [T]_1 \to [\tilde{T}]_1$$

constructed in Theorem A.3.4.
4. for any plane staged tree T decomposed into three plane staged forests by two horizontal cuts, an invertible modification

$[Y]_3 : \Big[$  $\Big]_2 \to [t^k]_2 : [T]_1 \to [t^n]_1.$

If two vertices i, j of a measured forest F' with k leaves are connected by an edge e, denote by F'' the measured forest F' with the edge e contracted and vertices i, j identified. The modifications $[Y]_3$ combine into an invertible modification

$$[Y_{i,j}]_3 : [F']_2 \to [F'']_2 : [T]_1 \to [t^n]_1;$$

such that
5. the following diagrams are commutative (for an arbitrary number of branches entering a vertex and an arbitrary choice of inner edge connecting the root with the previous level, etc.)

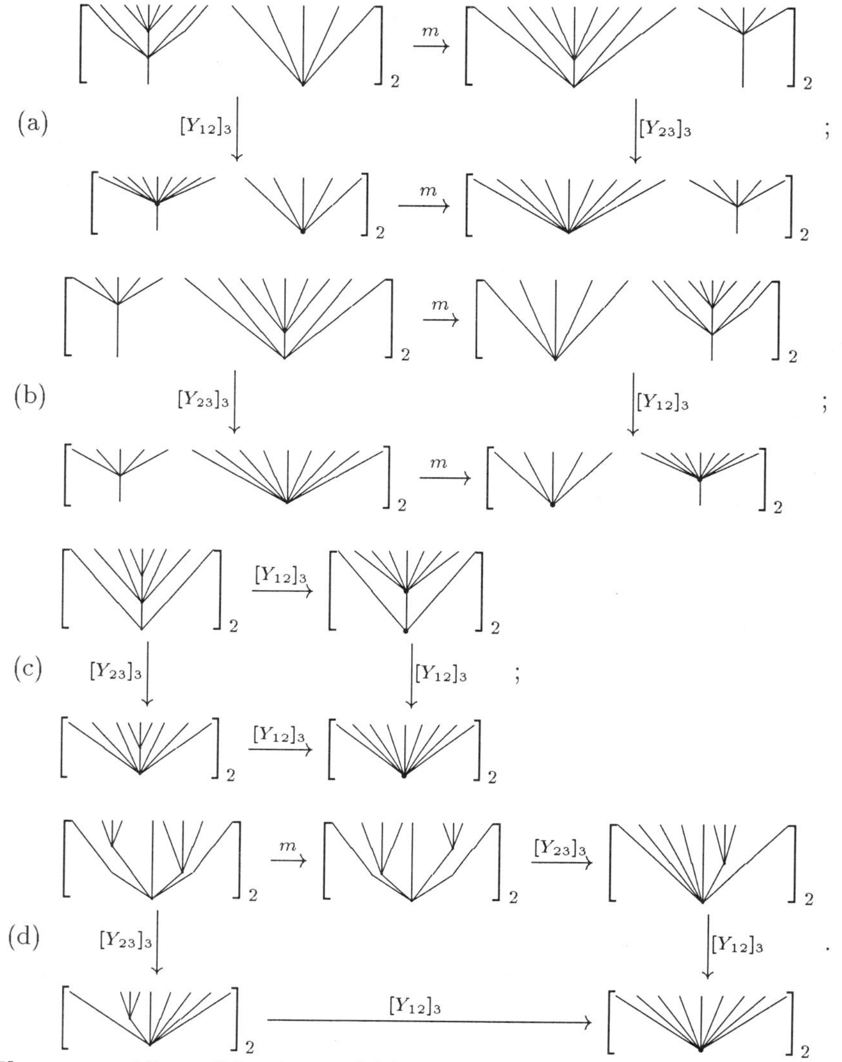

A.4.3. Theorem. *All modifications, which are products of $m_{F',F''}$ and $[Y_{i,j}]_3$ between given $[F_1]_2$ and $[F_2]_2$ are equal.*

PROOF. Follows from the local moves 5). □

A.4.4. Definition. A *monoidal 2-category* is a weak 3-category with one object.

A.5. A monoidal 2-category of abelian categories

Let $\Bbbk$ be an arbitrary field. Consider the 2-category $\mathfrak{A} = \mathfrak{Ab}^r_\Bbbk$ formed by
objects: essentially small $\Bbbk$-linear abelian categories $\mathcal{A} \in \mathfrak{A}_0$ with length,
1-morphisms: $\Bbbk$-linear right exact functors $F : \mathcal{A} \to \mathcal{B} \in \mathfrak{A}_1$,
2-morphisms: natural transformations (morphisms of functors) $\lambda \in \mathfrak{A}_2$.

When $\mathcal{A}_1, \dots, \mathcal{A}_n, \mathcal{B} \in \mathfrak{A}_0$ denote by $\mathrm{Hom}_{\text{m.l.,r.e.}}(\mathcal{A}_1 \times \cdots \times \mathcal{A}_n, \mathcal{B})$ the category of $\Bbbk$-multilinear right exact in each variable functors $F : \mathcal{A}_1 \times \cdots \times \mathcal{A}_n \to \mathcal{B}$. The following is the original form of Deligne's theorem valid without the assumption of perfectness of the field (cf. Proposition 1.1.5).

A.5.1. Theorem (Deligne [7]). *For any finite family of categories $\mathcal{A}_1, \dots, \mathcal{A}_n \in \mathfrak{A}_0$, there exists a category $\mathcal{A}$ and a $\Bbbk$-multilinear right exact in each variable functor $\boxtimes_{\mathcal{A}} : \mathcal{A}_1 \times \cdots \times \mathcal{A}_n \to \mathcal{A}$ such that for any $\mathcal{B} \in \mathfrak{A}_0$ the composition functor*

$$\operatorname{Hom}(\boxtimes_{\mathcal{A}}, \mathcal{B}) : \mathfrak{A}(\mathcal{A}, \mathcal{B}) \to \operatorname{Hom}_{m.l.,r.e.}(\mathcal{A}_1 \times \cdots \times \mathcal{A}_n, \mathcal{B})$$

is an equivalence of categories. □

A.5.2. Remark. For $n = 0$ by convention $\prod_1^n \mathcal{A}_k = \mathbf{1}$, the target category is $\operatorname{Hom}(\mathbf{1}, \mathcal{B}) \simeq \mathcal{B}$ – the category of all functors $\mathbf{1} \to \mathcal{B}$. We can take $\mathcal{A} = \Bbbk$-vect and the functor $\boxtimes_{\mathcal{A}}$ is $\mathbf{1} \to \Bbbk$-vect, object $\mapsto \Bbbk$.

A.5.3. Corollary. *For any functors $F, F', F'' \in \mathfrak{A}_1$ and isomorphisms of functors ϕ', ϕ'' as below*

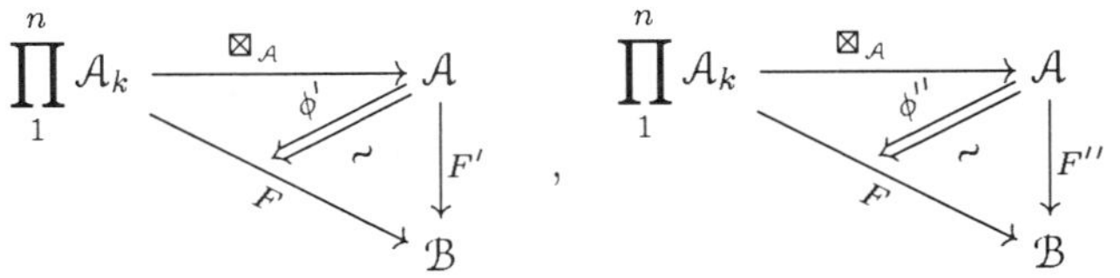

there is a unique isomorphism of functors $\phi : F' \to F''$, such that ϕ' equals the pasting

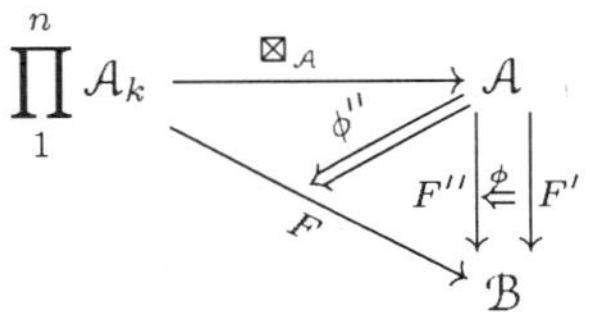

We want to show that using Deligne's theorem and some additional information on the relationship of categories $\mathcal{A}_1, \dots, \mathcal{A}_n$ with $\mathcal{A}$ one can construct a family of monoidal structures on the 2-category $\mathfrak{A}$.

The monoidal structure depends on several choices. The data for one of the monoidal structures follow.

0) As a weak 3-category, $\mathfrak{A}$ has one object.

1) A 2-category $\mathfrak{A}(\text{object,object}) = \mathfrak{A}$.

2) For any $n \geqslant 0$ a weak 2-functor

$$\boxtimes^n : \mathfrak{A}^n \to \mathfrak{A}$$

is given by the following data:

2.1) For each family $(\mathcal{A}_1, \dots, \mathcal{A}_n) \in \operatorname{Ob} \mathfrak{A}^n$, **choose** an object $\mathcal{A} \in \mathfrak{A}_0$ and a functor $\boxtimes_{\mathcal{A}} : \mathcal{A}_1 \times \cdots \times \mathcal{A}_n \to \mathcal{A}$ as in Theorem A.5.1. Set the first datum to

$$\boxtimes_1^n \mathcal{A}_k = \mathcal{A}_1 \boxtimes \cdots \boxtimes \mathcal{A}_n = \boxtimes^n (\mathcal{A}_1 \dots, \mathcal{A}_n) = \mathcal{A},$$

if $n > 1$. For $n = 1$ we choose

$$\boxtimes^1(\mathcal{A}) = \mathcal{A}, \quad \boxtimes_{\mathcal{A}} = \operatorname{Id}_{\mathcal{A}} : \mathcal{A} \to \mathcal{A} = \boxtimes^1(\mathcal{A}).$$

For $n = 0$ we choose

$$\boxtimes^0() = \Bbbk\text{-vect}, \quad \boxtimes : \mathbf{1} \to \Bbbk\text{-vect} = \boxtimes^0(), \quad \text{object} \mapsto \Bbbk.$$

2.2) For any $\Bbbk$-multilinear right exact in each variable functor $\Phi : \mathcal{A}_1 \times \cdots \times \mathcal{A}_n \to \mathcal{B}$, **choose** arbitrarily a functor $\tilde{\Phi} : \boxtimes_1^n \mathcal{A}_k \to \mathcal{B} \in \mathfrak{A}_1$ and an isomorphism $\overline{\Phi}$

$$\begin{array}{c} \mathcal{A}_1 \boxtimes \cdots \boxtimes \mathcal{A}_n \\ {}^{\boxtimes_\mathcal{A}} \nearrow \quad \Downarrow \overline{\Phi} \quad \searrow {}^{\tilde{\Phi}} \\ \mathcal{A}_1 \times \cdots \times \mathcal{A}_n \xrightarrow[\Phi]{} \mathcal{B} \end{array}$$

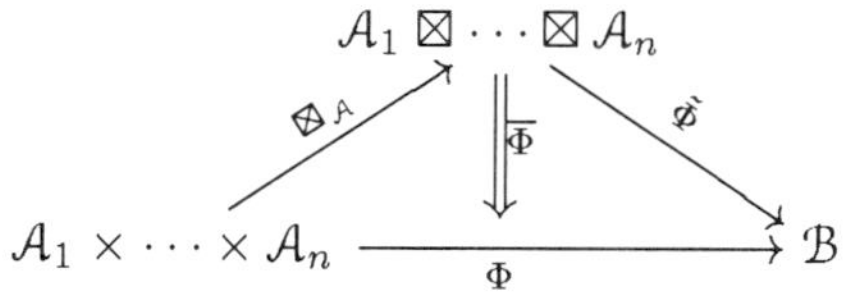

To simplify further considerations, assume that for $\Phi = \boxtimes_\mathcal{A} : \prod_1^n \mathcal{A}_k \to \boxtimes_1^n \mathcal{A}_k$ we choose $\tilde{\Phi} = \mathrm{Id}_{\boxtimes_1^n \mathcal{A}_k}$ and $\overline{\Phi} = \mathrm{id}_{\boxtimes_\mathcal{A}}$. These choices determine a functor quasi-inverse to $\mathrm{Hom}(\boxtimes_\mathcal{A}, \mathcal{B})$

$$D_{\mathcal{A}_1,\ldots,\mathcal{A}_n;\mathcal{B}} : \mathrm{Hom}_{\mathrm{m.l.,r.e.}}(\mathcal{A}_1 \times \cdots \times \mathcal{A}_n, \mathcal{B}) \to \mathfrak{A}(\mathcal{A}_1 \boxtimes \cdots \boxtimes \mathcal{A}_n, \mathcal{B}), \quad \Phi \mapsto \tilde{\Phi}$$
$$D_{\mathcal{A}_1,\ldots,\mathcal{A}_n;\mathcal{B}}(f : \Phi \to \Psi) = (g : \tilde{\Phi} \to \tilde{\Psi})$$

such that

$$\boxtimes_\mathcal{A} \bullet g = \left(\boxtimes_\mathcal{A} \bullet \tilde{\Phi} \xrightarrow{\overline{\Phi}} \Phi \xrightarrow{f} \Psi \xrightarrow{\overline{\Psi}^{-1}} \boxtimes_\mathcal{A} \bullet \tilde{\Psi} \right) : \prod_1^n \mathcal{A}_k \to \mathcal{B}.$$

Set the second datum, a functor $\boxtimes_{\mathcal{A};\mathcal{B}}$, to be

$$\boxtimes^n : \mathfrak{A}(\mathcal{A}_1, \mathcal{B}_1) \times \cdots \times \mathfrak{A}(\mathcal{A}_n, \mathcal{B}_n) \xrightarrow{\mathrm{Hom}(1, \boxtimes_\mathcal{B})} \mathrm{Hom}_{\mathrm{m.l.,r.e.}}(\prod_1^n \mathcal{A}_k, \boxtimes_1^n \mathcal{B}_l)$$
$$\xrightarrow{D_{\mathcal{A}_1,\ldots,\mathcal{A}_n; \boxtimes_1^n \mathcal{B}_l}} \mathfrak{A}(\boxtimes_1^n \mathcal{A}_k, \boxtimes_1^n \mathcal{B}_l).$$

We use the simplified notation

$$\boxtimes_1^n F_k = F_1 \boxtimes \cdots \boxtimes F_n = \boxtimes^n_{\mathcal{A};\mathcal{B}}(F_1, \ldots, F_n).$$

So we have an isomorphism

$$\begin{array}{ccc} \prod_1^n \mathcal{A}_k & \xrightarrow{\boxtimes_\mathcal{A}} & \boxtimes_1^n \mathcal{A}_k \\ {\scriptstyle \prod F_k} \downarrow & \overset{\widetilde{\prod F_k \bullet \boxtimes}}{\Rightarrow} & \downarrow {\scriptstyle \boxtimes F_k = \widetilde{\prod F_k \bullet \boxtimes}} \\ \prod_1^n \mathcal{B}_k & \xrightarrow{\boxtimes_\mathcal{B}} & \boxtimes_1^n \mathcal{B}_k \end{array}$$

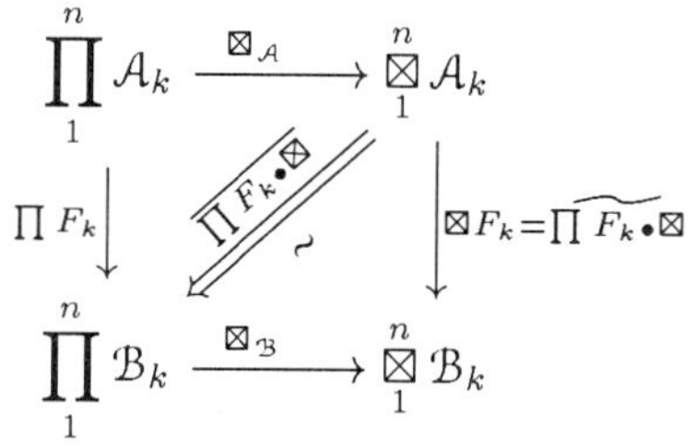

natural in F. That is, for any morphism $\mu = (\mu_k)$, $\mu_k : F_k \to G_k : \mathcal{A}_k \to \mathcal{B}_k$ there is a commutative cylinder

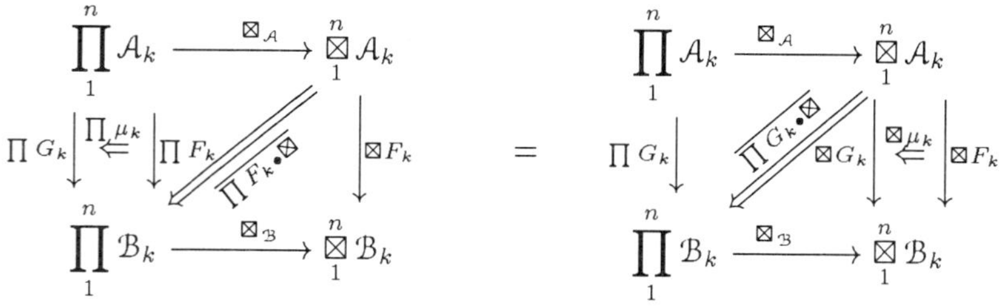

$$\begin{array}{ccc} \prod_1^n \mathcal{A}_k \xrightarrow{\boxtimes_\mathcal{A}} \boxtimes_1^n \mathcal{A}_k & & \prod_1^n \mathcal{A}_k \xrightarrow{\boxtimes_\mathcal{A}} \boxtimes_1^n \mathcal{A}_k \\ {\scriptstyle \prod G_k} \downarrow {\scriptstyle \overset{\prod \mu_k}{\Leftarrow}} \downarrow {\scriptstyle \prod F_k} \overset{\widetilde{\prod F_k \bullet \boxtimes}}{\Rightarrow} \downarrow {\scriptstyle \boxtimes F_k} & = & {\scriptstyle \prod G_k} \downarrow \overset{\widetilde{\prod G_k \bullet \boxtimes}}{\Rightarrow} \downarrow {\scriptstyle \boxtimes G_k} \overset{\boxtimes \mu_k}{\Leftarrow} \downarrow {\scriptstyle \boxtimes F_k} \quad (A.5.1) \\ \prod_1^n \mathcal{B}_k \xrightarrow{\boxtimes_\mathcal{B}} \boxtimes_1^n \mathcal{B}_k & & \prod_1^n \mathcal{B}_k \xrightarrow{\boxtimes_\mathcal{B}} \boxtimes_1^n \mathcal{B}_k \end{array}$$

which serves as a definition of $\boxtimes_1^n \mu_k : \boxtimes_1^n F_k \to \boxtimes_1^n G_k$.

2.3(a) In particular,
$$\mathrm{Id}_{\mathcal{A}_1} \boxtimes \cdots \boxtimes \mathrm{Id}_{\mathcal{A}_n} = \widetilde{\boxtimes_{\mathcal{A}}} = \mathrm{Id}_{\boxtimes_1^n \mathcal{A}_k}$$

due to the choice in 2.2). We **choose** the identity isomorphism
$$\phi^0 = \mathrm{id} : \mathrm{Id}_{\mathcal{A}_1 \boxtimes \cdots \boxtimes \mathcal{A}_n} \to \mathrm{Id}_{\mathcal{A}_1} \boxtimes \cdots \boxtimes \mathrm{Id}_{\mathcal{A}_n}$$

as the third datum.

2.3(b) One more datum is the isomorphism of functors

$$\begin{array}{ccc}
\prod_1^n \mathfrak{A}(\mathcal{A}_k, \mathcal{B}_k) \times \prod_1^n \mathfrak{A}(\mathcal{B}_k, \mathcal{C}_k) & \xrightarrow{\bullet^2} & \prod_1^n \mathfrak{A}(\mathcal{A}_k, \mathcal{C}_k) \\
\boxtimes^n \times \boxtimes^n \downarrow & \overset{\phi^2}{\Longrightarrow} & \downarrow \boxtimes^n \\
\mathfrak{A}(\boxtimes_1^n \mathcal{A}_k, \boxtimes_1^n \mathcal{B}_k) \times \mathfrak{A}(\boxtimes_1^n \mathcal{B}_k, \boxtimes_1^n \mathcal{C}_k) & \xrightarrow{\bullet^2} & \mathfrak{A}(\boxtimes_1^n \mathcal{A}_k, \boxtimes_1^n \mathcal{C}_k)
\end{array}$$

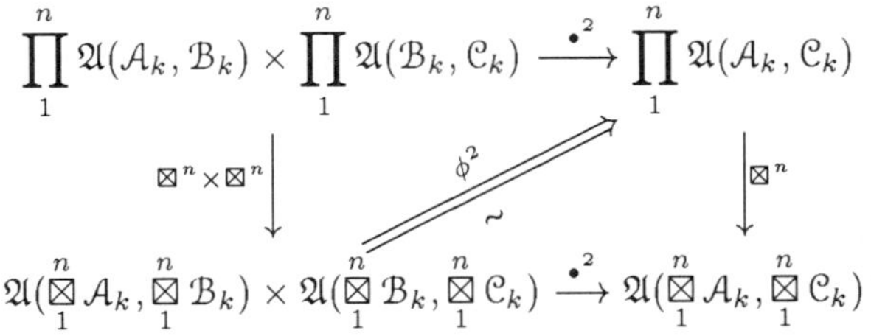

obtained from the diagram

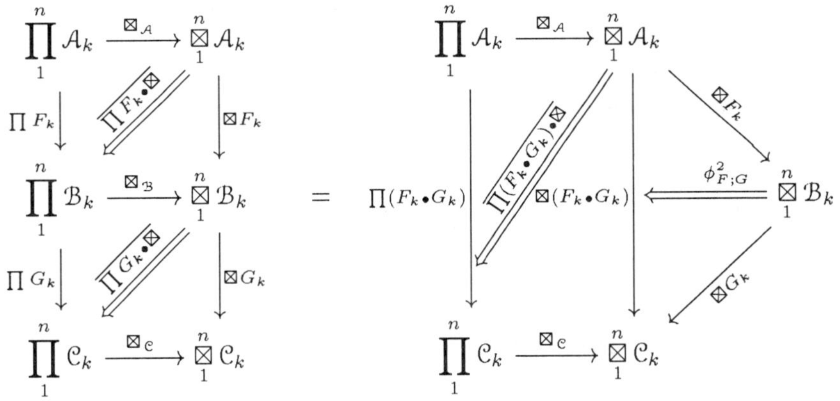

(A.5.2)

by Corollary A.5.3. We still have to check that
$$\phi^2_{F;G} : (\boxtimes_1^n F_k) \bullet (\boxtimes_1^n G_k) \to \boxtimes_1^n (F_k \bullet G_k)$$

is natural in F_k, G_k. If $\lambda_k : F_k \to F'_k$, $\mu_k : G_k \to G'_k$ are morphisms of functors, then the following string of equations holds

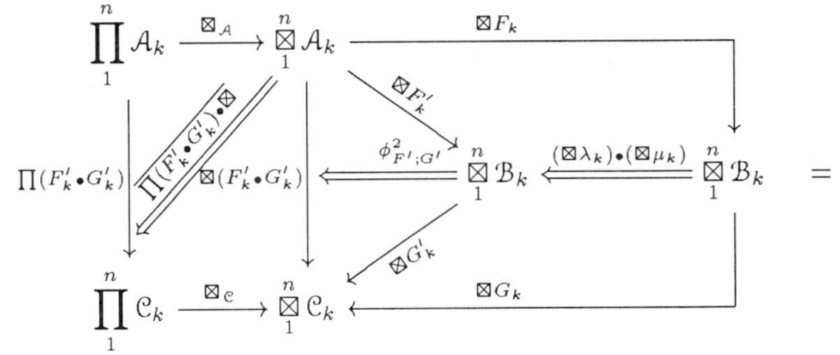

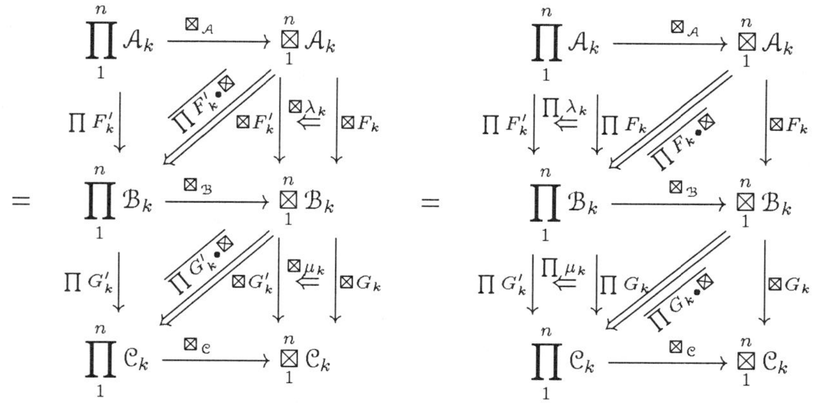

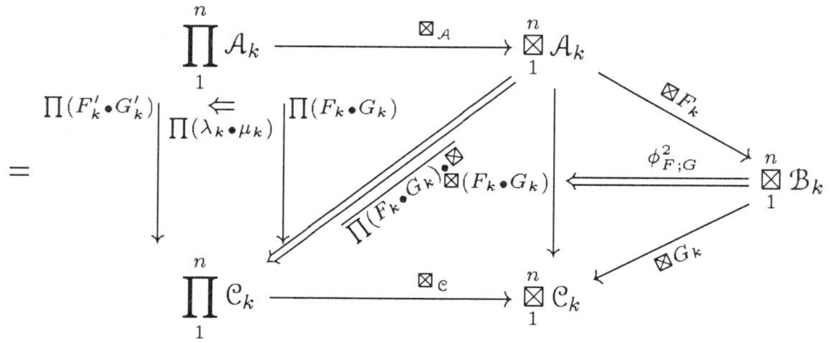

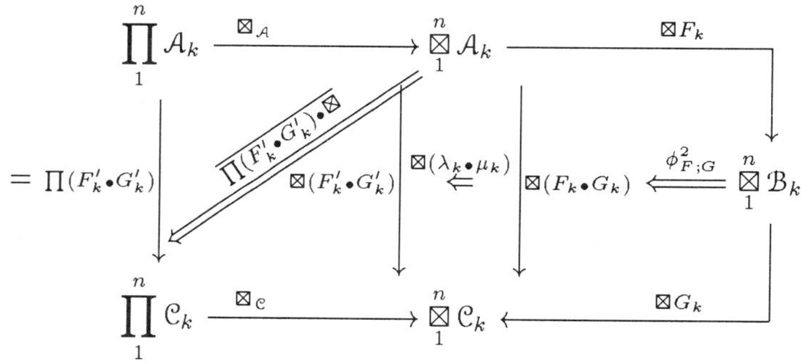

By Theorem A.5.1 the following diagram is commutative.

$$\begin{array}{ccc}
(\overset{n}{\underset{1}{\boxtimes}} F_k)\bullet(\overset{n}{\underset{1}{\boxtimes}} G_k) & \xrightarrow{\phi^2_{F;G}} & \overset{n}{\underset{1}{\boxtimes}}(F_k\bullet G_k) \\
{\scriptstyle (\boxtimes\lambda_k)\bullet(\boxtimes\mu_k)}\big\downarrow & & \big\downarrow{\scriptstyle \boxtimes(\lambda_k\bullet\mu_k)} \\
(\overset{n}{\underset{1}{\boxtimes}} F'_k)\bullet(\overset{n}{\underset{1}{\boxtimes}} G'_k) & \xrightarrow{\phi^2_{F';G'}} & \overset{n}{\underset{1}{\boxtimes}}(F'_k\bullet G'_k)
\end{array}$$

Thus all data for a weak 2-functor $\boxtimes^n$ are listed.

2.4) We have to verify some equations.

2.4(a) Since we have chosen $\phi_0 = \mathrm{id}$, equations (A.1.1) and (A.1.2) reduce to

$$\phi^2_{F_1,\dots,F_n;\mathrm{Id}_{\mathcal{B}_1},\dots,\mathrm{Id}_{\mathcal{B}_n}} = \mathrm{id}_{\boxtimes F_k}, \qquad \phi^2_{\mathrm{Id}_{\mathcal{A}_1},\dots,\mathrm{Id}_{\mathcal{A}_n};F_1,\dots,F_n} = \mathrm{id}_{\boxtimes F_k}$$

respectively for each family $F_k : \mathcal{A}_k \to \mathcal{B}_k \in \mathfrak{A}_1$. Verification of the first one follows from

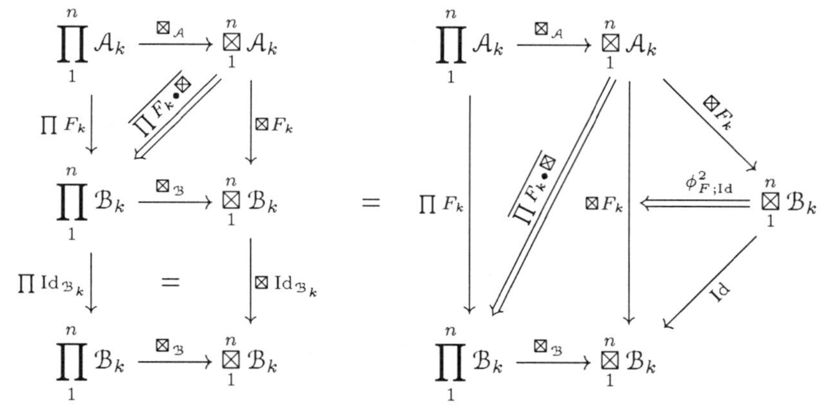

by Corollary A.5.3. The second equation is proved analogously.

2.4(b) For any system of 1-morphisms in $\mathfrak{A}_1$

$$\mathcal{A}_k \xrightarrow{F_k} \mathcal{B}_k \xrightarrow{G_k} \mathcal{C}_k \xrightarrow{H_k} \mathcal{D}_k,$$

equation A.1.2.4(b) follows from the equation

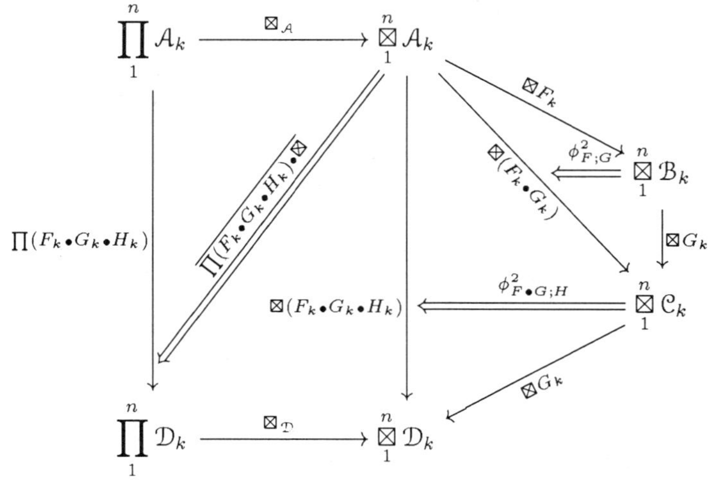

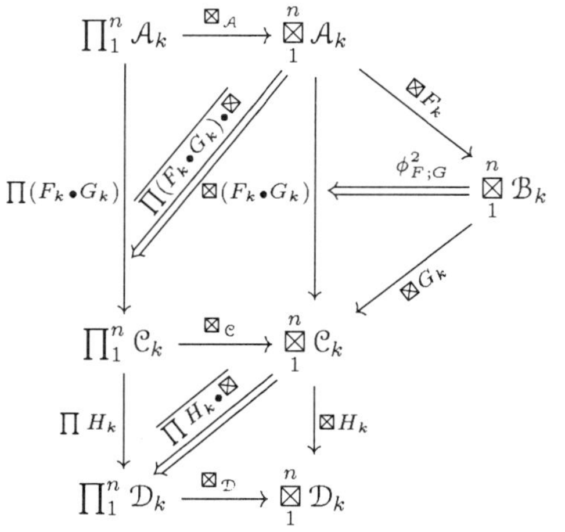

=

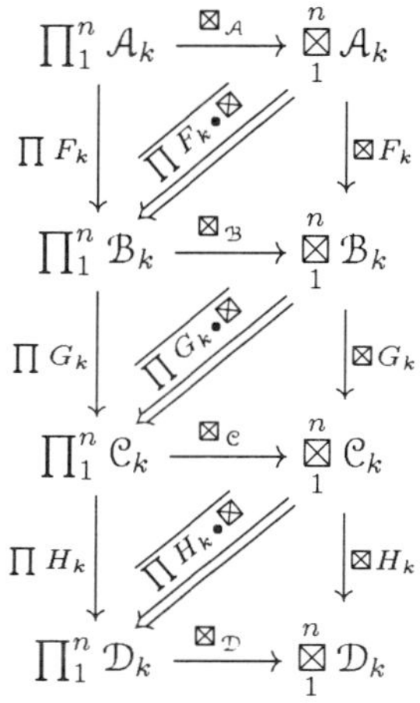

=

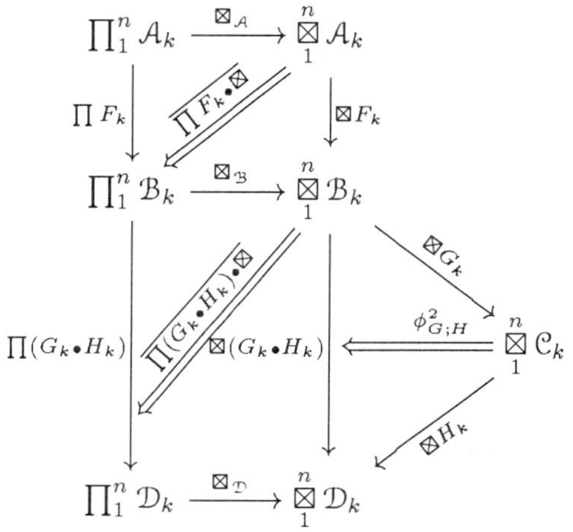

=

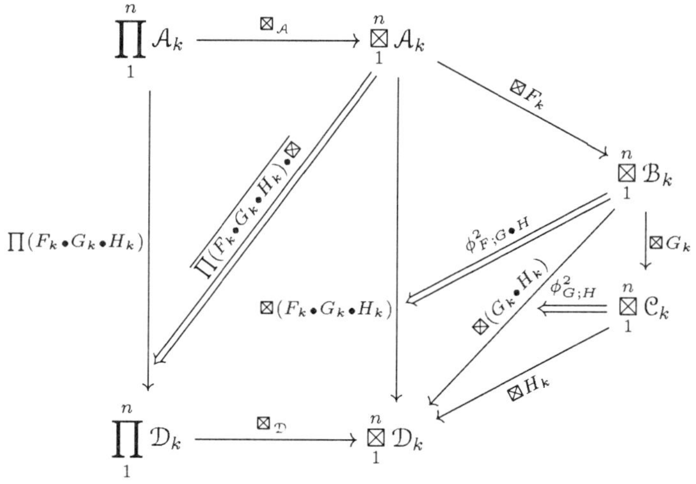

by Corollary A.5.3.

Therefore, $(\boxtimes^n, \phi^2, \mathrm{id})$ is a weak 2-functor.

3) To any plane staged tree T with n leaves (0-vertices are allowed) is associated a weak 2-functor $\boxtimes^T = [T]_1 : \mathfrak{A}^n \to \mathfrak{A}$. It is the composition of direct products

of 2-functors $\boxtimes^m$ corresponding to vertices with m entries, $m \geqslant 0$. It can be visualized as a word $\mathcal{A}_1 \dots \mathcal{A}_n$ with several $\Bbbk$-vect inserted, arguments separated by $\boxtimes$ signs with some parentheses. For any object $(\mathcal{A}_1, \dots, \mathcal{A}_n)$ of $\mathfrak{A}^n$, we have also a $\Bbbk$-multilinear right exact in each variable functor

$$\boxtimes^T_{\mathcal{A}} : \mathcal{A}_1 \times \dots \times \mathcal{A}_n \to \boxtimes^T(\mathcal{A}_1, \dots, \mathcal{A}_n),$$

the composition of direct products of chosen functors $\boxtimes_{\mathcal{B}} : \mathcal{B}_1 \times \dots \times \mathcal{B}_n \to \boxtimes^m_1 \mathcal{B}_k$.

We want to show that the categories $\boxtimes^T(\mathcal{A}_1, \dots, \mathcal{A}_n)$ are equivalent to $\mathcal{A}_1 \boxtimes \dots \boxtimes \mathcal{A}_n$. Clearly, it suffices to show that the functor $\boxtimes^T_{\mathcal{A}}$ possesses the universal properties of Theorem A.5.1.

A.5.4. Proposition. *For any plane staged forest T with n leaves and any $n+1$-tuple of categories $\mathcal{A}_1, \dots, \mathcal{A}_n, \mathcal{B} \in \mathfrak{A}_0$ the functor*

$$\mathrm{Hom}(\boxtimes^T_{\mathcal{A}}, \mathcal{B}) : \mathrm{Hom}_{m.l.,r.e.}(\boxtimes^T(\mathcal{A}_1, \dots, \mathcal{A}_n), \mathcal{B}) \to \mathrm{Hom}_{m.l.,r.e.}(\mathcal{A}_1 \times \dots \times \mathcal{A}_n, \mathcal{B})$$

is an equivalence.

PROOF. Consider first a particular case.

$T =$ 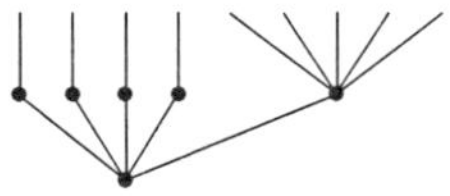

$$\boxtimes^T(\mathcal{A}_1, \dots, \mathcal{A}_n) = \mathcal{A}_1 \boxtimes \dots \boxtimes \mathcal{A}_{m-1} \boxtimes (\mathcal{A}_m \boxtimes \dots \boxtimes \mathcal{A}_n), \qquad 1 < m \leqslant n.$$

The functor $\mathrm{Hom}(\boxtimes^T_{\mathcal{A}}, \mathcal{B})$ is the composite of two functors

$$\mathfrak{A}(\boxtimes^T(\mathcal{A}_1, \dots, \mathcal{A}_n), \mathcal{B}) \xrightarrow{\mathrm{Hom}(\boxtimes_{\mathcal{A}_1, \dots, \mathcal{A}_m, \boxtimes^n_m \mathcal{A}_k}, \mathcal{B})} \mathrm{Hom}_{\mathrm{m.l.,r.e.}}\Big(\prod_1^{m-1} \mathcal{A}_k \times (\boxtimes_m^n \mathcal{A}_p), \mathcal{B}\Big)$$

$$\xrightarrow{\mathrm{Hom}(1 \times \boxtimes_{\mathcal{A}_m, \dots, \mathcal{A}_n}, \mathcal{B})} \mathrm{Hom}_{\mathrm{m.l.,r.e.}}\Big(\prod_1^n \mathcal{A}_k, \mathcal{B}\Big),$$

the first of which is an equivalence. So it remains to check that the second is an equivalence. In the commutative diagram of functors (where $\mathrm{Hom}_{\mathrm{m.l.}}$ denotes $\Bbbk$-multilinear functors)

$$\begin{array}{ccc}
\mathrm{Hom}_{\mathrm{m.l.,r.e.}}\Big(\prod_1^{m-1} \mathcal{A}_k \times (\boxtimes_m^n \mathcal{A}_p), \mathcal{B}\Big) & \hookrightarrow & \mathrm{Hom}_{\mathrm{m.l.}}\Big(\prod_1^{m-1} \mathcal{A}_k, \mathfrak{A}(\boxtimes_m^n \mathcal{A}_p, \mathcal{B})\Big) \\
\Big\downarrow{\scriptstyle \mathrm{Hom}(1 \times \boxtimes_{\mathcal{A}_m, \dots, \mathcal{A}_n}, \mathcal{B})} & & \Big\downarrow{\scriptstyle \mathrm{Hom}(1, \mathrm{Hom}(\boxtimes_{\mathcal{A}_m, \dots, \mathcal{A}_n}, \mathcal{B}))} \\
\mathrm{Hom}_{\mathrm{m.l.,r.e.}}\Big(\prod_1^n \mathcal{A}_k, \mathcal{B}\Big) & \hookrightarrow & \mathrm{Hom}_{\mathrm{m.l.}}\Big(\prod_1^{m-1} \mathcal{A}_k, \mathrm{Hom}_{\mathrm{m.l.,r.e.}}\big(\prod_m^n \mathcal{A}_p, \mathcal{B}\big)\Big)
\end{array}$$

the rows are full and faithful and the right vertical arrow is an equivalence. Hence, the left vertical arrow is full and faithful.

Any $\Bbbk$-multilinear right exact in each variable functor $F : \prod_1^n \mathcal{A}_k \to \mathcal{B}$ viewed as a functor $\prod_1^{m-1} \mathcal{A}_k \to \mathrm{Hom}_{\mathrm{m.l.,r.e.}}(\prod_m^n \mathcal{A}_p, \mathcal{B})$ is isomorphic to the composition of some $G' : \prod_1^{m-1} \mathcal{A}_k \to \mathfrak{A}(\boxtimes_m^n \mathcal{A}_p, \mathcal{B})$ with $\mathrm{Hom}(\boxtimes_{\mathcal{A}_m, \dots, \mathcal{A}_n}, \mathcal{B})$. In other words, there is a $\Bbbk$-multilinear functor $G : \prod_1^{m-1} \mathcal{A}_k \times (\boxtimes_m^n \mathcal{A}_p) \to \mathcal{B}$, right exact in the argument $M \in \boxtimes_m^n \mathcal{A}_p$, and an isomorphism

$$F \simeq F' = \big(\prod_1^n \mathcal{A}_k \xrightarrow{1\times\boxtimes_{\mathcal{A}_m,\dots,\mathcal{A}_n}} \prod_1^{m-1} \mathcal{A}_k \times (\boxtimes_m^n \mathcal{A}_p) \xrightarrow{G} \mathcal{B}\big).$$

Since F is $\Bbbk$-multilinear right exact in each variable, so is F'. We have to show that G is exact in arguments $X_k \in \mathcal{A}_k$, $1 \leqslant k \leqslant m-1$. By Deligne's theory [7] (see Propositions 1.1.2, 1.1.6) for any object $M \in \boxtimes_m^n \mathcal{A}_p$ there is an exact sequence in $\boxtimes_m^n \mathcal{A}_p$

$$Y_m \boxtimes \cdots \boxtimes Y_n \xrightarrow{\sum_r f_1^r \boxtimes \cdots \boxtimes f_n^r} Z_m \boxtimes \cdots \boxtimes Z_n \to M \to 0$$

with $f_p^r \in \mathcal{A}_p$. Fix k, $1 \leqslant k \leqslant m-1$ and an exact sequence in $\mathcal{A}_k$

$$X_k'' \to X_k' \to X_k \to 0.$$

The diagram with $A = X_1 \times \cdots \times X_{k-1}$, $B = X_{k+1} \times \cdots \times X_{m-1}$ has exact columns.

$$\begin{array}{ccccccc}
G(A\times X_k''\times B\times\boxtimes_m^n Y_p) & \longrightarrow & G(A\times X_k'\times B\times\boxtimes_m^n Y_p) & \longrightarrow & G(A\times X_k\times B\times\boxtimes_m^n Y_p) & \longrightarrow & 0 \\
\downarrow & & \downarrow & & \downarrow & & \\
G(A\times X_k''\times B\times\boxtimes_m^n Z_p) & \longrightarrow & G(A\times X_k'\times B\times\boxtimes_m^n Z_p) & \longrightarrow & G(A\times X_k\times B\times\boxtimes_m^n Z_p) & \longrightarrow & 0 \\
\downarrow & & \downarrow & & \downarrow & & \\
G(A\times X_k''\times B\times M) & \longrightarrow & G(A\times X_k'\times B\times M) & \longrightarrow & G(A\times X_k\times B\times M) & \longrightarrow & 0 \\
\downarrow & & \downarrow & & \downarrow & & \\
0 & & 0 & & 0 & &
\end{array}$$

The first and the second rows are also exact, since G can be replaced with F' in them. One half of the 3×3-Lemma implies that the third row is also exact. Hence, $\mathrm{Hom}(1\times\boxtimes_{\mathcal{A}_m,\dots,\mathcal{A}_n}, \mathcal{B})$ is essentially surjective on objects, therefore, an equivalence.

The case $m = n + 1$

$$\boxtimes^T(\mathcal{A}_1, \dots, \mathcal{A}_n) = \mathcal{A}_1 \boxtimes \cdots \boxtimes \mathcal{A}_n \boxtimes \Bbbk\text{-vect}$$

is easy to consider separately.

The case of any plane staged tree T with 2 levels with all vertices having 1 entry except two of them is completely analogous.

In the general case $\mathrm{Hom}(\boxtimes_{\mathcal{A}}^T, \mathcal{B})$ is a product of equivalences considered above. □

A.5.5. Corollary. *For any plane staged tree T with n leaves and any $n+1$-tuple of categories $\mathcal{A}_1, \dots, \mathcal{A}_n, \mathcal{B} \in \mathfrak{A}_0$ the functor*

$$\mathrm{Hom}(\boxtimes_{\mathcal{A}}^T, \mathcal{B}) : \mathfrak{A}(\boxtimes^T(\mathcal{A}_1, \dots, \mathcal{A}_n), \mathcal{B}) \to \mathrm{Hom}_{m.l.,r.e.}(\mathcal{A}_1 \times \cdots \times \mathcal{A}_n, \mathcal{B})$$

is an equivalence.

3.1) Universality of Deligne's tensor product (Theorem A.5.1) implies that for any plane staged tree T we can (and we will) **choose** a natural equivalence $\lambda^T_{\mathcal{A}}$ and

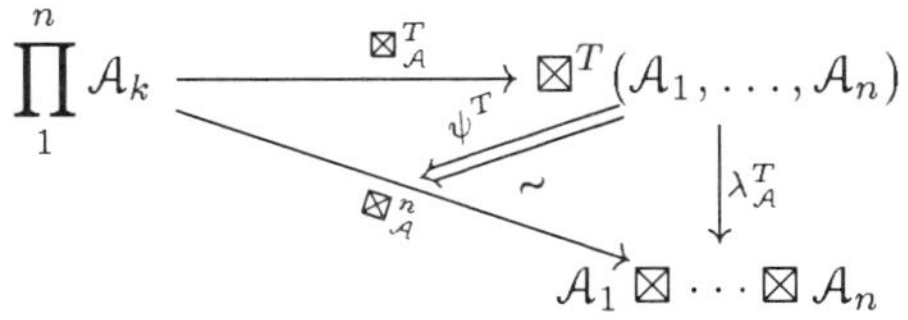

an isomorphism of functors ψ^T. The family of equivalences

$$\lambda^T_{\mathcal{A}} : \boxtimes^T(\mathcal{A}_1, \ldots, \mathcal{A}_n) \to \mathcal{A}_1 \boxtimes \cdots \boxtimes \mathcal{A}_n$$

is the *first datum* of a weak 2-transformation

$$\lambda^T = [t^k]_2 : [T]_1 \to [t^n]_1 : \mathfrak{A}^n \to \mathfrak{A}.$$

3.2) The *second datum* of $[t^k]_2$ is an isomorphism λ^T_F constructed for a family of functors $F_k : \mathcal{A}_k \to \mathcal{B}_k \in \mathfrak{A}_1$

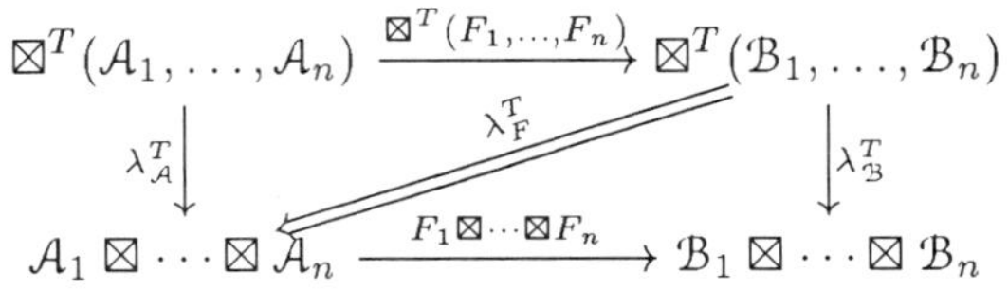

from the equation

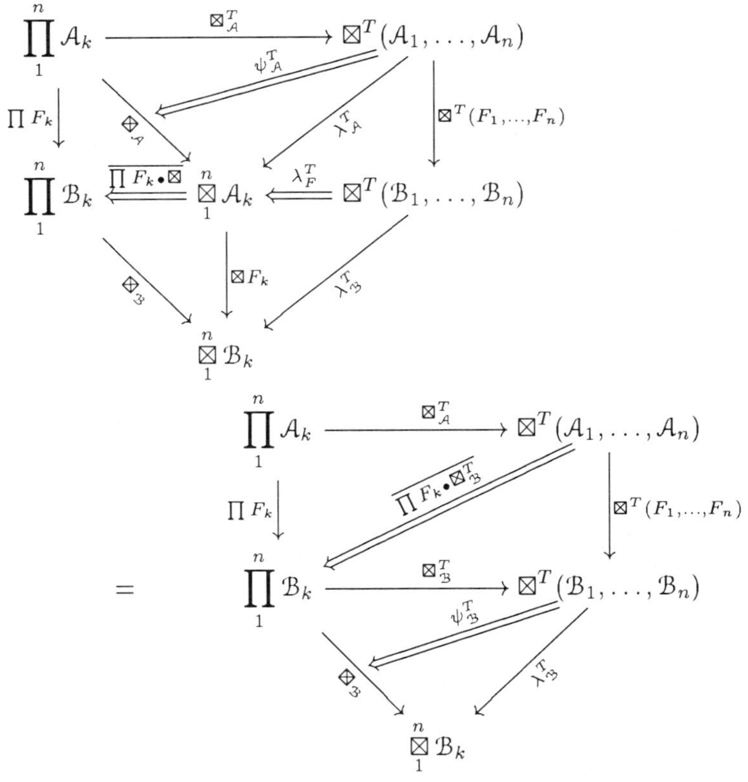

(A.5.3)

by Corollary A.5.3. Here the isomorphism $\overline{\prod F_k \bullet \boxtimes^T_{\mathcal{B}}}$ is the pasting of isomorphisms corresponding to multiplicands of $\boxtimes^T_{\mathcal{B}}$.

We have to show that $\lambda^T_{F_1,\ldots,F_n}$ is natural in $F_1, \ldots, F_n$. Let $\mu_k : F_k \to G_k$, $1 \leqslant k \leqslant n$, be natural transformations. Then

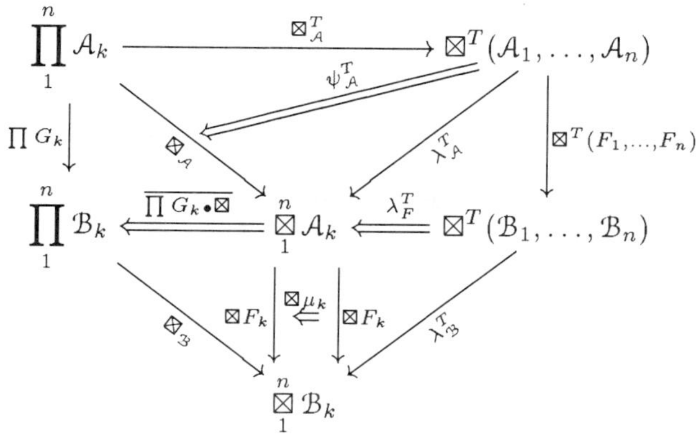

=

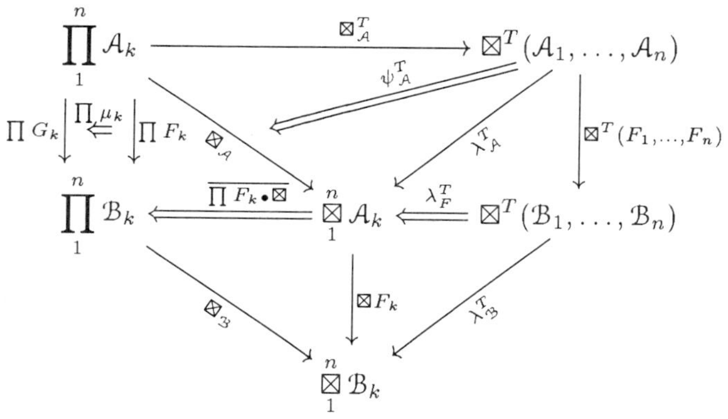

=

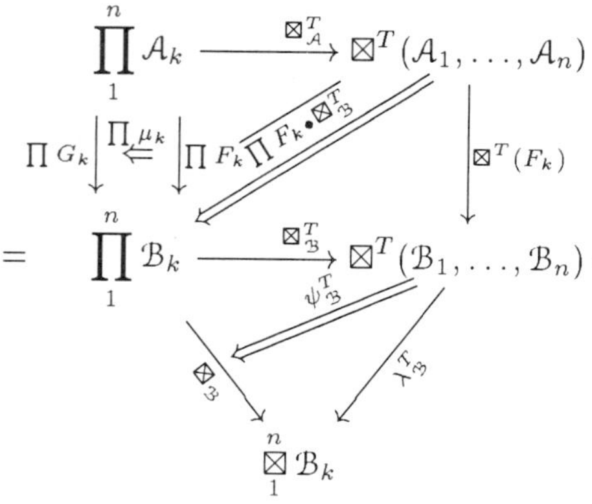

=

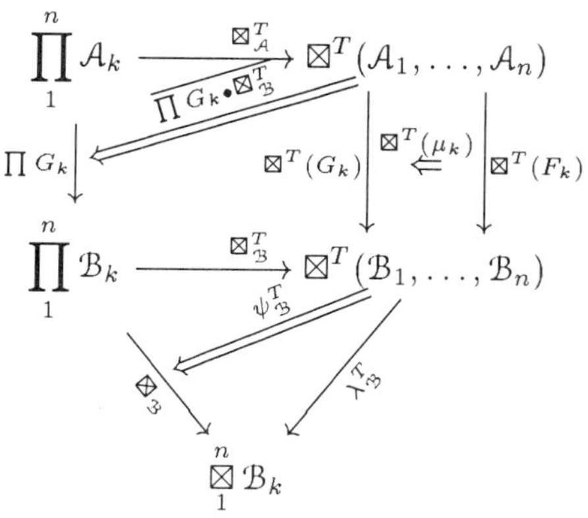

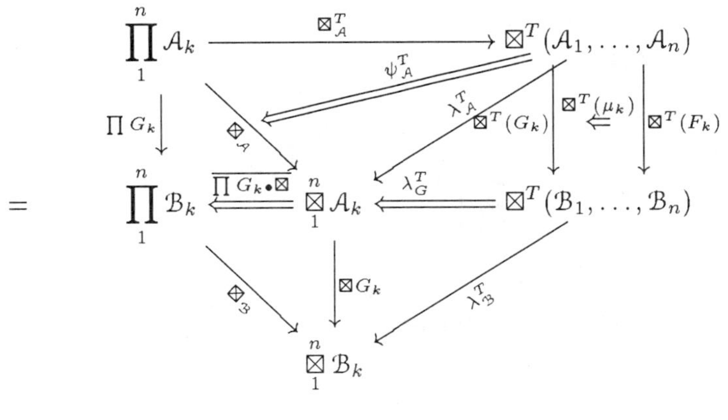

By Theorem A.5.1 we deduce that

$$
\begin{array}{c}
\boxtimes_1^n \mathcal{A}_k \xleftarrow{\lambda_{\mathcal{A}}^T} \boxtimes^T(\mathcal{A}_1,\dots,\mathcal{A}_n) \\
\boxtimes G_k \downarrow \boxtimes \mu_k \Downarrow \boxtimes F_k \quad \boxtimes^T(F_1,\dots,F_n) \downarrow \\
\boxtimes_1^n \mathcal{B}_k \xleftarrow{\lambda_{\mathcal{B}}^T} \boxtimes^T(\mathcal{B}_1,\dots,\mathcal{B}_n)
\end{array}
\quad = \quad
\begin{array}{c}
\boxtimes_1^n \mathcal{A}_k \xleftarrow{\lambda_{\mathcal{A}}^T} \boxtimes^T(\mathcal{A}_1,\dots,\mathcal{A}_n) \\
\boxtimes G_k \downarrow \quad \lambda_G^T \quad \boxtimes^T(G_k) \; \boxtimes^T(\mu_k) \Leftarrow \; \boxtimes^T(F_k) \downarrow \\
\boxtimes_1^n \mathcal{B}_k \xleftarrow{\lambda_{\mathcal{B}}^T} \boxtimes^T(\mathcal{B}_1,\dots,\mathcal{B}_n)
\end{array}
$$

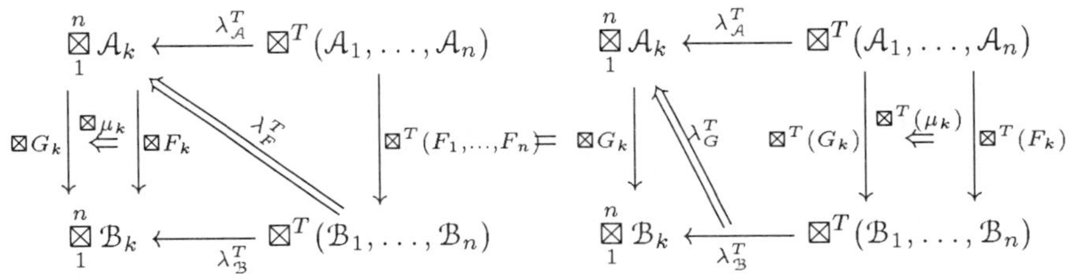

This is the functoriality of $\lambda_{F_1,\dots,F_n}^T$ in $F_1,\dots,F_n$.

3.3(a) We have to check that

$$
\begin{array}{c}
\boxtimes^T(\mathcal{A}_1,\dots,\mathcal{A}_n) \xrightarrow[\boxtimes^T(\mathrm{Id}_{\mathcal{A}_1},\dots,\mathrm{Id}_{\mathcal{A}_n})]{\mathrm{Id} \;\Downarrow \phi_0^T} \boxtimes^T(\mathcal{A}_1,\dots,\mathcal{A}_n) \\
\lambda_{\mathcal{A}}^T \downarrow \quad \lambda_{\mathrm{Id},\dots,\mathrm{Id}} \quad \downarrow \lambda_{\mathcal{A}}^T \\
\mathcal{A}_1 \boxtimes \dots \boxtimes \mathcal{A}_n \xrightarrow{\mathrm{Id}_{\mathcal{A}_1} \boxtimes \dots \boxtimes \mathrm{Id}_{\mathcal{A}_n}} \mathcal{A}_1 \boxtimes \dots \boxtimes \mathcal{A}_n
\end{array}
$$

$$
= \quad
\begin{array}{c}
\boxtimes^T(\mathcal{A}_1,\dots,\mathcal{A}_n) \xrightarrow{\mathrm{Id}} \boxtimes^T(\mathcal{A}_1,\dots,\mathcal{A}_n) \\
\lambda_{\mathcal{A}}^T \downarrow \quad = \quad \downarrow \lambda_{\mathcal{A}}^T \\
\mathcal{A}_1 \boxtimes \dots \boxtimes \mathcal{A}_n \xrightarrow[\mathrm{Id}_{\mathcal{A}_1}\boxtimes\dots\boxtimes\mathrm{Id}_{\mathcal{A}_n}]{\mathrm{Id}_{\mathcal{A}_1\boxtimes\dots\boxtimes\mathcal{A}_n} \;\Downarrow \phi_0} \mathcal{A}_1 \boxtimes \dots \boxtimes \mathcal{A}_n
\end{array}
$$

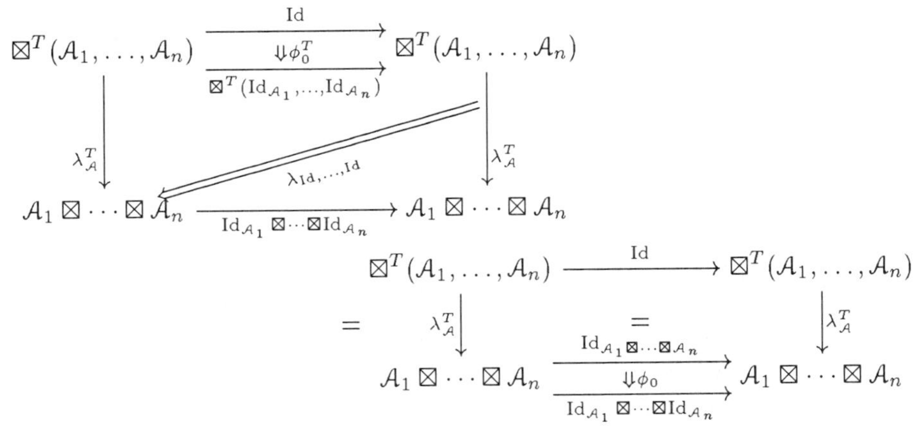

We have chosen $\phi_0 = \mathrm{id}_{\mathrm{Id}}$. Since ϕ_0^T is a composition of several ϕ_0 it also equals $\phi_0^T = \mathrm{id}_{\mathrm{Id}}$. Hence the equation reduces to

$$\lambda_{\mathrm{Id}_{\mathcal{A}_1}\boxtimes\dots\boxtimes\mathrm{Id}_{\mathcal{A}_n}} = \mathrm{id}_{\lambda_{\mathcal{A}}^T},$$

which we verify now. Indeed,

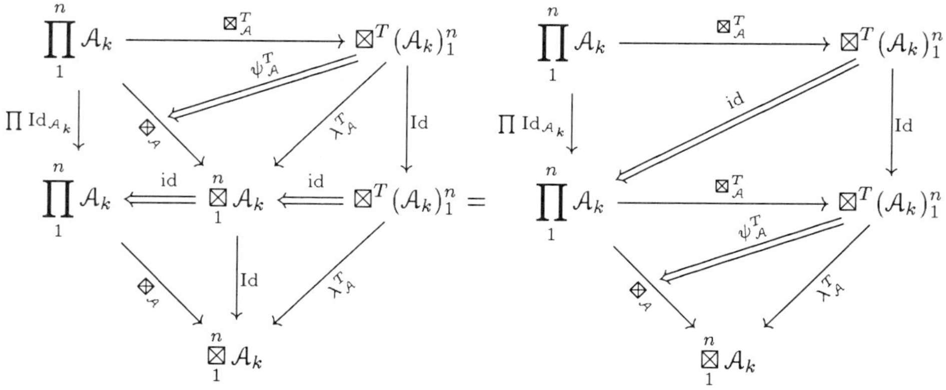

3.3(b) Consider a family of functors from $\mathfrak{A}_1$ $H_k = \left(\mathcal{A}_k \xrightarrow{F_k} \mathcal{B}_k \xrightarrow{G_k} \mathcal{C}_k\right)$. We have the following string of equations

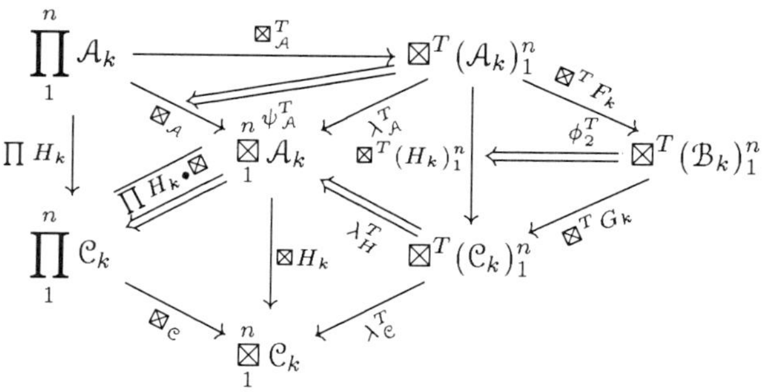

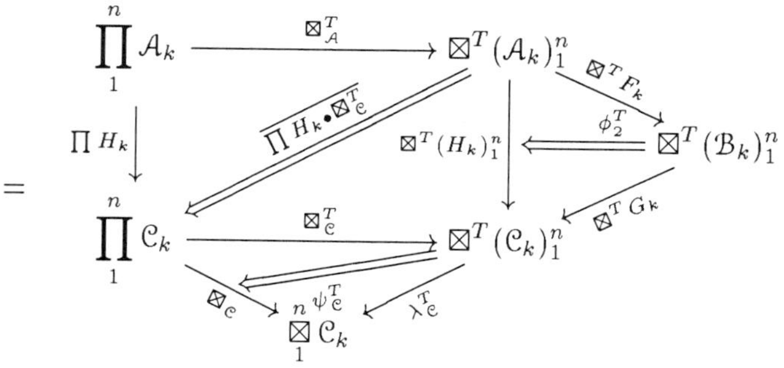

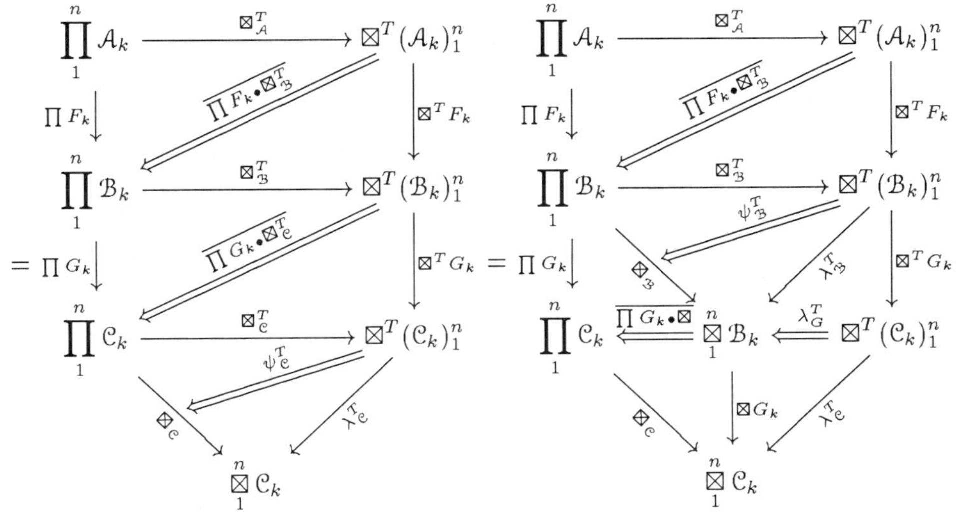

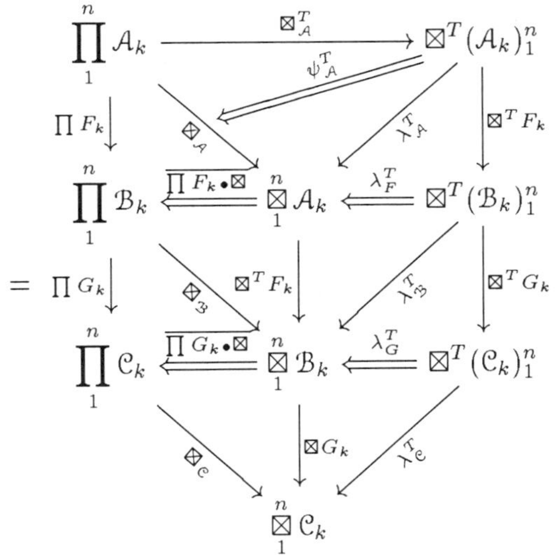

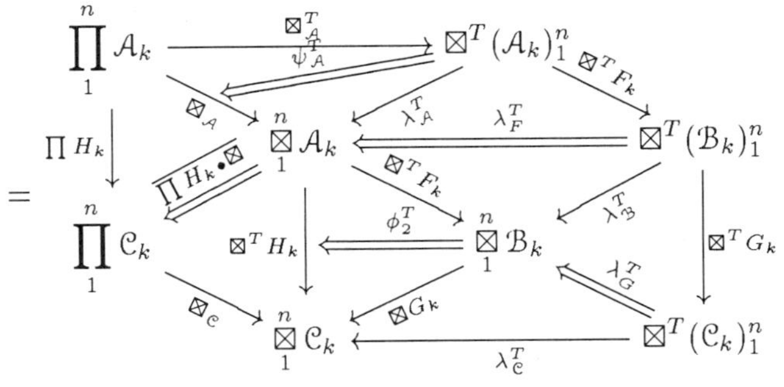

We deduce from the above by Corollary A.5.3 that

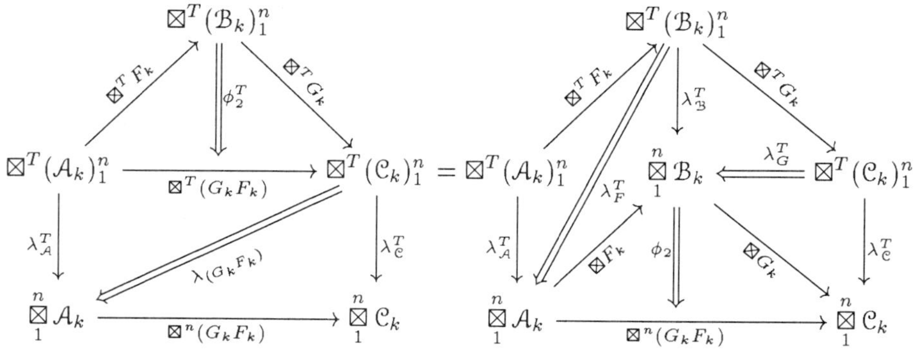

This finishes the proof that λ^T is a weak 2-transformation.

4) Consider a plane staged tree T decomposed into three plane staged forests by two horizontal cuts,

$$T = \begin{matrix} g \\ f \\ t \end{matrix} .$$

The lowest forest t is a tree. Denote by $\bar{f}$ the forest f with all inner edges (not leaves) shrunk to a point. Each tree in f is replaced with a 1-level tree in $\bar{f}$. Let

$$\overline{T} = \begin{matrix} g \\ \bar{f} \\ t \end{matrix} ,$$

which is T with several levels compressed. We want to construct

$[Y_{12}]_3 :$  $\to [t^k]_2 : [T]_1 \to [t^n]_1 : \mathfrak{A}^n \to \mathfrak{A}.$

The first 2-transformation is the composition of two vertical arrows in

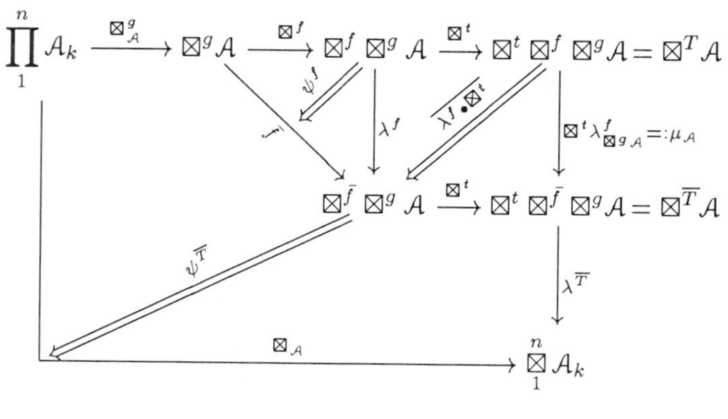

(A.5.4)

where we define $\mu_{\mathcal{A}} = \boxtimes^t \lambda^f_{\boxtimes^g \mathcal{A}}$. The morphism $Y_{\mathcal{A}} = [Y_{12}]_3$ is found by Corollary A.5.3 so that the above pasting equals

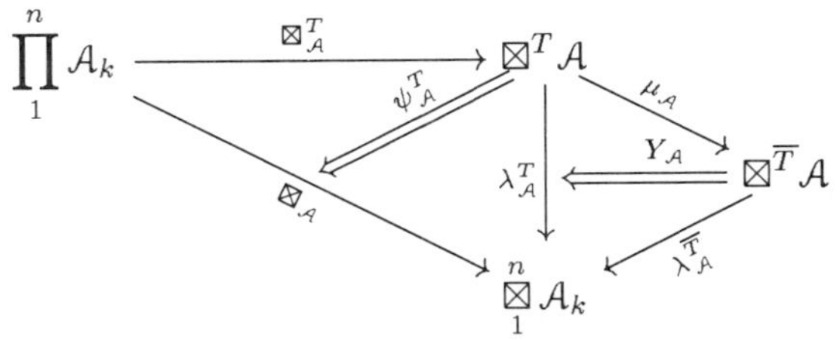

(A.5.5)

We want to show that $[Y_{12}]_3 = (Y_{\mathcal{A}})_{\mathcal{A}}$ is a modification. To do this we remark that the following prism with two triangles added is commutative in the sense that there is a unique morphism between any two functors corresponding to two arbitrary paths from $\prod_1^n \mathcal{A}_k$ to $\boxtimes_1^n \mathcal{B}_k$.

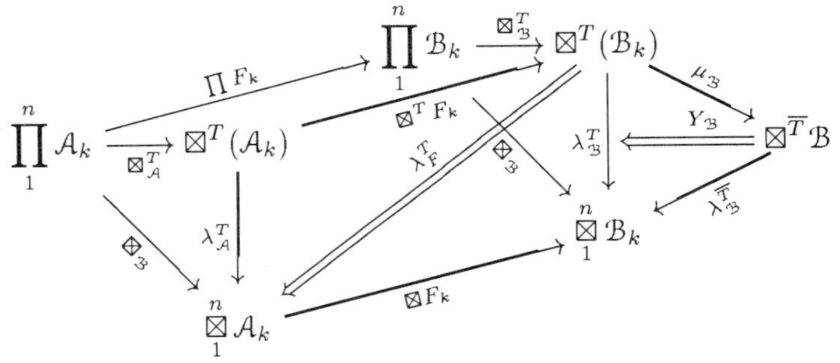

Commutativity follows by the definition of λ^T_F. On the other hand, the prism

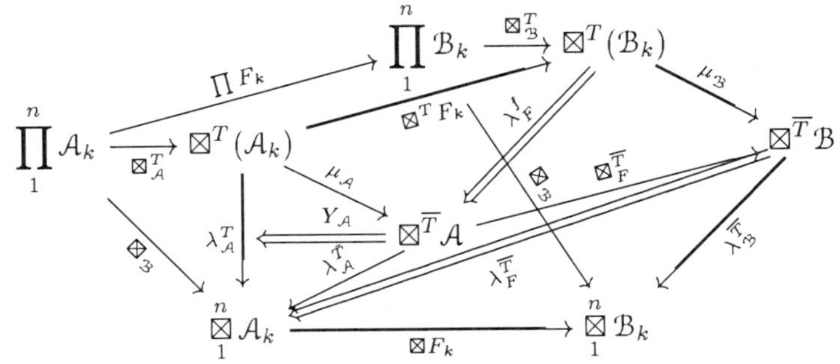

also commutes. One finds this replacing the faces (A.5.5) with the faces (A.5.4). The obtained prism consists of two prisms with a triangular base, each commutative by the same reason as above, and one cube

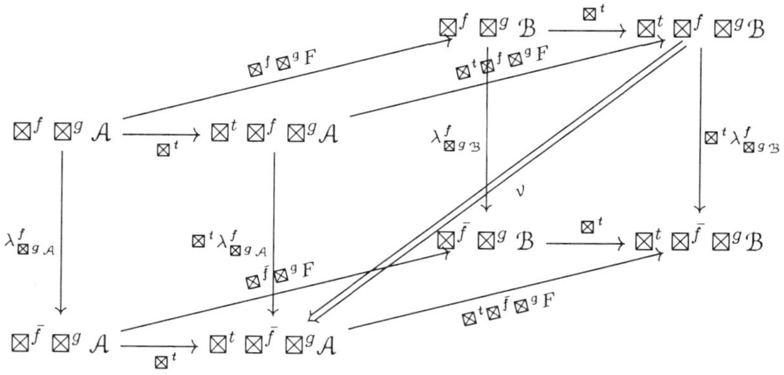

where the only explicitly marked 2-cell is

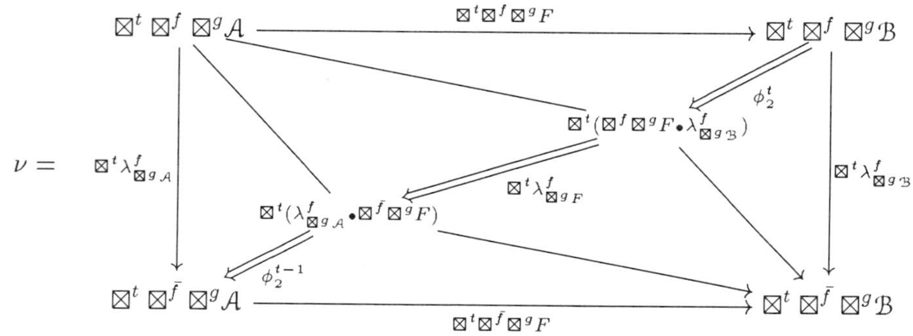

Commutativity of the cube follows from commutativity of two prisms with triangular base (A.5.2) and one cylinder (A.5.1).

By Corollary A.5.3 we deduce that

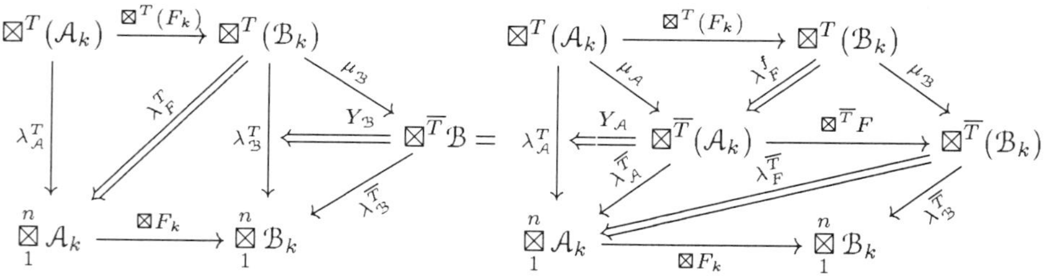

Taking direct products of the constructed modifications and inserting them into compositions of weak 2-transformations, we get invertible modifications

$$[Y_{ij}]_3 : [T']_2 \to [T'']_2 : [T]_1 \to [t^n]_1 : \mathfrak{A}^n \to \mathfrak{A},$$

where T'' is obtained from T' by contracting an edge $(i, j) \subset T'$.

5) We have to check equations A.4.2.5. Equations 5(a) and 5(b) follow from Lemmas A.3.3 and A.3.2, namely, from equations (A.3.5) with $l = k = [Y]_3$.

5(c) Let us prove now A.4.2.5(c), where weak 2-transformations are of the form $[T]_1 \to [t^n]_1 : \mathfrak{A}^n \to \mathfrak{A}$. Decompose the plane staged tree T into the product of five forests corresponding to the choice of T' (at the illustration below the forests f_1,

f_5 have 1 level, the forests f_2, f_4 have 2 levels, and the forest f_3 has 3 levels):

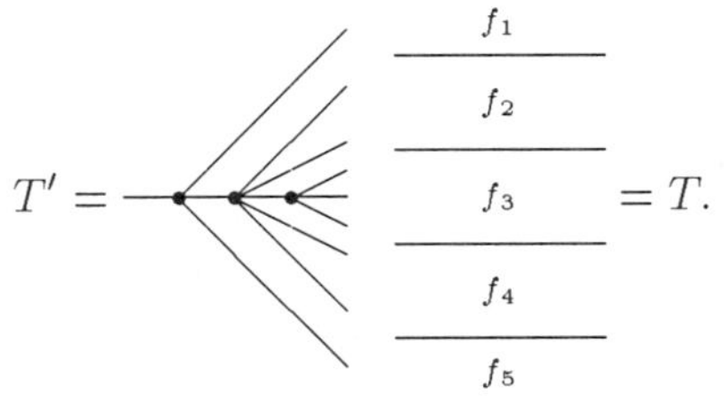

The forest f_5 is a tree. Correspondingly $\boxtimes^T_{\mathcal{A}}$ decomposes as the upper row in the following diagram:

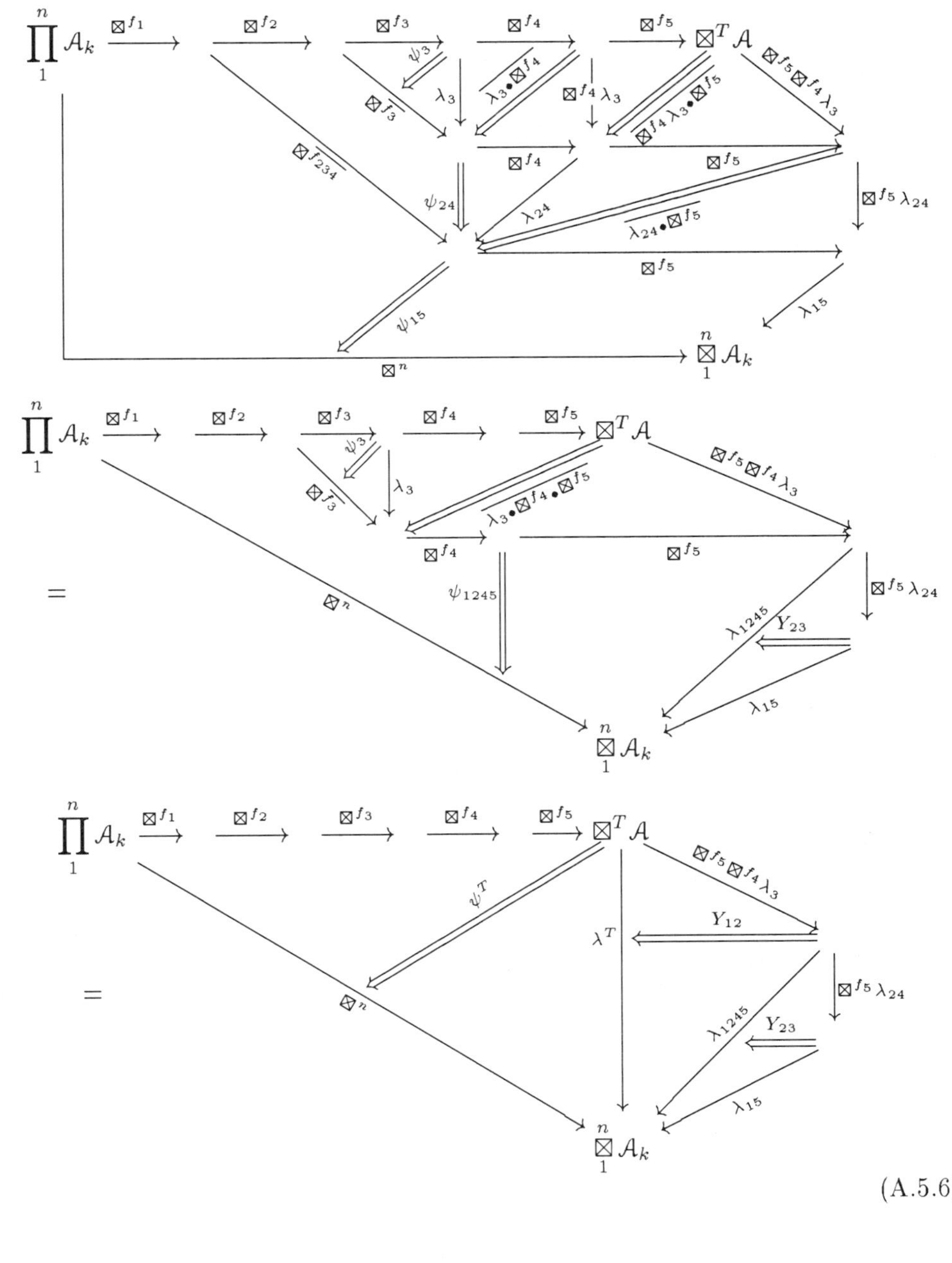

(A.5.6)

On the other hand, pasting of the first diagram also equals

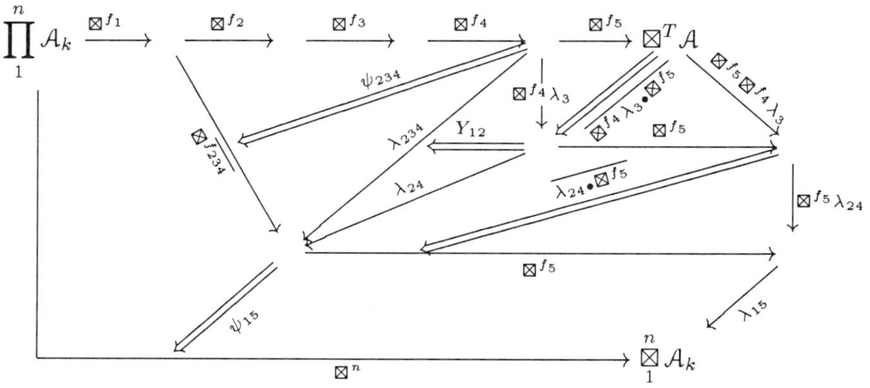

=

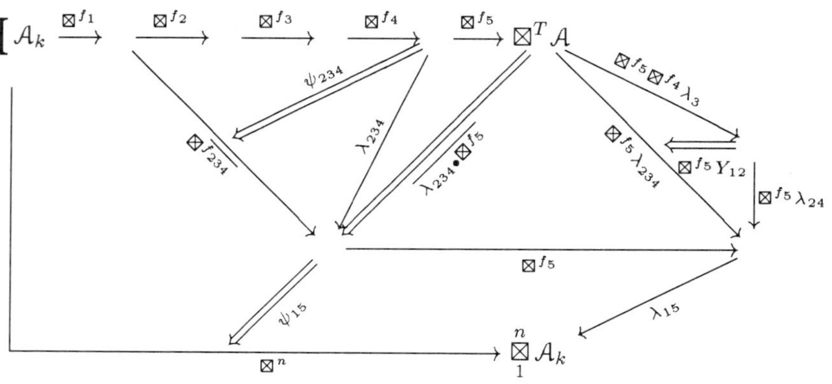

=

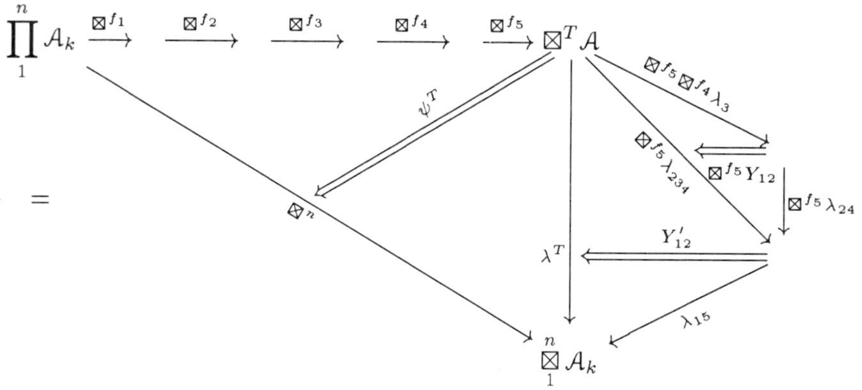

Comparing this diagram with (A.5.6) we get by Corollary A.5.3

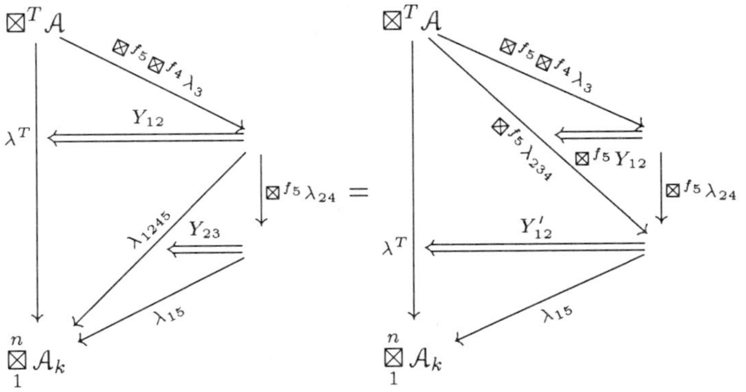

which is the equation A.4.2.5(c).

5(d) To prove A.4.2.5(d) decompose the plane staged tree T into the composition of five forests determined by the choice of T' as in the picture

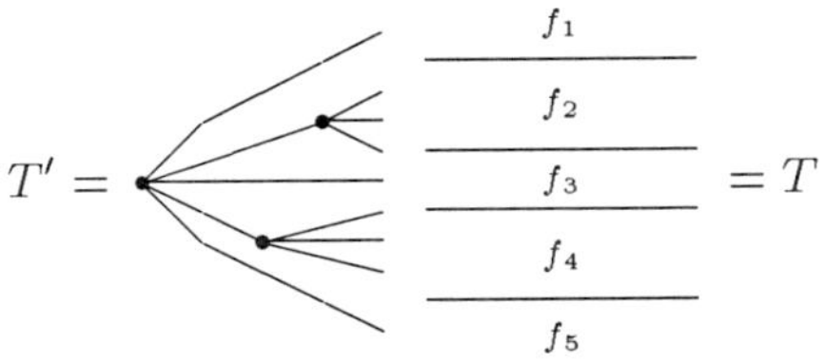

where f_5 is a tree. The composition

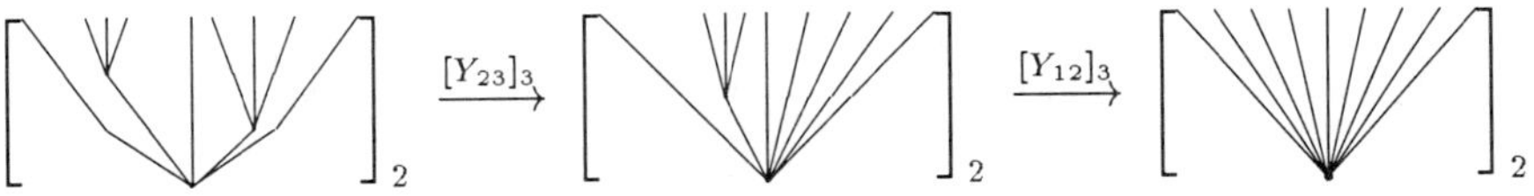

is obtained from the equation

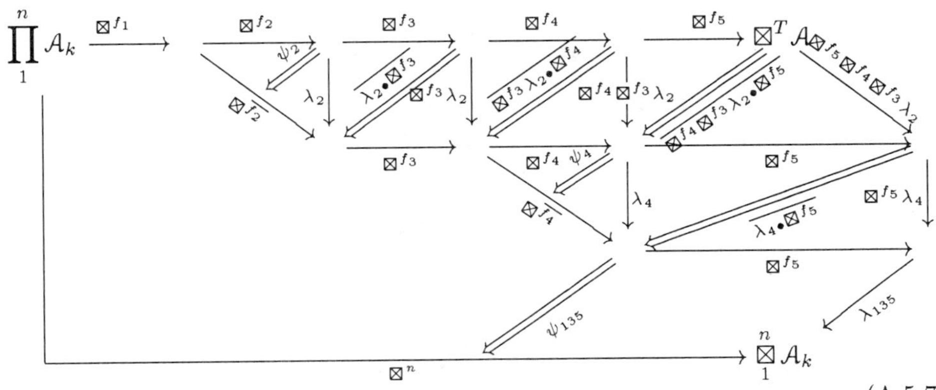

(A.5.7)

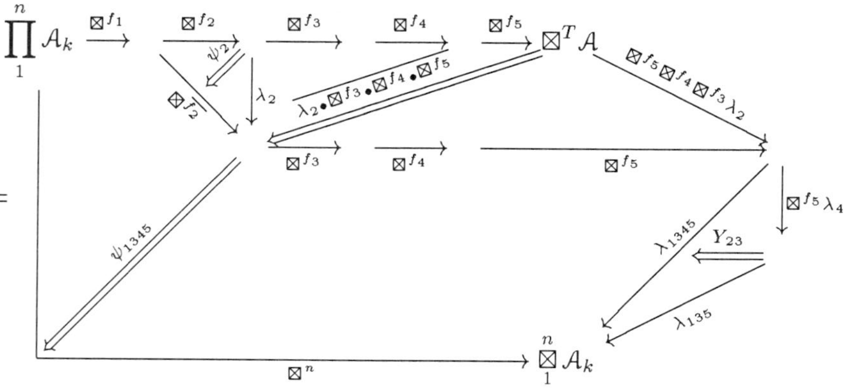

$=$

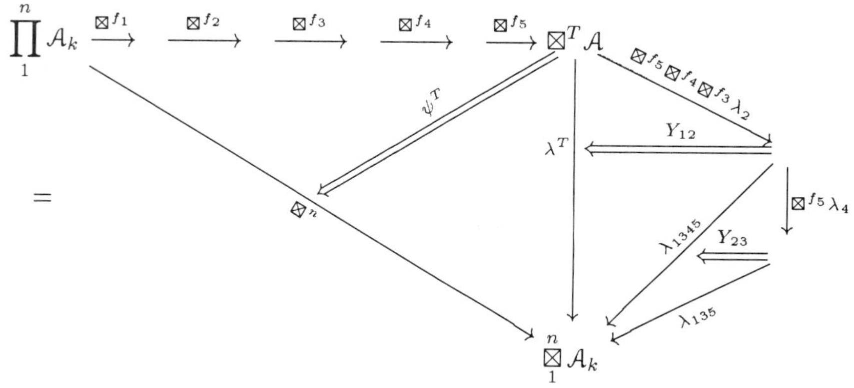

The composition

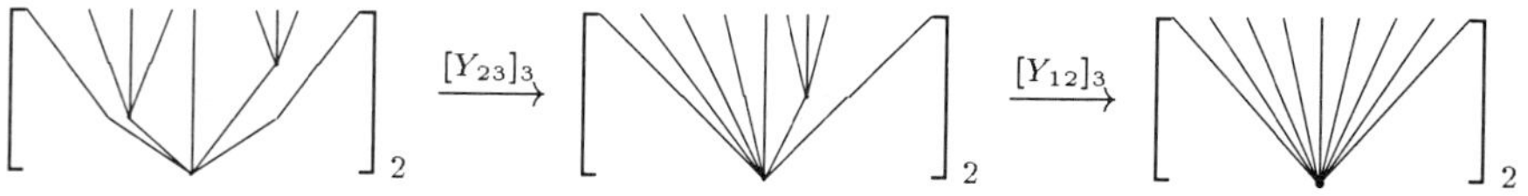

is obtained from the equation

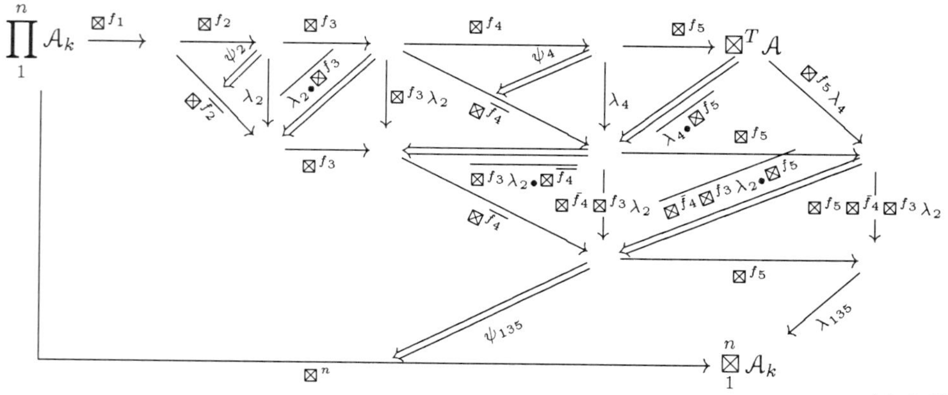

(A.5.8)

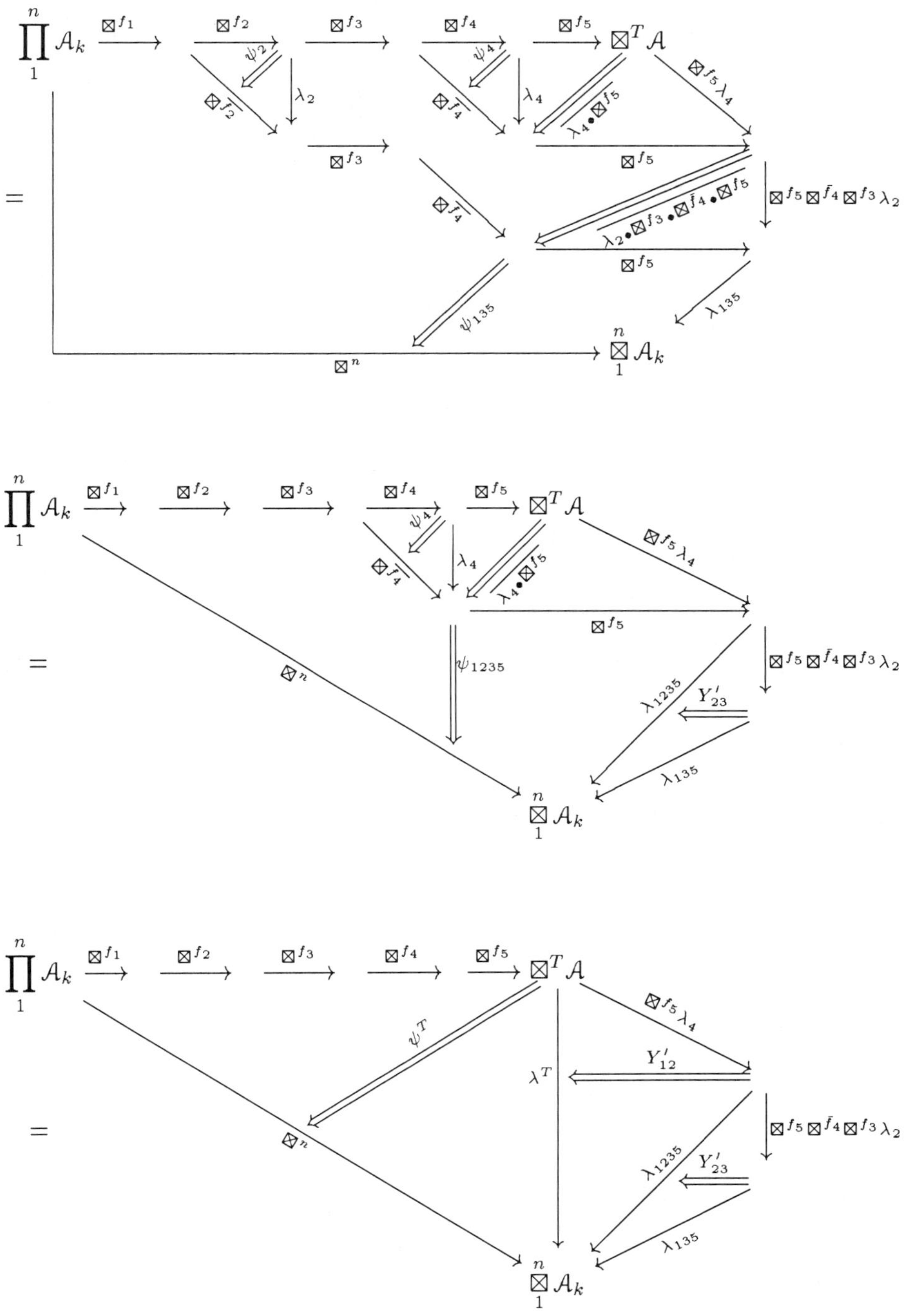

The homomorphism

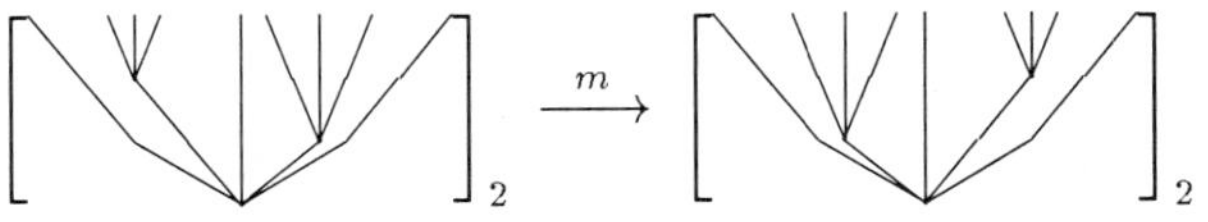

is given by

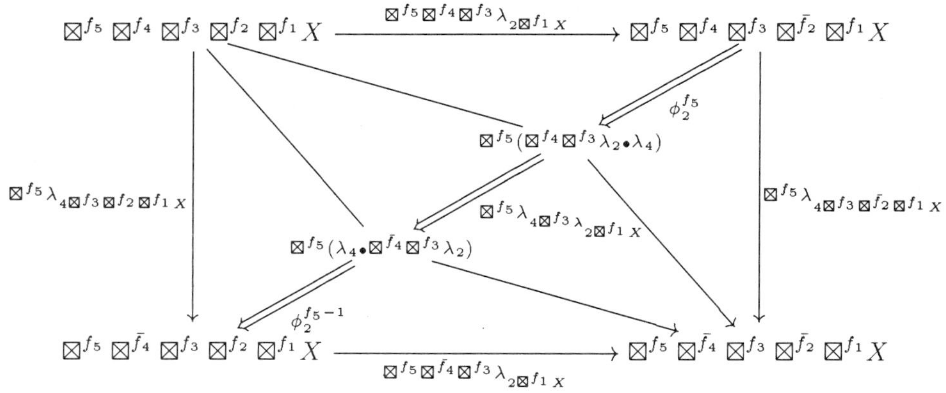

by (A.3.4) and Section A.1.5. We want to show that pasting (A.5.8) with the above square gives (A.5.7). That would imply A.4.2.5(d).

It remains to prove the equation

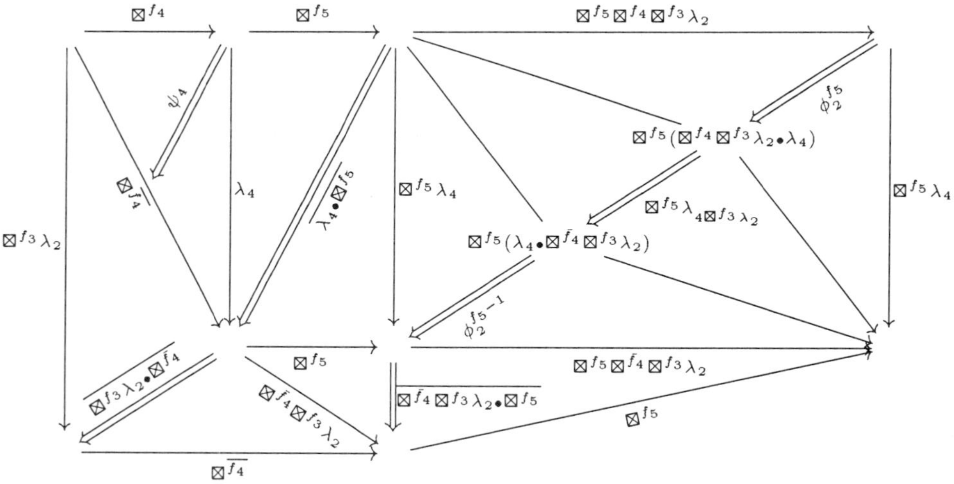

=

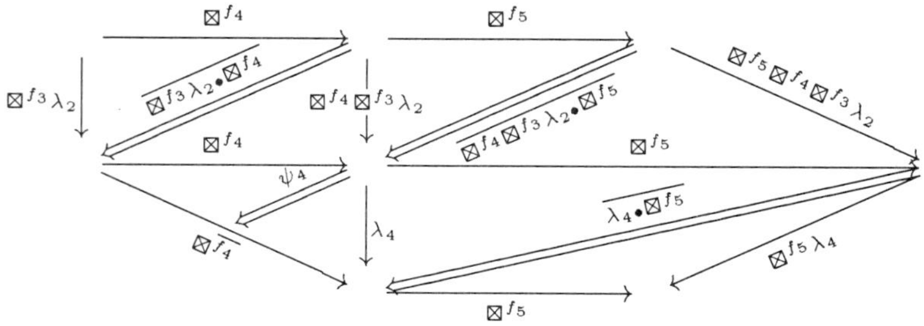

Using (A.5.2) twice we reduce the above to an equivalent equation

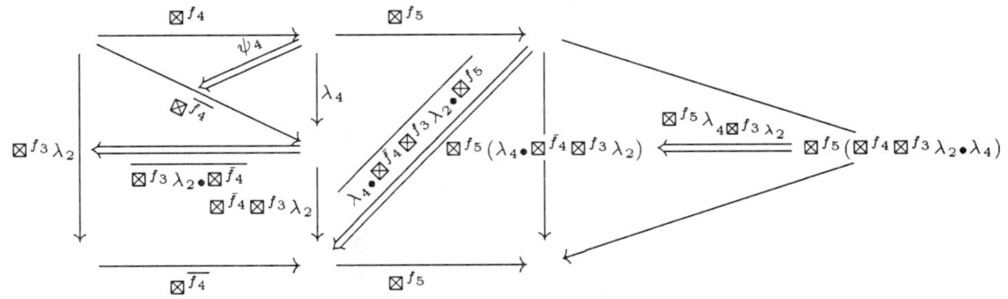

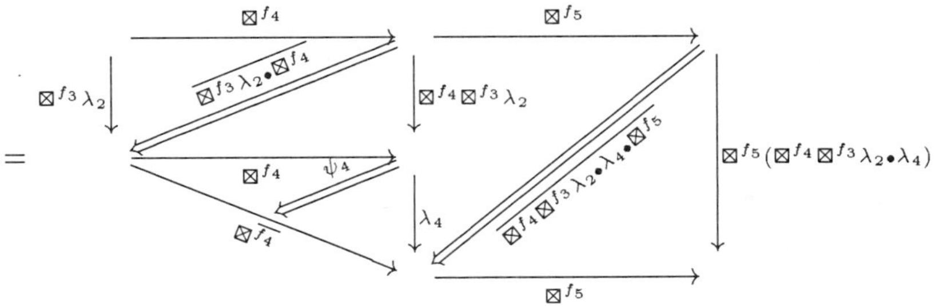

However, the left hand side of the above equation equals to

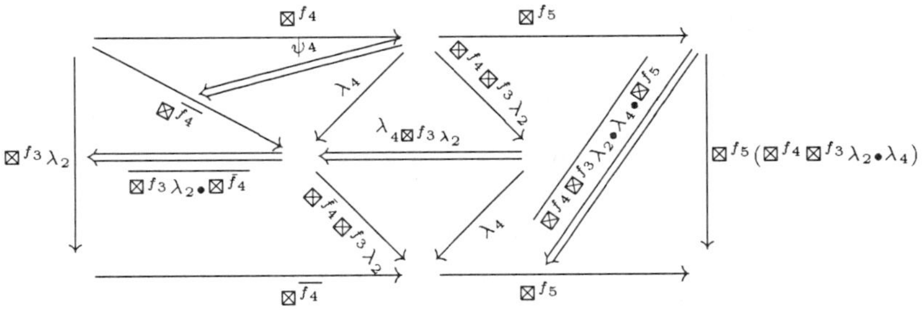

by definition of $\boxtimes^{f_5}\lambda_4$. The right hand side equals to the same diagram by definition (A.5.3) of $\lambda_{4\boxtimes^{f_3}\lambda_2}$.

Therefore, $\mathfrak{A}$ is a monoidal 2-category. □

A.6. Symmetric monoidal 2-categories

Let $\mathcal{S}$ be the category of finite linearly ordered sets and arbitrary maps (which are not required to preserve the order) from Section 1.2. We define a *staged forest* as a functor $f : [n] \to \mathcal{S}$ (cf. Section A.4.1). We say that f has n levels. A *staged tree* is a staged forest $t : [n] \to \mathcal{S}$ with $t(n) \simeq [0]$. The set of *leaves* of a staged forest f is $f(0)$. The set of *roots* of $f : [n] \to \mathcal{S}$ is $f(n)$.

The following definition practically repeats Definition A.4.4 of monoidal 2-categories with plane staged forests replaced with staged forests. We give it in complete form for the convenience of the reader.

A.6.1. Definition. A *symmetric monoidal 2-category* $\mathcal{C}$ consists of the following data

1. a 2-category $\mathcal{C}$;
2. for any $n \geqslant 0$ a weak 2-functor

$$\boxtimes^n : \mathcal{C}^n \to \mathcal{C}$$

called tensor product. For $n = 1$ the 2-functor $\boxtimes^1$ is the identity functor. For $n = 0$ we assume that $\boxtimes^0 : \mathbf{1} \to \mathcal{C}$ is a strict 2-functor. $\mathbf{1}$ is a 2-category with one object, one 1-morphism and one 2-morphism. $\boxtimes^0$(object) is denoted $\mathcal{I} \in \operatorname{Ob} \mathcal{C}$; it is called the unit object;

Since the composition of weak 2-functors is strictly associative, to any staged tree T with n leaves is associated a weak 2-functor

$$[T]_1 : \mathcal{C}^n \to \mathcal{C},$$

so that 2-categories are appropriately permuted and each vertex of the tree with k entries is replaced with $\boxtimes^k$. This 2-functor is a product of functors $\boxtimes^\phi_*$ defined for $\phi : I \to J \in \mathcal{S}$ similarly to $\otimes^\phi_*$ from Definition 1.2.2. To a function $X : I \to \mathcal{S}_s$, $i \mapsto X_i$, $s = 0, 1, 2$, the 2-functor $\boxtimes^\phi_*$ assigns the function

$$j \longmapsto \boxtimes^{f^{-1}(j)}\big(f^{-1}(j) \xrightarrow{e_j} I \xrightarrow{X} \mathcal{C}\big) = \boxtimes^{f^{-1}(j)}(X_{i_1}, \dots, X_{i_m}),$$

where $\{i_1, \dots, i_m\} = \{i \in I \mid f(i) = j\}$, $i_1 < \cdots < i_m$. In particular, for a 1-level tree t^n with n leaves we have $[t^n]_1 = \boxtimes^n$;

3. for any staged tree T with k levels and n leaves, a weak 2-natural equivalence, denoted

$$[t^k]_2 : [T]_1 \to [t^n]_1 : \mathcal{C}^n \to \mathcal{C}.$$

Consider a measured tree T' with k leaves. To T' corresponds a unique 2-natural equivalence

$$[T']_2 : [T]_1 \to [t^n]_1 : \mathcal{C}^n \to \mathcal{C}$$

where an l-vertex of T' is replaced with $[t^l]_2$ and 2-transformations for higher vertices precede the lower ones. More generally, to a measured forest F with k leaves corresponds a unique 2-natural equivalence

$$[F]_2 : [T]_1 \to [\tilde{T}]_1 : \mathcal{C}^n \to \mathcal{C}$$

where the number of levels in $\tilde{T}$ is the number of trees in F. For any two measured forests F' and F'' isomorphic as rooted forests, which differ only by the linear order of vertices, there is a unique invertible modification

$$m_{F',F''} : [F']_2 \to [F'']_2 : [T]_1 \to [\tilde{T}]_1 : \mathcal{C}^n \to \mathcal{C}$$

constructed in Theorem A.3.4.

4. for any staged tree T decomposed into three staged forests by two horizontal cuts, an invertible modification

$[Y]_3 :$ 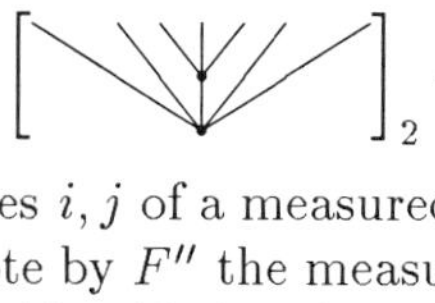 $\to [t^k]_2 : [T]_1 \to [t^n]_1 : \mathcal{C}^n \to \mathcal{C}.$

If two vertices i, j of a measured forest F' with k leaves are connected by an edge e, denote by F'' the measured forest F' with the edge e contracted and vertices i, j identified. The modifications $[Y]_3$ combine into an invertible modification

$$[Y_{i,j}]_3 : [F']_2 \to [F'']_2 : [T]_1 \to [t^n]_1 : \mathcal{C}^n \to \mathcal{C};$$

such that

5. equations 5(a)–5(d) from Definition A.4.2 hold.

A.7. A symmetric monoidal 2-category of abelian categories

Let $\Bbbk$ be an arbitrary field. We have proved in Section A.5 that the 2-category $\mathfrak{A} = \mathfrak{Ab}^r_\Bbbk$ is monoidal. Using staged forests instead of plane staged forests we immediately generalise this result to the following one.

A.7.1. Theorem. *$\mathfrak{Ab}^r_\Bbbk$ is a symmetric monoidal 2-category.*

For the convenience of the reader we summarize here the choices which specify a symmetric monoidal structure on the 2-category $\mathfrak{A} = \mathfrak{Ab}^r_\Bbbk$.

1) A 2-category $\mathfrak{A} = \mathfrak{Ab}^r_\Bbbk$.

2) For any $n \geqslant 0$ a weak 2-functor

$$\boxtimes^n : \mathfrak{A}^n \to \mathfrak{A}$$

is given by the following data:

2.1) For each family $(\mathcal{A}_1, \dots, \mathcal{A}_n) \in \operatorname{Ob}\mathfrak{A}^n$, **choose** an object $\mathcal{A} \in \mathfrak{A}_0$ and a functor $\boxtimes_\mathcal{A} : \mathcal{A}_1 \times \cdots \times \mathcal{A}_n \to \mathcal{A}$ as in Theorem A.5.1. Set the first datum to

$$\boxtimes_1^n \mathcal{A}_k = \mathcal{A}_1 \boxtimes \cdots \boxtimes \mathcal{A}_n = \boxtimes^n(\mathcal{A}_1 \dots, \mathcal{A}_n) = \mathcal{A},$$

if $n > 1$. For $n = 1$ we choose

$$\boxtimes^1(\mathcal{A}) = \mathcal{A}, \quad \boxtimes_\mathcal{A} = \mathrm{Id}_\mathcal{A} : \mathcal{A} \to \mathcal{A} = \boxtimes^1(\mathcal{A}).$$

For $n = 0$ we choose

$$\boxtimes^0() = \Bbbk\text{-vect}, \quad \boxtimes : \mathbf{1} \to \Bbbk\text{-vect} = \boxtimes^0(), \quad \text{object} \mapsto \Bbbk.$$

2.2) For any $\Bbbk$-multilinear right exact in each variable functor $\Phi : \mathcal{A}_1 \times \cdots \times \mathcal{A}_n \to \mathcal{B}$, **choose** arbitrarily a functor $\tilde{\Phi} : \boxtimes_1^n \mathcal{A}_k \to \mathcal{B} \in \mathfrak{A}_1$ and an isomorphism $\overline{\Phi}$

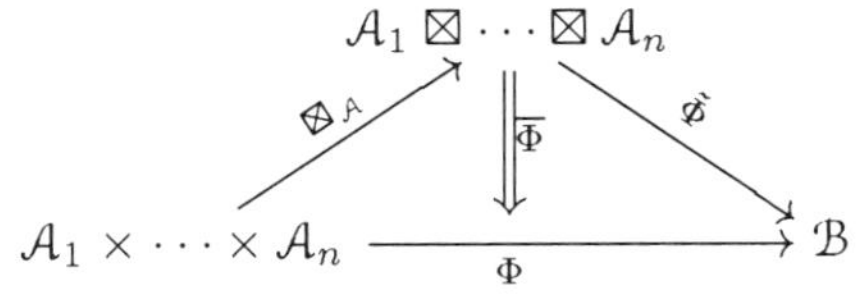

For $\Phi = \boxtimes_\mathcal{A} : \prod_1^n \mathcal{A}_k \to \boxtimes_1^n \mathcal{A}_k$ we choose $\tilde{\Phi} = \mathrm{Id}_{\boxtimes_1^n \mathcal{A}_k}$ and $\overline{\Phi} = \mathrm{id}_{\boxtimes_\mathcal{A}}$. These choices determine a functor quasi-inverse to $\mathrm{Hom}(\boxtimes_\mathcal{A}, \mathcal{B})$

$$D_{\mathcal{A}_1,\dots,\mathcal{A}_n;\mathcal{B}} : \mathrm{Hom}_{\text{m.l.,r.e.}}(\mathcal{A}_1 \times \cdots \times \mathcal{A}_n, \mathcal{B}) \to \mathfrak{A}(\mathcal{A}_1 \boxtimes \cdots \boxtimes \mathcal{A}_n, \mathcal{B}), \quad \Phi \mapsto \tilde{\Phi}$$
$$D_{\mathcal{A}_1,\dots,\mathcal{A}_n;\mathcal{B}}(f : \Phi \to \Psi) = (g : \tilde{\Phi} \to \tilde{\Psi}).$$

Set the second datum, a functor $\boxtimes_{\mathcal{A};\mathcal{B}}$, to be

$$\boxtimes^n : \mathfrak{A}(\mathcal{A}_1, \mathcal{B}_1) \times \cdots \times \mathfrak{A}(\mathcal{A}_n, \mathcal{B}_n) \xrightarrow{\mathrm{Hom}(1, \boxtimes_\mathcal{B})} \mathrm{Hom}_{\text{m.l.,r.e.}}(\prod_1^n \mathcal{A}_k, \boxtimes_1^n \mathcal{B}_l)$$
$$\xrightarrow{D_{\mathcal{A}_1,\dots,\mathcal{A}_n;\boxtimes_1^n \mathcal{B}_l}} \mathfrak{A}(\boxtimes_1^n \mathcal{A}_k, \boxtimes_1^n \mathcal{B}_l).$$

2.3(a) In particular,

$$\mathrm{Id}_{\mathcal{A}_1} \boxtimes \cdots \boxtimes \mathrm{Id}_{\mathcal{A}_n} = \widetilde{\boxtimes_\mathcal{A}} = \mathrm{Id}_{\boxtimes_1^n \mathcal{A}_k}$$

due to the choice in 2.2). We **choose** the identity isomorphism

$$\phi^0 = \mathrm{id} : \mathrm{Id}_{\mathcal{A}_1 \boxtimes \cdots \boxtimes \mathcal{A}_n} \to \mathrm{Id}_{\mathcal{A}_1} \boxtimes \cdots \boxtimes \mathrm{Id}_{\mathcal{A}_n}$$

as the third datum.

2.3(b) One more datum is the isomorphism of functors

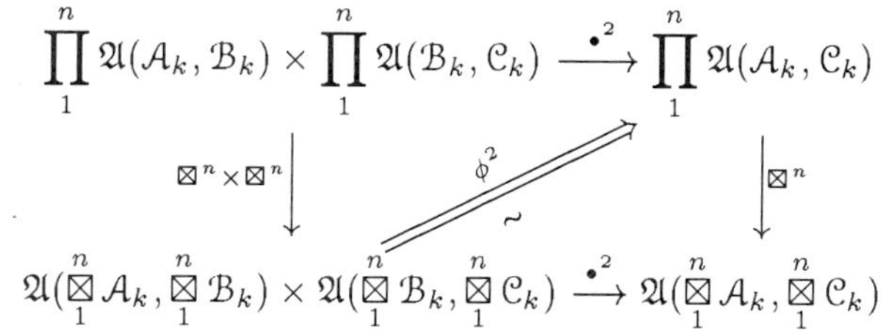

obtained from the diagram

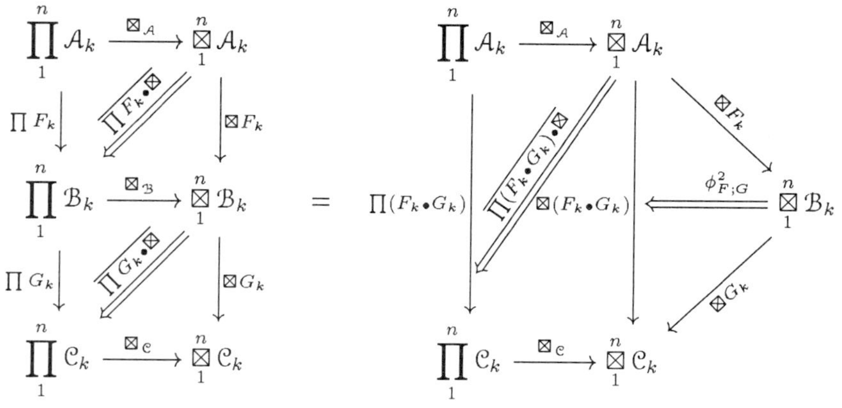

by Corollary A.5.3.

Therefore, $(\boxtimes^n, \phi^2, \mathrm{id})$ is a weak 2-functor.

3) With any staged tree T with n leaves (0-vertices are allowed) is associated a weak 2-functor $\boxtimes^T = [T]_1 : \mathfrak{A}^n \to \mathfrak{A}$. It can be visualized as a word made of permuted arguments $\mathcal{A}_1, \dots, \mathcal{A}_n$ with several $\Bbbk$-vect inserted, arguments separated by $\boxtimes$ signs with some parentheses. For any n-tuple $(\mathcal{A}_1, \dots, \mathcal{A}_n) \in \mathrm{Ob}\,\mathfrak{A}^n$ we have also a $\Bbbk$-multilinear right exact in each variable functor

$$\boxtimes^T_{\mathcal{A}} : \mathcal{A}_1 \times \cdots \times \mathcal{A}_n \to \boxtimes^T(\mathcal{A}_1, \dots, \mathcal{A}_n)$$

obtained from the chosen functors $\boxtimes_{\mathcal{B}} : \mathcal{B}_1 \times \cdots \times \mathcal{B}_n \to \boxtimes^m_1 \mathcal{B}_k$.

We want to show that categories $\boxtimes^T(\mathcal{A}_1, \dots, \mathcal{A}_n)$ are equivalent to $\mathcal{A}_1 \boxtimes \cdots \boxtimes \mathcal{A}_n$. Clearly, it suffices to show that the functor $\boxtimes^T_{\mathcal{A}}$ possesses the universal properties of Theorem A.5.1.

A.7.2. Proposition. *For any staged forest T with n leaves and any $n+1$-tuple of categories $\mathcal{A}_1, \dots, \mathcal{A}_n, \mathcal{B} \in \mathfrak{A}_0$ the functor*

$$\mathrm{Hom}(\boxtimes^T_{\mathcal{A}}, \mathcal{B}) : \mathrm{Hom}_{m.l.,r.e.}(\boxtimes^T(\mathcal{A}_1, \dots, \mathcal{A}_n), \mathcal{B}) \to \mathrm{Hom}_{m.l.,r.e.}(\mathcal{A}_1 \times \cdots \times \mathcal{A}_n, \mathcal{B})$$

is an equivalence.

3.1) Universality of the Deligne's tensor product (Theorem A.5.1) implies that for any plane staged tree T we can (and we will) **choose** a natural equivalence $\lambda^T_{\mathcal{A}}$

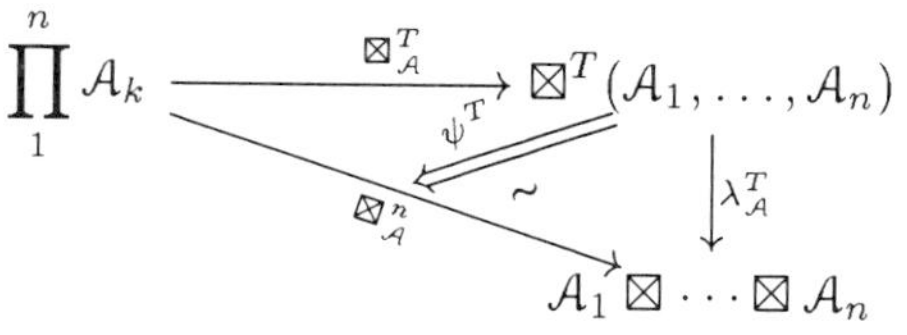

and an isomorphism of functors ψ^T. The family of equivalences

$$\lambda^T_{\mathcal{A}} : \boxtimes^T(\mathcal{A}_1, \dots, \mathcal{A}_n) \to \mathcal{A}_1 \boxtimes \cdots \boxtimes \mathcal{A}_n$$

is the *first datum* of a weak 2-transformation

$$\lambda^T = [t^k]_2 : [T]_1 \to [t^n]_1 : \mathfrak{A}^n \to \mathfrak{A}.$$

3.2) The *second datum* of $[t^k]_2$ is an isomorphism λ^T_F constructed for a family of functors $F_k : \mathcal{A}_k \to \mathcal{B}_k \in \mathfrak{A}_1$

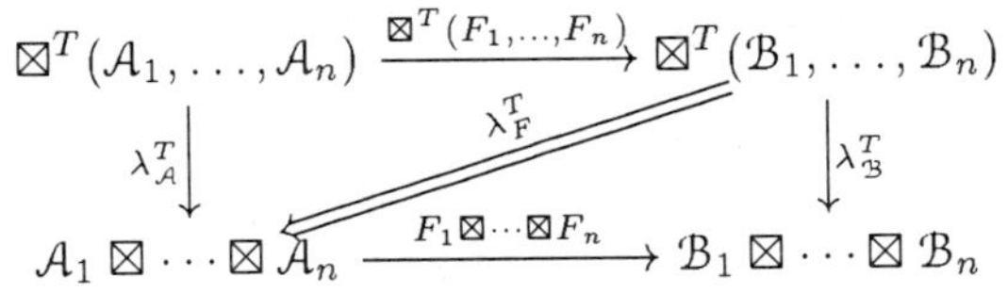

from the equation

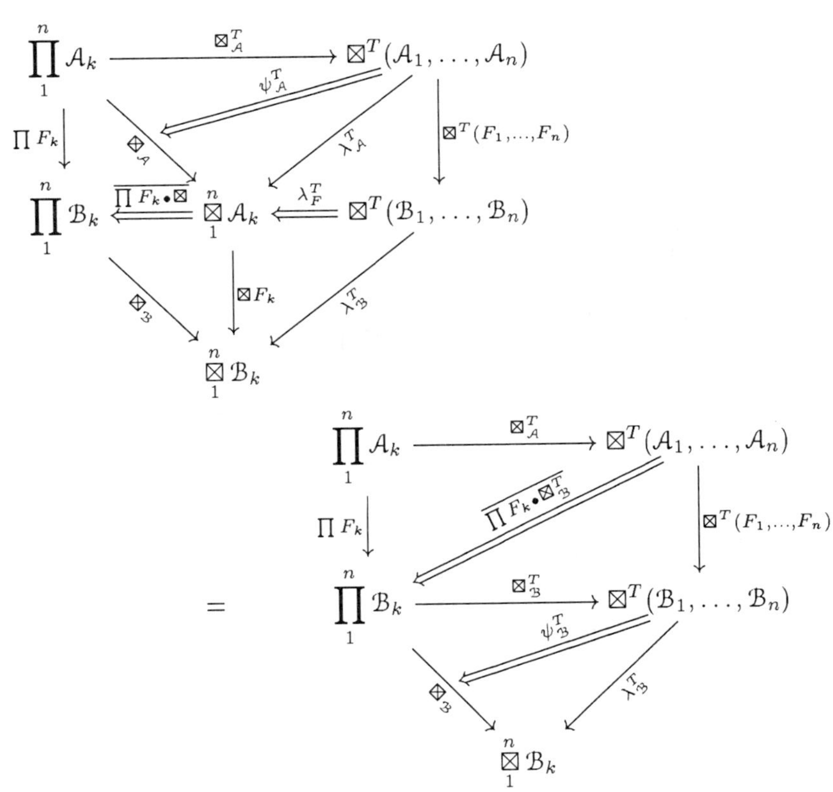

by Corollary A.5.3. Here the isomorphism $\overline{\prod F_k \bullet \boxtimes^T_{\mathcal{B}}}$ is the pasting of isomorphisms corresponding to multiplicands of $\boxtimes^T_{\mathcal{B}}$.

4) Consider a plane staged tree T decomposed into three plane staged forests by two horizontal cuts,

$$T = \begin{matrix} g \\ f \\ t \end{matrix}\,.$$

The lowest forest t is a tree. Denote by $\bar{f}$ the forest f with all inner edges (not leaves) shrunk to a point. Each tree in f is replaced with a 1-level tree in $\bar{f}$. Let

$$\overline{T} = \begin{matrix} g \\ \bar{f} \\ t \end{matrix},$$

which is T with several levels compressed. We want to construct

$[Y_{12}]_3 :$ 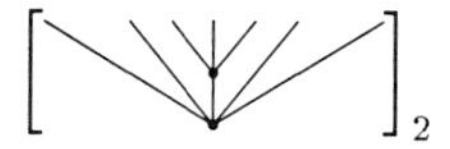 $\to [t^k]_2 : [T]_1 \to [t^n]_1 : \mathfrak{A}^n \to \mathfrak{A}.$

The first 2-transformation is the composition of two vertical arrows in

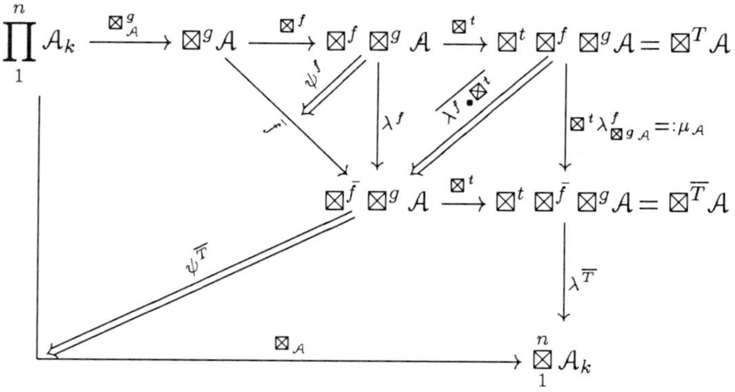

(A.7.1)

where we define $\mu_{\mathcal{A}} = \boxtimes^t \lambda^f_{\boxtimes^g \mathcal{A}}$. The morphism $Y_{\mathcal{A}} = [Y_{12}]_3$ is found by Corollary A.5.3 so that the above pasting equals to

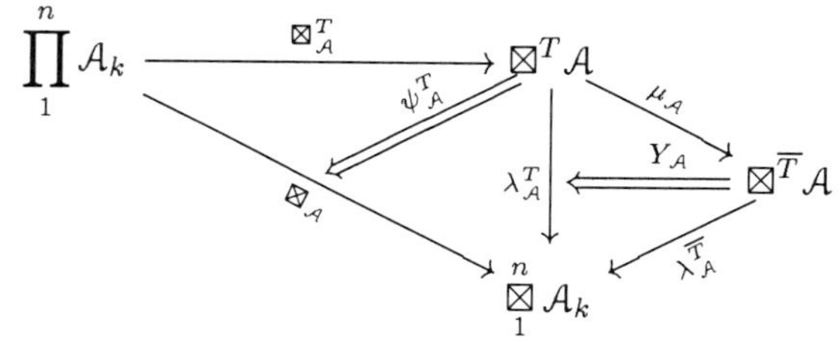

(A.7.2)

Bibliography

1. E. Abe, *Hopf algebras*, Cambridge Tracts in Math., vol. 74, Cambridge Univ. Press, Cambridge, 1980.
2. J. C. Baez and M. Neuchl, *Higher-Dimensional Algebra I: Braided Monoidal 2-Categories*, Adv. Math. **121** (1996), no. 2, 196–244.
3. M. A. Batanin, *Monoidal globular categories as a natural environment for the theory of weak n-categories*, preprint of Macquarie University, 1997.
4. J. Bénabou, *Introduction to Bicategories*, Lect. Notes Math. **47**, Berlin, Heidelberg, New York: Springer-Verlag, 1967, 1–77.
5. L. Breen, *On the classification of 2-gerbes and 2-stacks*, Astérisque **225**, Soc. Math. de France, 1994.
6. B. Day and R. Street, *Monoidal bicategories and Hopf algebroids*, Advances in Math. **129** (1997), no. 1, 99–157.
7. P. Deligne, *Catégories tannakiennes*, in: The Grothendieck Festschrift, Vol. II, Progress in Math. **87**, Boston, Basel, Berlin: Birkhäuser, 1991, 111–195.
8. P. Deligne, *Action du groupe des tresses sur une catégorie*, Invent. Math. **128** (1997), no. 1, 159–175.
9. P. Deligne and J. S. Milne, *Tannakian Categories*, Lect. Notes Math. **900**, Berlin, Heidelberg, New York: Springer-Verlag, 1982, 101–228.
10. V. G. Drinfeld, *Quantum groups*, Proceedings of the ICM, AMS, Providence, R.I. **1** (1987), 798–820.
11. R. Gordon, A. J. Power and R. Street, *Coherence for Tricategories*, Memoirs of the AMS, **117** (1995) No 558, Amer. Math. Soc., Providence, Rhode Island.
12. J. Gray, *Formal category theory: Adjointness for 2-categories*, Lecture Notes in Mathematics, vol. 391, Springer, Berlin, 1974.
13. J. Gray, *Coherence for tensor product of 2-categories and braid groups*, Algebra, Topology, and Category Theory (a collection of papers in honor of Samuel Eilenberg), Academic Press, 1976, 63–76.
14. A. Grothendieck and J. L. Verdier, *Préfaisceaux*, in: Théorie des topos et cohomologie étale des schémas (SGA 4), Lecture Notes Math. **269**, Berlin, Heidelberg, New York: Springer-Verlag, 1972, 1–217.
15. L. Hlavaty, *Quantized braided groups*, J. Math. Phys. **35** (1994), no. 5, 2560–2569.
16. A. Joyal and R. Street, *The geometry of tensor calculus I*, Advances in Math. **88** (1991), 55–112.
17. A. Joyal and R. Street, *An introduction to Tannaka duality and quantum groups*, in: Category Theory, Proceedings, Como 1990 (A. Carboni, M. C. Pedicchio, and G. Rosolini, eds.), Lecture Notes in Math. 1488, Springer-Verlag, Berlin, Heidelberg, (1991) 411–492.
18. M. M. Kapranov and V. A. Voevodsky, *2-categories and Zamolodchikov Tetrahedra Equations*, Proc. Symposia Pure Math., AMS, **56** (1994), Part 2, 177–259.
19. G. M. Kelly, *Basic Concepts of Enriched Category Theory*, London Math. Soc. Lect. Notes Ser. **64**, Cambridge Univ. Press, 1982.
20. V. Lyubashenko, *Tangles and Hopf algebras in braided categories*, J. Pure Appl. Algebra **98** (1995) n. 3, 245–278.
21. S. Mac Lane, *Natural associativity and commutativity*, Rice Univ. Studies **49** (1963), 28–46.
22. S. MacLane, *Categories for the Working Mathematician*, Springer-Verlag, 1971.
23. B. Pareigis, *Reconstruction of hidden symmetries*, J. Algebra **183** (1996), no. 1, 90–154.
24. A. J. Power, *A 2-categorical pasting theorem*, J. of Algebra **129** (1990), 439–445.

25. D. Quillen, *Higher algebraic K-theory*, Lecture Notes Math., **341**, Berlin, Heidelberg, New York: Springer-Verlag, (1973) 85–147.
26. N. Saavedra Rivano, *Catégories Tannakiennes*, Lecture Notes Math. **265**, Berlin, Heidelberg, New York: Springer-Verlag, 1972.
27. P. Schauenburg, *Tannaka duality for Arbitrary Hopf Algebras*, Algebra-Berichte **66**, München: R. Fisher, 1992.

Author adress:

Institute of Mathematics, Ukrainian National Academy of Sciences,
3 Tereshchenkivska st., Kyiv 252601, Ukraine
E-mail address: lub@ukma.kiev.ua

Editorial Information

To be published in the *Memoirs*, a paper must be correct, new, nontrivial, and significant. Further, it must be well written and of interest to a substantial number of mathematicians. Piecemeal results, such as an inconclusive step toward an unproved major theorem or a minor variation on a known result, are in general not acceptable for publication. *Transactions* Editors shall solicit and encourage publication of worthy papers. Papers appearing in *Memoirs* are generally longer than those appearing in *Transactions* with which it shares an editorial committee.

As of July 31, 1999, the backlog for this journal was approximately 5 volumes. This estimate is the result of dividing the number of manuscripts for this journal in the Providence office that have not yet gone to the printer on the above date by the average number of monographs per volume over the previous twelve months, reduced by the number of issues published in four months (the time necessary for preparing an issue for the printer). (There are 6 volumes per year, each containing at least 4 numbers.)

A Copyright Transfer Agreement is required before a paper will be published in this journal. By submitting a paper to this journal, authors certify that the manuscript has not been submitted to nor is it under consideration for publication by another journal, conference proceedings, or similar publication.

Information for Authors and Editors

Memoirs are printed by photo-offset from camera copy fully prepared by the author. This means that the finished book will look exactly like the copy submitted.

The paper must contain a *descriptive title* and an *abstract* that summarizes the article in language suitable for workers in the general field (algebra, analysis, etc.). The *descriptive title* should be short, but informative; useless or vague phrases such as "some remarks about" or "concerning" should be avoided. The *abstract* should be at least one complete sentence, and at most 300 words. Included with the footnotes to the paper, there should be the 1991 *Mathematics Subject Classification* representing the primary and secondary subjects of the article. This may be followed by a list of *key words and phrases* describing the subject matter of the article and taken from it. A list of the numbers may be found in the annual index of *Mathematical Reviews*, published with the December issue starting in 1990, as well as from the electronic service e-MATH [**telnet e-MATH.ams.org** (or **telnet 130.44.1.100**). Login and password are **e-math**]. For journal abbreviations used in bibliographies, see the list of serials in the latest *Mathematical Reviews* annual index. When the manuscript is submitted, authors should supply the editor with electronic addresses if available. These will be printed after the postal address at the end of each article.

Electronically prepared papers. The AMS encourages submission of electronically prepared papers in $\mathcal{AMS}$-TeX or $\mathcal{AMS}$-LaTeX. The Society has prepared author packages for each AMS publication. Author packages include instructions for preparing electronic papers, the *AMS Author Handbook*, samples, and a style file that generates the particular design specifications of that publication series for both $\mathcal{AMS}$-TeX and $\mathcal{AMS}$-LaTeX.

Authors with FTP access may retrieve an author package from the Society's Internet node `e-MATH.ams.org` (130.44.1.100). For those without FTP access, the author package can be obtained free of charge by sending e-mail to `pub@ams.org` (Internet) or from the Publication Division, American Mathematical Society, P.O. Box 6248, Providence, RI 02940-6248. When requesting an author package, please specify $\mathcal{A}\mathcal{M}\mathcal{S}$-TeX or $\mathcal{A}\mathcal{M}\mathcal{S}$-LaTeX, Macintosh or IBM (3.5) format, and the publication in which your paper will appear. Please be sure to include your complete mailing address.

Submission of electronic files. At the time of submission, the source file(s) should be sent to the Providence office (this includes any TeX source file, any graphics files, and the DVI or PostScript file).

Before sending the source file, be sure you have proofread your paper carefully. The files you send must be the EXACT files used to generate the proof copy that was accepted for publication. For all publications, authors are required to send a printed copy of their paper, which exactly matches the copy approved for publication, along with any graphics that will appear in the paper.

TeX files may be submitted by email, FTP, or on diskette. The DVI file(s) and PostScript files should be submitted only by FTP or on diskette unless they are encoded properly to submit through e-mail. (DVI files are binary and PostScript files tend to be very large.)

Files sent by electronic mail should be addressed to the Internet address `pub-submit@ams.org`. The subject line of the message should include the publication code to identify it as a Memoir. TeX source files, DVI files, and PostScript files can be transferred over the Internet by FTP to the Internet node `e-math.ams.org` (130.44.1.100).

Electronic graphics. Figures may be submitted to the AMS in an electronic format. The AMS recommends that graphics created electronically be saved in Encapsulated PostScript (EPS) format. This includes graphics originated via a graphics application as well as scanned photographs or other computer-generated images.

If the graphics package used does not support EPS output, the graphics file should be saved in one of the standard graphics formats—such as TIFF, PICT, GIF, etc.—rather than in an application-dependent format. Graphics files submitted in an application-dependent format are not likely to be used. No matter what method was used to produce the graphic, it is necessary to provide a paper copy to the AMS.

Authors using graphics packages for the creation of electronic art should also avoid the use of any lines thinner than 0.5 points in width. Many graphics packages allow the user to specify a "hairline" for a very thin line. Hairlines often look acceptable when proofed on a typical laser printer. However, when produced on a high-resolution laser imagesetter, hairlines become nearly invisible and will be lost entirely in the final printing process.

Screens should be set to values between 15% and 85%. Screens which fall outside of this range are too light or too dark to print correctly.

Any inquiries concerning a paper that has been accepted for publication should be sent directly to the Editorial Department, American Mathematical Society, P. O. Box 6248, Providence, RI 02940-6248.

Editors

This journal is designed particularly for long research papers (and groups of cognate papers) in pure and applied mathematics. Papers intended for publication in the *Memoirs* should be addressed to one of the following editors:

Ordinary differential equations, partial differential equations, and applied mathematics to JOHN MALLET-PARET, Division of Applied Mathematics, Brown University, Providence, RI 02912-9000; electronic mail: `jmp@cfm.brown.edu.`

Harmonic analysis, representation theory, and Lie theory to ROBERT J. STANTON, Department of Mathematics, The Ohio State University, 231 West 18th Avenue, Columbus, OH 43210-1174; electronic mail: `stanton@math.ohio-state.edu.`

Ergodic theory and dynamical systems to ROBERT F. WILLIAMS, Department of Mathematics, University of Texas at Austin, Austin, TX 78712-1082; e-mail: `bob@math.utexas.edu`

Real and harmonic analysis and geometric partial differential equations to WILLIAM BECKNER, Department of Mathematics, University of Texas at Austin, Austin, TX 78712-1082; e-mail: `beckner@math.utexas.edu.`

Algebra to CHARLES CURTIS, Department of Mathematics, University of Oregon, Eugene, OR 97403-1222 e-mail: `cwc@darkwing.uoregon.edu`

Algebraic topology and cohomology of groups to STEWART PRIDDY, Department of Mathematics, Northwestern University, 2033 Sheridan Road, Evanston, IL 60208-2730; e-mail: `s_priddy@math.nwu.edu.`

Differential geometry and global analysis to CHUU-LIAN TERNG, Department of Mathematics, Northeastern University, Huntington Avenue, Boston, MA 02115-5096; e-mail: `terng@neu.edu.`

Probability and statistics to RODRIGO BAÑUELOS, Department of Mathematics, Purdue University, West Lafayette, IN 47907-1968; e-mail: `banuelos@math.purdue.edu.`

Combinatorics and Lie theory to PHILIP J. HANLON, Department of Mathematics, University of Michigan, Ann Arbor, MI 48109-1003; e-mail: `hanlon@math.lsa.umich.edu.`

Logic to THEODORE SLAMAN, Department of Mathematics, University of California at Berkeley, Berkeley, CA 94720-3840; e-mail: `slaman@math.berkeley.edu.`

Number theory and arithmetic algebraic geometry to ALICE SILVERBERG, MSRI, 1000 Centennial Dr., Berkeley, CA 94720; e-mail: `silver@math.ohio-state.edu.`

Complex analysis and complex geometry to DANIEL M. BURNS, Department of Mathematics, University of Michigan, Ann Arbor, MI 48109-1003; e-mail: `dburns@math.lsa.umich.edu.`

Algebraic geometry and commutative algebra to LAWRENCE EIN, Department of Mathematics, University of Illinois, 851 S. Morgan (M/C 249), Chicago, IL 60607-7045; e-mail: `ein@uic.edu.`

Geometric topology, knot theory, hyperbolic geometry, and general topoogy to JOHN LUECKE, Department of Mathematics, University of Texas at Austin, Austin, TX 78712-1082; e-mail: `luecke@math.utexas.edu.`

Partial differential equations and applied mathematics to BARBARA LEE KEYFITZ, Department of Mathematics, University of Houston, 4800 Calhoun, Houston, TX 77204-3476; e-mail: `keyfitz@uh.edu`

Operator algebras and functional analysis to BRUCE E. BLACKADAR, Department of Mathematics, University of Nevada, Reno, NV 89557; e-mail: `bruceb@math.unr.edu`

All other communications to the editors should be addressed to the Managing Editor, PETER SHALEN, Department of Mathematics, University of Illinois, 851 S. Morgan (M/C 249), Chicago, IL 60607-7045; e-mail: `shalen@math.uic.edu.`

Selected Titles in This Series

(Continued from the front of this publication)